Biochemistry and Genetic Regulation of Commercially Important Antibiotics

BIOTECHNOLOGY

BIOTECHNOLOGY SERIES

1. R. Saliwanchik	*Legal Protection for Microbiological and Genetic Engineering Inventions*
2. L. Vining (editor)	*Biochemistry and Genetic Regulation of Commercially Important Antibiotics*

Other volumes in preparation

K. Herrmann and R. Somerville (editors)	*Amino Acid Biosynthesis and Genetic Regulation*
D. Wise (editor)	*Organic Chemicals from Biomass*
A. Demain (editor)	*Biology of Industrial Microorganisms*
A. Laskin (editor)	*Applications of Isolated Enzymes and Immobilized Cells to Biotechnology*
S. Dagley (editor)	*Biodegradation*
W. Thilly (editor)	*Mammalian Cell Technology*
V. Sato (editor)	*Monoclonal Antibodies: A Technology for Industry and Medicine*

Biochemistry and Genetic Regulation of Commercially Important Antibiotics

Edited by

Leo C. Vining
Dalhousie University
Halifax, Nova Scotia, Canada

1983

ADDISON-WESLEY PUBLISHING COMPANY
Advanced Book Program/World Science Division
Reading, Massachusetts

London • Amsterdam • Don Mills, Ontario • Sydney • Tokyo

Library of Congress Cataloging in Publication Data

Main entry under title:

Biochemistry and genetic regulation of commercially important antibiotics.

(Biotechnology ; 2)
Includes bibliographical references and index.
1. Antibiotics. 2. Microbial genetics.
3. Chemistry, Pharmaceutical. I. Vining, Leo C.
II. Series. [DNLM: 1. Biochemistry. 2. Antibiotics.
3. Genetics, Biochemical. W1 BI918W no. 2 /
QV 350 B6154]
RS431.A6B5 1983 615′.329 83-2492
ISBN 0-201-10984-0

Published simultaneously in Canada.

MANUFACTURED IN THE UNITED STATES OF AMERICA

ABCDEFGHIJ-MA-89876543

CONTRIBUTORS

Yair Aharonowitz
Department of Microbiology
Tel Aviv University
Tel Aviv, Israel

Christopher Ball
The Squibb Institute for Medical Research
Princeton, New Jersey

Vladislav Běhal
Institute of Microbiology
Czechoslovak Academy of Sciences
Prague, Czechoslovakia

Michal Bučko
Biotika
Slovenská Lupča, Czechoslovakia

Julian E. Davies
Biogen, S.A.
Geneva, Switzerland

Arnold L. Demain
Department of Nutrition and Food Science
Massachusetts Institute of Technology
Cambridge, Massachusetts

David A. Hopwood
John Innes Institute
Norwich, England

Zdeněk Hošťálek
Institute of Microbiology
Czechoslovak Academy of Sciences
Prague, Czechoslovakia

Horst Kleinkauf
Institut für Biochemie und Molekulare
Biologie der Technischen Universität
Berlin, Federal Republic of Germany

Giancarlo Lancini
Gruppo Lepetit S.p.A.
Milano, Italy

Kenneth D. Macdonald
Chemical Defence Establishment
Salisbury, England

Vedpal S. Malik
Philip Morris, Research Center
Richmond, Virginia

Juan-Francisco Martín
Departamento de Microbiologia
Universidad de Leon
Leon, Spain

Yasushi Okumura
Sanraku-Ocean Company, Ltd.
Fujisawa, Japan

Satoshi Ōmura
School of Pharmaceutical Sciences
Kitasato University
Tokyo, Japan
and
The Kitasato Institute
Tokyo, Japan

Joseph O'Sullivan
The Squibb Institute for Medical Research
Princeton, New Jersey

Yoshitake Tanaka
School of Pharmaceutical Sciences
Kitasato University
Tokyo, Japan
and
The Kitasato Institute
Tokyo, Japan

Hans von Döhren
Institut für Biochemie und Molekulare
Biologie der Technischen Universität
Berlin, Federal Republic of Germany

Richard J. White
Lederle Laboratories
Pearl River, New York

Jeffrey L.C. Wright
National Research Council of Canada
Halifax, Canada

Morimasa Yagisawa
Japan Antibiotics Research Association
Tokyo, Japan

CONTENTS

PREFACE

Commercial production of antibiotics has been an established area of biotechnology for about 40 years. Over this period, it has undergone intensive development, and recent progress in the manipulation of genetic material promises even more striking advances in the years ahead. Genetic engineering in the antibiotic field depends on a sound knowledge of the way in which the drugs are biosynthesized. The genetic systems in producing microorganisms must also be understood. We need to know their scope and complexity, but we also require information about the location and function of genes that specify or control the enzymes for antibiotic biosynthesis. As secondary metabolites, antibiotics share a wide literature on biochemical and physiological aspects of the fermentation process. It, too, is a rich source of information, pointing the way to future development of antibiotic biotechnology.

This book surveys current knowledge in these areas. The first three chapters deal broadly with the genetics of the most important antibiotic producers and with the range of metabolic controls that affect antibiotic production. In each subsequent chapter, the biosynthesis of a commercially important group of antibiotics is related to genetic and metabolic control systems in the producing organisms. For some antibiotic groups, our knowledge extends little further than an appreciation of biosynthetic relationships, and, in general, much remains to be done before the genetic control of antibiotic production is well understood.

The aim of this book, in bringing the information together for the first time, is to provide an authoritative reference for those in industry, universities, and other institutions concerned with microbial biosynthesis of useful products. The book should be of particular value to those seeking to apply the new techniques of genetic engineering in this field. Because it not only considers the most recent advances but also provides the background for an overall appreciation of the subject, the book should also meet the needs of advanced students in applied microbiology and biochemistry classes.

LEO C. VINING

CHAPTER

1

Actinomycete Genetics and Antibiotic Production

David A. Hopwood

I. INTRODUCTION

There are several recent reviews that deal with the genetics of antibiotic production by actinomycetes (Hopwood and Merrick, 1977; Hopwood, 1979) and the possible involvement of plasmid-borne genes in some cases (Hopwood, 1978; Okanishi, 1979). In the few years since these summaries were compiled, the fragmentary knowledge of the number and organization of the genes concerned with the synthesis of even the most important antibiotics has not grown very greatly. However, there has been a significant increase in the versatility and scope of genetic analysis in antibiotic-producing actinomycetes (Chater and Hopwood, 1982), so a much more complete and incisive study of the subject is now possible. Most of this chapter is concerned with these methodological advances. In later sections, I review the progress that has been made, largely by use of the older genetic techniques, in understanding genetic control of antibiotic biosynthesis by certain actinomycetes. From this, we see that progress has been made in analyzing some examples where antibiotic biosynthesis is determined by clusters of chromosomal genes. As for plasmid involvement, methylenomycin A in *Streptomyces coelicolor* A3(2) (and now also in a related strain of *S. violaceus-*

ruber) remains the only well-established example of an antibiotic whose biosynthesis is determined by plasmid-borne structural genes, although tylosin in *Streptomyces fradiae* may belong in this category. The last couple of years have seen some further preliminary suggestions of plasmid involvement in antibiotic production but a larger number of cases in which such suggestions have later been questioned by the authors of the original claims. The subject is still open. The genetic and molecular tools are available for decisive analysis of specific examples of antibiotic synthesis; where they have not been used, conclusions should be viewed with caution.

This chapter is not primarily concerned with industrial applications of actinomycete genetics; its main preoccupation is with the *analysis* of genetic control of antibiotic biosynthesis rather than with strain construction. However, in the final section, some examples of antibiotic discovery through genetic recombination are discussed, partly to emphasize the latent potential of this approach and to show how understanding genetic control of antibiotic biosynthesis would enable such work to proceed in a much more certain fashion. There is no doubt that genetics has a considerable part to play in the rational or semirational development of higher-yielding strains of industrial antibiotic producers. This topic has been covered elsewhere (Hopwood and Chater, 1980) and I will not discuss it here, except to point out that empirical gene amplification by shotgun self-cloning on multicopy plasmid vectors, made possible by recent advances in recombinant DNA technology, is a realistic additional approach to this problem (Chater et al., 1982).

II. ROUTES TO GENETIC ANALYSIS IN ACTINOMYCETES

The actinomycetes, and particularly certain members of the genus *Streptomyces,* might now be regarded as the most versatile group of microorganisms in terms of the range of tools available for use in genetic analysis. In the following sections, some of the features of these genetic tools are outlined.

A. Natural Systems of Genetic Recombination

1. Plasmid-mediated Conjugation. Intrastrain genetic recombination in mixed cultures of auxotrophic mutants has been reported in many streptomycetes and in *Nocardia* (formerly *Streptomyces*) *mediterranea* and *Micromonospora* spp. (reviewed by Hopwood and Merrick, 1977). Recombination between strains of other *Nocardia* spp., now classified as *Rhodococcus,* has also been studied (Adams and Brownell, 1976; Brownell et al., 1981), but these organisms, like the related strains of *Mycobacterium* in which recombination has also been reported, are not known as antibiotic producers. Evidence that such recombination in *Streptomyces* and *N. mediterranea* is due to conjugation between hyphae rests primarily on genetic analysis, which showed that groups of distantly linked genes from each parent are frequently inherited together by recombinants; this result is expected for a conjugation process but not for recombination brought about by transduction or transformation. That this conjugation is (at least in part) plasmid-mediated has been demonstrated in *S. coelicolor* A3(2), *Streptomyces lividans* 66, and *Streptomyces rimosus*.

The wild-type *S. coelicolor* A3(2) harbors two autonomous sex plasmids, SCP1 and SCP2, which are responsible for yielding recombinants at frequencies up to about 10^{-5} of the total progeny of a mixed culture of two genetically marked strains (Bibb and Hopwood, 1981), since recombination between pairs of SCP1$^-$ SCP2$^-$ strains occurs only at frequencies of 10^{-7} or lower. Direct interaction of SCP1 with the chromosome, leading to chromosomal transfer, is shown by the behavior of strains carrying SCP1 integrated into the chromosome (NF strains and other donor types), which give extremely high levels (greater than 10^{-1}) of recombinants in crosses with SCP1$^-$ strains (Hopwood et al., 1973) as well as by the isolation of SCP1′ strains (Hopwood and Wright, 1976). SCP2 has been directly implicated in gene exchange through the discovery of SCP2* variants, which, though still present autonomously, enhance chromosomal recombination by a factor of at least 10^3 (Bibb et al., 1977; Bibb and Hopwood, 1981). It is possible that the "residual" recombination manifested in crosses between pairs of SCP1$^-$ SCP2$^-$ strains is also plasmid-mediated, since there are at least two further plasmids in *S. coelicolor* A3(2). SLP1 occurs as a DNA sequence integrated into the chromosome close to the *strA* locus; it becomes excised from the chromosome, together with contiguous chromosomal segments of variable length, and can be found as autonomous ccc DNA molecules of a series of sizes (SLP1.1, SLP1.2, etc.) after interspecific matings (or by in vitro DNA transfer) into *S. lividans* 66 (Bibb et al., 1981). SLP4 was also detected in *S. lividans* after matings between *S. coelicolor* A3(2) and *S. lividans* 66 derivatives. In this case, plasmid DNA was not detected; the plasmid was revealed by the fact that putative SLP4$^+$ *S. lividans* transconjugants inhibited the parent *S. lividans* culture by the so-called "lethal zygosis" reaction, a phenotype (see later) characteristic of conjugative *Streptomyces* plasmids (Bibb et al., 1977). Both SLP1 and SLP4 promote recombination of chromosomal genes in *S. lividans*. Whether or not one or both of these sex plasmids is responsible for the low-level recombination seen in SCP1$^-$ SCP2$^-$ matings of *S. coelicolor* A3(2) is not yet known, since strains lacking SLP1 or SLP4 have not been isolated.

In *S. lividans* 66 itself, the picture with respect to plasmid involvement in chromosomal recombination is somewhat simpler (T. Kieser, D. A. Hopwood, and H. M. Wright, unpublished results). Marked derivatives of the wild-type undergo recombination at a level of about 10^{-6}. This has been shown to be due to the presence of one, or perhaps two, plasmids. These were detected by the isolation, following protoplast formation and regeneration (see later), of strains sensitive to lethal zygosis by the parent strain. Two independent plasmids, SLP2 and SLP3, were revealed by the isolation of SLP2$^-$ SLP3$^+$ strains (sensitive to lethal zygosis by SLP2$^+$ but not by SLP3$^+$ strains), SLP2$^+$ SLP3$^-$ strains (sensitive to lethal zygosis by SLP3$^+$ but not SLP2$^+$ strains), and SLP2$^-$ SLP3$^-$ strains (sensitive to lethal zygosis by both plasmids). SLP2 certainly promotes chromosomal recombination, which occurs at a level of around 5×10^{-5} in SLP2$^+$ $\times$ SLP2$^-$ crosses. However, it is not certain that SLP3 can do so. In any event, the important finding is that pairs of SLP2$^-$ SLP3$^-$ strains yield no detectable recombinants (only occasional nonparental genotypes of classes explicable by reversion of single markers have been detected, rather than the more complex genotypes expected for most recombinants). It follows that all detectable recombination in *S. lividans* 66 is plasmid-mediated. SLP2$^-$ SLP3$^-$ strains therefore provide a useful background in which the effects of other plasmids can be studied (Kieser et al., 1982).

In *S. rimosus,* part of the recombination observed in pairwise crosses was shown to be mediated by a plasmid (SRP1), but a residual low level of recombination was observed in crosses between pairs of SRP1$^-$ strains (Friend et al., 1978). Recently, a second sex plasmid, SRP2, has been identified in these strains (E. J. Friend, personal communication); this is responsible for part of the recombination observed in SRP1$^-$ $\times$ SRP1$^-$ crosses.

Apart from plasmids native to *S. coelicolor* A3(2) and *S. lividans* 66, some further plasmids, originally detected in other strains, have been transferred to these two strains and found to promote chromosomal recombination. Notable is pIJ101, a multicopy, broad-host-range plasmid from *S. lividans* ISP5434, which is a very efficient sex plasmid, giving 1% or more of non-selected recombinants when present in either *S. coelicolor* A3(2) or *S. lividans* 66 (Kieser et al., 1982). This finding may be particularly significant because it suggests that this, or other broad-host-range plasmids, might be used to introduce efficient chromosomal recombination, and so a versatile genetic mapping capability, into strains that are originally found to give a few or no detectable recombinants.

The nature of the conjugation process in *Streptomyces* is almost unknown, although it is clear that it involves few plasmid-coded functions (many fewer than in gram-negative bacteria; Willetts, 1980) since small plasmids are conjugative. SLP2, SLP3, SLP4, SRP1, and SRP2 have not been detected physically and so are of unknown size; SCP1 is large, probably in excess of 170 kb (Westpheling, 1980), like a related sex plasmid, pSV1, which resembles SCP1 in coding for methylenomycin production and resistance (Aguilar and Hopwood, 1982). SCP2 is about 30 kb in size (Schrempf et al., 1975; Bibb et al., 1977; Schrempf and Goebel, 1977), but no more than half of it is required for conjugation (Bibb et al., 1980a). The size of the smallest SLP1 plasmid so far isolated, SLP1.6, is only 9.4 kb (Bibb et al., 1981), and pIJ101 is only 8.9 kb in size (Kieser et al., 1982). Moreover, derivatives of pIJ101 that are nonconjugative have deletions or insertions of DNA affecting only about 2 kb of the plasmid, suggesting that very few plasmid-borne genes are needed for conjugation.

2. Generalized Transduction. This is well documented in *Streptomyces venezuelae,* in which a phage called ϕSV1 has been found to promote transduction of a number of auxotrophic mutations to prototrophy (Stuttard, 1979).

3. Transformation of Competent Mycelia. Strains of *Streptomyces virginiae* and *Streptomyces kasugaensis,* were found to be unusual in going through a stage of competence for the uptake of linear fragments of DNA (Roelants et al., 1976), but these experiments involved radioactively labeled DNA, not genetic markers, and no system of genetic transformation was developed. In *Thermoactinomyces vulgaris,* a classical competence for transformation was found, and generalized transformation of chromosomal markers was studied (Hopwood and Wright, 1972). This thermophilic strain is probably not closely related to the *Streptomyces, Micromonospora,* and *Nocardia* spp. (conceivably, it represents a parallel development of the mycelial habit from a different eubacterial stock), and so does not provide a model for genetic manipulation of these mesophilic organisms, although it could perhaps be harnessed to analyze antibiotic production by *Thermoactinomyces* itself (Pirali et al., 1974).

B. Artificial Systems of Genetic Analysis

1. Generalized Recombination through Protoplast Fusion. Protoplast fusion is a means of bringing about intrastrain chromosomal recombination at very high frequency (Hopwood et al., 1977; Baltz, 1978; Ochi and Katz, 1978). Up to 20% of the total progeny arising from nonselective regeneration of fused protoplasts can be recombinant genotypes when fusion is carried out under optimal conditions in the presence of about 50% polyethyleneglycol (PEG) (reviewed by Hopwood, 1981a). In fusions between multiply marked strains, when crossing-over around the whole chromosome could be monitored, it was deduced that crossing-over can occur in widely separated map intervals, indicating a probable complete diploidy of the fusion bodies (Hopwood and Wright, 1978; Hopwood, 1981a); this contrasts with the situation in the mero-diploids generated by conjugation (Hopwood, 1967). Also, several rounds of recombination occur between fusion and recovery of the haploid progeny of regeneration (Hopwood and Wright, 1978). Both these factors lead to a loosening of linkage, making the determination of map locations for distant markers more difficult, though not impossible (Baltz, 1980).

2. Plasmid "Curing" by Protoplasting and Regeneration. As in *Staphylococcus aureus* (Novick et al., 1980), so in *Streptomyces* (Hopwood, 1981b; Aguilar and Hopwood, 1982), with some plasmids the simple expedient of forming and regenerating protoplasts can lead to a significant frequency of plasmid-free individuals. This effect has been suggested to occur at protoplast regeneration in *S. aureus* (Novick et al., 1980); in *Streptomyces,* loss at protoplast formation and regeneration have not been discriminated (Hopwood, 1981b).

3. Transformation/Transfection of Protoplasts by Plasmid or Phage DNA. When *Streptomyces* protoplasts are treated with plasmid or phage DNA in the presence of 20–25% PEG and are subsequently allowed to regenerate, transformants or transfectants are obtained (Bibb et al., 1978; Kruegel et al., 1980; Suarez and Chater, 1980a). At saturating DNA concentrations, up to 80% of regenerating protoplast colonies can be transformed with plasmid DNA (Bibb et al., 1980b). Phage/host systems vary in this respect; in some, over 1% of protoplasts are transfected, while in others a "competent" fraction, representing only about 10^{-4} of protoplasts, become transfected (Suarez and Chater, 1980a). At nonsaturating DNA concentrations, up to 10^7 transformants can be achieved per microgram of plasmid DNA (Thompson et al., 1982a), but only about 5×10^5 transfectants can be achieved per microgram of phage DNA. However, if positively charged DNA-free liposomes are added to the mixture of phage DNA and protoplasts and the PEG concentration is raised to that optimal for protoplast fusion (about 50%), at least 10^7 transfectants can be obtained per microgram of phage DNA (Rodicio and Chater, 1982). This treatment appears also to aid reproducible recovery of the maximum frequency of plasmid transformants, or may even raise it.

4. Gene Cloning on Plasmid or Phage Vectors. Use of the above transformation/transfection system allowed the development of DNA cloning procedures on endogenous *Streptomyces* plasmid and phage vectors (Bibb et al., 1980a; Suarez and Chater, 1980b; Thompson et al., 1980). SLP1.2 has provided the best low-copy-number plasmid-cloning vectors so far developed—namely, pIJ41 and pIJ61, which carry two antibiotic resistance markers, one for vector selection and the other allowing clone recognition by insertional inactivation (Thompson et al., 1982b). However, the possibly rather restricted host range of the SLP1 plasmids (*S. lividans* 66 and *Streptomyces reticuli* to date) may limit their utility as vectors for streptomycetes in general. SCP2 may have a slightly wider host range and has been used for cloning in *S. coelicolor* A3(2) (Bibb et al., 1980a; Thompson et al., 1982a), where SLP1-based vectors cannot be used because of incompatibility with the resident chromosomal SLP1 sequences, as well as in *S. lividans* 66 and *Streptomyces parvulus* ATCC 12434. The development of SCP2-based vectors is less advanced so far, but there is no insuperable obstacle to their refinement. Being multicopy (40–300 per chromosome) and of broad host range, pIJ101-based vectors are particularly useful as generalized vectors (Hopwood et al., 1981; Kieser et al., 1982). ϕC31 is being developed into a series of comparatively wide-host-range phage vectors (Chater et al., 1981a, b). Both this phage and the plasmids are available as bifunctional vectors for *Escherichia coli* and *Streptomyces*. The development and use of vectors for gene cloning in *Streptomyces* has recently been reviewed (Chater et al., 1982; Hopwood and Chater, 1982).

5. Transformation of Protoplasts by Chromosomal DNA Entrapped in Liposomes. The PEG-mediated transformation of protoplasts with naked DNA, which is so effective for plasmid and phage DNA, does not yield recombinants at detectable levels with chromosomal DNA (Makins and Holt, 1981). However, when the DNA is entrapped in liposomes and these are mixed with protoplasts and treated with PEG under conditions optimal for protoplast fusion, high frequencies of transformants are found for individual chromosomal markers. Since up to 10% transformants were found for each of a number of distantly linked markers and each marker was transformed independently, presumably nearly the whole of the recipient population was transformed for at least one marker (Makins and Holt, 1981).

6. Artificial Feeding of Metabolic Precursors Using Liposomes. Potential antibiotic precursors to which mycelia or protoplasts are impermeable might be introduced into *Streptomyces* by inclusion in liposomes followed by fusion with protoplasts (Hopwood, 1981b). This approach has not yet been reported for actinomycetes, but it was successfully used in the filamentous molds *Penicillium chrysogenum* and *Aspergillus nidulans* (Makins and Holt, 1982), in which penicillin nonproducing mutants were fed with the ACV tripeptide penicillin precursor.

III. USES OF GENETIC ANALYSIS IN THE STUDY OF ANTIBIOTIC PRODUCTION

A. Mapping of Chromosomal Genes

The best evidence that genes controlling the biosynthesis of an antibiotic are chromosomal rather than extrachromosomal is to map blocked mutants on a chromosomal linkage group. Mapping studies are also needed to locate genes controlling different biosynthetic steps, or regulatory genes, relative to each other. Thus, ideally, the means to carry out both gross mapping and fine mapping are required in any particular antibiotic producer of interest.

Gross mapping can be particularly straightforward in some streptomycetes. This is because crossing-over is relatively infrequent; that is, the linkage group is short in recombination units (only 260 centimorgans in *S. coelicolor* A3(2) compared with at least four times that value in *E. coli;* Hopwood, 1967), so a single cross involving as few as four markers can span the whole of the circular map. This has allowed a system of genetic analysis (the four-on-four method) to be developed in which crosses between two strains, each carrying two selectable alleles, are analyzed by selection of recombinant genotypes on four different media, each selecting one of the combinations of one selective marker from each parent (Hopwood, 1959). When this system is used, a single cross can produce an unambiguous circular map location for the four markers (Hopwood, 1972); thereafter, simpler crosses can be used to add further markers to the map (Hopwood, 1967). Although this system, using the natural recombination available in particular strains, has been employed for many years and was instrumental in the development of linkage maps in *S. rimosus* (Friend and Hopwood, 1971) *Streptomyces glaucescens* (Baumann et al., 1974) and *N. mediterranea* (Schupp et al., 1975), it can now be aided by two recent developments: the use of identified sex plasmids and protoplast fusion.

The utility of specific sex plasmids as mapping tools is illustrated by the case of *S. lividans* 66. Recombination occurs at low frequency in crosses between most normal derivatives of this organism (that is, in SLP2^{+} SLP3^{+} × SLP2^{+} SLP3^{+} crosses; see above), so the rare recombinants tend to be obscured by parental background growth on the necessarily heavily inoculated plates. In crosses of any of the types SLP2^{+} × SLP2^{-}, SLP1^{+} × SLP1^{-}, SCP2* × SCP2^{-}, or pIJ101^{+} × pIJ101^{-}, selective genetic analysis was straightforward (D. A. Hopwood and H. M. Wright, unpublished results). As was mentioned above, the use of pIJ101 as a very efficient sex plasmid is particularly significant, since its broad host range within *Streptomyces* may make it applicable in a variety of different strains.

Protoplast fusion has disadvantages for gross mapping because of the higher crossover frequencies observed compared with matings, which may lead to some ambiguity in map locations. The data of Hopwood and Wright (1978) for a fusion cross involving six markers of known map location can be analyzed retrospectively; the correct sequence of markers is revealed by minimizing the total number of crossovers, but the choice of sequence depends on small differences in the numbers of critical genotypes. In *S. fradiae,* a map for a dozen markers was constructed from the data of protoplast fusions; here, there was no map derived from conjugational crosses for comparison (Baltz, 1980). The biggest benefit of protoplast fusion in relation to gross mapping is probably in the development of multiply marked strains for subsequent conjugational analysis. Because of the very high recombinant frequencies attainable, nonselectable markers (typically auxotrophic mutations) can be combined by fusion (Hopwood et al., 1977). Thus a single successful auxotroph isolation, yielding a number of different classes of mutations, can be followed by several fusions to yield double or multiple auxotrophic strains, which can be used in conjugational analysis. By this strategy, it was possible rapidly to build up for *S. lividans* 66, a circular linkage map carrying about a dozen auxotrophic and resistance markers (D. A. Hopwood, T. Kieser, and H. M. Wright, unpublished results).

For mapping closely linked mutations, conjugational matings can be used, but they have poor resolving power (Hopwood and Chater, 1974). Protoplast fusion might be expected to be somewhat more efficient, because of the reduction in linkage resulting from multiple rounds of mating, but this has not been tested. In principle, the use of liposome-mediated transformation should be beneficial. Generalized transduction would also be useful, should it become widely available.

Localized mutagenesis would be a valuable adjunct to fine mapping. Several systems might be applicable for this purpose, of which co-mutation is one (Randazzo et al., 1976). Mutagenesis of transducing phages is another possibility (Stuttard, 1981), but there is no immediate prospect that systems will be widely available within the actinomycetes. Other, untried, possibilities involve the UV irradiation of mutagenized protoplasts of one parent before fusion with unirradiated protoplasts of another (Hopwood and Wright, 1981), or the use of liposome-mediated transformation with in vitro mutagenesis of the donor DNA (Makins and Holt, 1981). However, the most potentially versatile approach, to both fine structure analysis (including physical mapping) and localized mutagenesis, will become possible through gene cloning. Once a DNA segment "complementing" a mapped chromosomal mutation has been cloned, adjacent segments of the chromosome can be isolated and manipulated. For example, existing clones can be used as hybridization probes to find longer DNA sequences by the Southern (1975) technique. Once isolated, such cloned DNA can be subjected to in vitro mutagenesis or deletion and reintroduced into host organisms to study the phenotypic effects of such changes.

With the availability, actual or potential, of both efficient plasmid-mediated conjugation and versatile gene cloning systems for many streptomycetes of interest, the best approach to the analysis of chromosomal genes for antibiotic production would therefore seem to be gross mapping of a sample of mutations by conjugational matings (aided by protoplast fusion in the development of suitable parent strains), followed by gene cloning and its use to discover and characterize other nearby genes. The existing examples described in a later section (actinorhodin, rifamycin, oxytetracycline, and undecylprodigiosin) suggest very strongly that different steps in the same pathway of antibiotic synthesis will often turn out to be controlled by clusters of chromosomal genes; thus the above approach has a high probability of success.

B. The Implication of Plasmids in the Genetic Determination of Antibiotic Synthesis

The literature of the last few years reinforces the need for critical evidence before concluding that production of a particular antibiotic is controlled by plasmid-borne genes. Of the various classes of acceptable evidence discussed by Hopwood (1978) and Okanishi (1980), segregational data remain the most central; a demonstration that antibiotic production segregates in a cross independently of known chromosomal markers, and is therefore by definition extrachromosomal, fulfills the original criterion of a plasmid (Lederberg, 1952). Thus there is rarely an effective substitute for developing a system of chromosomal mapping in any antibiotic producer of interest; either some or all of the genes controlling antibiotic production will turn out to be chromosomal, in which case mapping becomes a prerequisite for further progress (see above), or the mapping attempts will reveal such genes to be extrachromosomal. Exceptions arise when, by infectious conjugal transfer or, better, by transformation, a plasmid coding for antibiotic production can be transferred to a strain already possessing a well-marked linkage map, which can then provide a background against which to study the phenotypic properties conferred by the foreign plasmid. An example is provided by pSV1, which occurs naturally in an unmarked strain of *Streptomyces violaceus-ruber* but can be transferred, by transformation or conjugation, to marked derivatives of *S. coelicolor* A3(2) or *S. lividans* 66 (Aguilar and Hopwood, 1982).

The widespread availability of protoplast transformation in streptomycetes means that transformation of an antibiotic nonproducing variant by ccc DNA of a plasmid thought to be implicated in antibiotic production in the parent strain is now a realistic direct route to the proof of such a hypothesis. Methods of plasmid isolation from streptomycetes are also being improved, so negative evidence, based on failure to detect plasmid DNA in antibiotic nonproducing variants, may have more significance. However, there are pitfalls, partly arising from the fact that nonproducing variants often differ pleiotropically from their producing parents and such changes may include factors affecting the successful isolation of ccc DNA molecules. Two recent examples illustrate this. Yagisawa et al. (1978) detected plasmid DNA in a neomycin-producing strain of *S. fradiae* but failed to find it in a neomycin-nonproducing variant (A083) and used this as evidence that plasmids might be implicated in neomycin biosynthesis. Later (Komatsu et al., 1981), plasmid DNA was found in A083. In *Streptomyces ambofaciens,* a plasmid was isolated from the spiramycin-producing parental strain but was not at first detected in nonproducing variants (Ōmura et al., 1979). Later, it was discovered that identical plasmid DNA could be extracted from the variants by an alternative procedure. In the variants, there appeared to be only about one copy of the plasmid DNA per chromosome, and this was associated with the chromosomal and membrane fractions of the mycelium, whereas the parental strain contained rather more plasmid DNA, some of it in a free intracellular form that could be recovered by a normal cleared lysate procedure (Ikeda et al., 1982).

In view of the ever-present doubt that must exist as to whether a putative plasmid-negative variant has really lost the plasmid or harbors it in a state that renders it harder to visualize than in the parent strain, the Southern hybridization technique should be extremely useful whenever enough plasmid DNA (or a fragment of it) can be obtained from the parent strain for in vitro labeling (either directly or after cloning) and used as a radioactive probe to detect the presence or absence of homologous sequences in the variant strain. Investigation of the origin of SLP1 plasmids from the *S. coelicolor* A3(2) chromosome (Bibb et al., 1981) provides a good example of the utility of the Southern technique. In difficult cases, it may be possible to make a useful probe by cloning a specific fragment of the plasmid in question on to a multicopy vector in *E. coli,* as in the example of pSV1 in *S. violaceus-ruber* SANK 95570 (Aguilar and Hopwood, 1982) (see later).

Infectious transfer of antibiotic production, at a frequency orders of magnitude higher than the frequency of chromosomal recombination between antibiotic-producing and nonproducing strains, can provide evidence of plasmid involvement in antibiotic production, but what is the significance of a negative result? Many naturally occurring *Streptomyces* plasmids do appear to be self-transmissible, including of course those mentioned above that have been shown to be capable of promoting chromosomal recombination. However, examples of non-self-transmissible plasmids have recently been provided by deletion derivatives of pIJ101 (Kieser et al., 1982). Three of these (pIJ102, pIJ103, and pIJ104) occur naturally in *S. lividans* ISP5434 along with pIJ101 itself, and others have been produced in vitro by deletion of (or insertion into) the short segment of pIJ101 needed for self-transfer (see above). These transfer-defective variants can be mobilized by unrelated conjugative plasmids present in the same strain (or in the recipient), but the frequency of mobilization is about the same as that of chromosomal genes in the same cross (Kieser et al., 1982); no *infectious* transfer of characters determined by the mobilized plasmid is therefore apparent, although lack of linkage of these characters with chromosomal genes is observed.

Recent experience has also emphasized the unreliability of "curing" evidence as an indication of plasmid involvement in antibiotic production. It is clear that "curing agents," such as acridines and ethidium bromide, are often more likely to cause loss or rearrangement of DNA segments than simple plasmid loss (Schrempf and Goebel, 1979; Komatsu et al., 1981). On the other hand, protoplast formation and regeneration, suggested as a "mild" means of effecting plasmid loss in the diagnosis of plasmid involvement in a phenotype (Hopwood, 1981b), does seem to be a useful diagnostic tool in some situations, for example in *S. violaceus-ruber* SANK 95570, in which many regenerant colonies had lost methylenomycin production and resistance together with the detectable DNA of the pSV1 plasmid (Aguilar and Hopwood, 1982).

Another useful discovery is that a very large number of *Streptomyces* conjugative plasmids elicit the reaction that has been called lethal zygosis (Bibb et al., 1977) by analogy (perhaps superficial) with the phenomenon of this name in *E. coli.* On an agar surface, a colony of a plasmid-carrying strain, surrounded by a confluent culture of the corresponding plasmid-free host, gives rise to a narrow zone of inhibition of this plasmid-free strain as the plasmid is transferred to and inhibits development of the latter. When individual plasmid-carrying spores or protoplasts develop in a plasmid-free lawn, they give rise to small circular inhibition zones called "pocks" (Bibb et al., 1978; Bibb and Hopwood, 1981). This pock reaction can provide good presumptive evidence that a phenotypic variant differs from its parent by plasmid loss, or at least by deletion of that part of the plasmid carrying the determinant(s) of resistance to lethal zygosis (Kieser et al., 1982). The lethal zygosis reaction can also be used to detect the presence of a hitherto unknown plasmid in a strain. In this case, individual colonies (perhaps derived by protoplasting and regeneration) are tested for sensitivity to lethal zygosis by the parent. This method led to the discovery of SLP2 and SLP3 in *S. lividans* 66 (see above).

IV. CHROMOSOMAL CLUSTERS OF BIOSYNTHETIC STRUCTURAL GENES

A. Rifamycin Genes in *N. mediterranea*

N. mediterranea ATCC 13685, derivatives of which are used in commercial production of rifamycin, shows the general features of recombination seen in *S. coelicolor* A3(2), and gives a similar circular linkage map (Schupp et al., 1975; Schupp, 1976). No direct evidence for plasmid involvement in the conjugation system has been obtained, nor are there strains displaying the markedly different fertility properties that might indicate different states in respect to one or more sex factors; most crosses yield recombinants at a frequency of about 10^{-5}. No progress in producing or regenerating protoplasts appears to have been reported, so the new methods of genetic manipulation used in *Streptomyces* are not yet available in the rifamycin producer.

Genetic analysis has revealed that six mutations (*ans-1* to *ans-6*) blocking the final step in rifamycin B synthesis, addition of glycolic acid via an ether linkage to the aromatic ansamycin nucleus of rifamycin S, map to the chromosome between *pro-1* and *str-2* (Schupp and Nüesch, 1979). In later work, mutations blocking two earlier steps, conversion of protorifamycin I to rifamycin W (*ans-13*) and conversion of rifamycin W to rifamycin S (*ans-11*), were also mapped to this same region of the linkage map, suggesting the existence of a cluster of chromosomal genes controlling successive steps in rifamycin biosynthesis (Ghisalba et al., 1982).

B. Oxytetracycline Genes in *S. rimosus*

Earlier reports (summarized by Hopwood, 1978) suggested that genetic control of oxytetracycline (OTC) biosynthesis by *S. rimosus* might differ in the three strains that have been studied in Moscow, Zagreb, and Sandwich (Pfizer). Although in each strain there was clear evidence of chromosomal linkage of many (presumably structural) gene mutations blocking OTC biosynthesis, there were some indications that a plasmid might be required for expression of the structural genes in the Russian strain, perhaps indirectly through an effect on OTC resistance; OTC^- OTC-sensitive variant strains frequently arose after treatment with "curing agents," and OTC resistance showed infectious transfer in $OTC^R \times OTC^S$ crosses. In the Pfizer strain, a similar class of variants occurred at high frequency, but they were found to be due to a chromosomal change. It was recently reported that extrachromosomal DNA of unknown size was present in the OTC^+ Russian strain and absent from the OTC^S variants (Borisoglebskaya et al., 1979), but in view of experience with other systems discussed earlier, this negative result may not be significant. The situation is best regarded as an example of genetic instability which could be elucidated by the techniques of DNA manipulation now available.

There has, however, been significant progress in the biochemical genetics of OTC^- blocked mutants in the Pfizer strain of *S. rimosus* (Rhodes et al., 1981). Fifty-seven nonleaky mutants, isolated from 40,000 mutagenized colonies, fell into nine groups on the basis of the cosynthesis of OTC when pairs of mutants were grown in mixed culture. In most cases the secretor or converter status of the mutations could not be determined, since the cosynthetic reactions were too weak when the mutants were separated even by a short distance on an agar surface. Nevertheless, it was clear that three representative mutations (*otc-19, -56,* and *-155*) called class 1 acted as converters when grown close to secretor mutations of the other six types, called class 2 (*otc-4, -20, -25, -75, -90,* and *-151*). Class 1 mutations blocked synthesis of a cofactor (CSF1) required catalytically for the last step in the OTC pathway, from 5a,11a-dehydro OTC to OTC, while class 2 mutations represented blocks in the main pathway of OTC synthesis itself. CSF1 was the material responsible for the strong cosynthetic reaction observed between class 1 and class 2 mutants.

By feeding precursors of OTC to the blocked mutants, the biochemical lesions in three class 2 mutations—*otcA25, otcC75,* and *otcD4*—were related to three of the four final steps of the biosynthesis (Fig. 1.1); mutations in the fourth step, a hydroxylation controlled by the hypothetical *otcB* gene, would not have been expected amongst *otc* mutations, since lack of this hydroxylation would not destroy the antibiotic activity of the molecule. The other three class-2 mutations, *otcX20, otcY90,* and *otcZ151,* were blocked at three of the five known early steps between malonyl-CoA and 4-aminodimethylaminoanhydrotetracycline, but the particular steps could not be identified.

Genetic mapping, using the plasmid-mediated conjugation system (Friend and Hopwood, 1971; Friend et al., 1978), showed that each of the nine classes of OTC^- mutations mapped to the chromosome; moreover, they

Malonyl-CoA

↓

Nonaketideamide

↓ methionine

otcX20 — 6-Methylpretetramid (6MPT)

otcY90 — ↓

otcZ151 — 4-Hydroxy-6-methylpretetramid (4-hydroxy-6MPT)

↓

4-Ketodedimethylaminoanhydrotetracycline (4-keto-ATC)

↓

4-Aminodedimethylaminoanhydrotetracycline (4-amino-ATC)

otcD4 — ↓ methionine

Anhydrotetracycline (ATC)

otcC75 — ↓

5a,11a-Dehydrotetracycline (DHTC)

(otcB) — ↓ O_2, NADPH

5a, 11a-Dehydrooxytetracycline (DHOTC)

otcA25 — ↓ CSF1, NADPH

Oxytetracycline (OTC)

FIGURE 1.1 Steps in the biosynthesis of oxytetracycline controlled by *otc* genes *A-D* and *X-Z; otcB* has not been identified by the isolation of a mutation (from Rhodes et al., 1981).

fell into two clusters, diametrically opposed on the circular linkage group. One cluster contained genes controlling the earliest four steps in the biosynthetic pathway (*otcD, X, Y,* and *Z*), while the other contained the later blocked mutations (*otcA* and *C*), together with the three types of class-1 mutations, called *csfA, B,* and *C*.

C. Actinorhodin and Undecylprodigiosin Genes in *S. coelicolor* A3(2)

Genetic studies of two chemically unrelated pigmented antibiotics synthesized by *S. coelicolor* A3(2), by the isolation and phenotypic characterization of blocked mutants followed by genetic mapping of representative mutations, showed unambiguously that structural genes for each antibiotic are chromosomally located and presumptively linked in a cluster. For actinorhodin, an isochromanequinone antibiotic, at least seven classes of mutations (*actI*-

VII) were mapped to the short segment between *hisD* and *guaA* (Rudd and Hopwood, 1979), while for the "red" antibiotic, five classes (*redA-E*) were mapped in the somewhat larger region between *cysD* and *leuB* (Rudd and Hopwood, 1980).

Attempts to define the biosynthetic steps blocked in these various mutants have so far made limited progress, although a sequence (not defined in terms of chemical compounds) accommodating most of the *act* classes was suggested by the results of cosynthesis studies (Rudd and Hopwood, 1979). Attempts to characterize the precursors responsible for such cosynthesis are underway (S. Prasatik and H. G. Floss, personal communication); they are hampered by the (not unexpected) finding that the major pigments accumulated by the mutants appear to be shunt products rather than true precursors of actinorhodin (B. A. M. Rudd, personal communication). Meanwhile, the point of dimerization of the two identical half molecules of actinorhodin, long unknown, is now known to be the C-10 carbon rather than C-9 (Gorst-Allman et al., 1981). Actinorhodin belongs to the opposite enantiomeric series from that to which similar isochromanequinone antibiotics such as granaticin (Brufani and Dobler, 1968; Zeeck and Christiansen, 1969), belong making detailed elucidation of the genetic and enzymological control of their synthesis a fascinating prospect.

The "red" antibiotic has been tentatively identified as undecylprodigiosin (S.-W. Lee, C.-J. Chang, and H. G. Floss, personal communication); a comparative study of the biochemical genetics of such compounds in *Streptomyces* and *Serratia* (prodigiosin itself is the well-known red pigment of *Serratia marcescens*) might be rewarding.

V. PLASMID-DETERMINED ANTIBIOTIC PRODUCTION

A. Plasmid-linked Biosynthetic Genes

1. Methylenomycin in *S. coelicolor* A3(2) and *S. violaceus-ruber* SANK 95570. Evidence that structural genes for methylenomycin A biosynthesis in *S. coelicolor* A3(2) are located on the SCP1 plasmid has been summarized elsewhere (Kirby and Hopwood, 1977; Hornemann and Hopwood, 1981). Westpheling (1980) obtained some physical information on SCP1, which had previously been characterized genetically. Visualization of the plasmid on agarose gels was poorly reproducible, but a size in excess of 120 Md was suggested. SCP1-specific DNA sequences were also detected in a very different approach involving either a positive selection for homologous DNA sequences in SCP1$^+$ *S. coelicolor* A3(2) and SCP1$^+$ *S. parvulus* ATCC 12434 (the general base sequence homology between their chromosomes is very low) or a negative selection for nonhomologous sequences between SCP1$^+$ *S. coelicolor* and SCP1$^-$ *S. coelicolor* (Hopwood et al., 1979). However, it was not possible to isolate SCP1 DNA, by either of these procedures, in a form suitable for DNA cloning. Recently, more progress has been made with another plasmid, pSV1. This was detected, by electron microscopy only (Okanishi et al., 1980), in the strain of *S. violaceus-ruber* SANK 95570 from which methylenomycin was first isolated (Haneishi et al., 1974). We found that pSV1 could regularly be revealed on agarose gels using a slightly modified version of Westpheling's lysis procedure. Its size was estimated as about 110 Md. That pSV1, like SCP1, codes for methylenomycin production and resistance was shown as follows.

1. After protoplast regeneration, about 15% of colonies of SANK 95570 had become methylenomycin-sensitive nonproducers and failed to show pSV1 DNA on gels.
2. A radioactive DNA probe, consisting of the *E. coli* vector pBR322 carrying the methylenomycin-resistance determinant cloned by Bibb et al. (1980a) from *S. coelicolor* A3(2), hybridized by the Southern technique specifically to pSV1.
3. *S. lividans* 66 and SCP1$^-$ strains of *S. coelicolor* A3(2) could be converted to methylenomycin production and resistance by transformation with DNA preparations containing pSV1 or by conjugation with SANK 95570; pSV1 could be detected in the transformants or transconjugants.

It is hoped that this successful isolation of pSV1 will soon lead to the cloning of some or all of the genes involved in methylenomycin biosynthesis.

2. Tylosin in *S. fradiae*. A detailed biochemical study of *S. fradiae* mutants blocked in the biosynthesis of this macrolide has recently been made (Baltz and Seno, 1981; Seno and Baltz, 1981). Out of 72 mutants, 59 were blocked in synthesis of the 16-membered lactone (tylactone); these were all designated *tylG* mutants, although it is very likely that they represented mutations in DNA controlling several different functions in a multienzyme complex (or multifunctional enzyme) analogous to a fatty acid synthetase. The remaining 13 mutants fell into eight classes (*tylA-F, tylH, I*), three of which were each defective in the synthesis or addition of one of the three sugar moieties to the lactone; one failed to add all three sugars; two lacked O-methyltransferases active at separate positions on the 6-deoxy-D-allose sugar residue after its addition to the lactone; and two each lacked the ability to oxidize a methyl group at a different position on the lactone.

Unfortunately, there is not yet available a detailed chromosomal linkage map for this strain of *S. fradiae,* though some linkage data were obtained by protoplast fusion (Baltz, 1980). Available evidence, summarized by Baltz et al. (1981), suggests that many structural genes for tylosin biosynthesis may be carried on a self-transmissible plasmid that also controls tylosin resistance, a situation analogous to that for methylenomycin involving SCP1 and pSV1. A tylosin nonproducing (Tyl$^-$), tylosin-sensitive (Tyl^S) variant

(JS85), obtained by protoplast regeneration, could be converted to tylosin production and resistance at high frequency (8%) by mixed growth with a Tyl^+ Tyl^R strain, under conditions in which chromosomal mutations did not recombine; infectious transfer thus occurred. The critical finding was that JS85 could be converted to Tyl^R by conjugation with blocked mutants (Tyl^R Tyl^-), in which case the presumptive transconjugants remained Tyl^-, as expected, and often had the specific phenotype of the blocked mutant used as donor of the Tyl^R character (*tylA, tylB,* etc., as appropriate). However, they sometimes differed from the donor in failing to secrete the expected tylosin precursor, contradicting a hypothesis of simple plasmid transfer and necessitating the postulation of features such as plasmid dissociation. Clearly, more work is needed to clarify this interesting situation.

B. More Doubtful Cases of Plasmid Control

Numerous publications dealing with the possible or probable involvement of plasmids (in some cases the plasmids themselves not unambiguously defined) were summarized in earlier reviews (Hopwood, 1978, 1979; Chater and Hopwood, 1982) and need not be cited again.

1. Aureothricin in *S. kasugaensis* An earlier suggestion that a plasmid might control the synthesis of the pyrrothine moiety of aureothricin in strain M338 of *S. kasugaensis,* with chromosomal genes controlling propionylation of pyrrothine, has been disproved (Okanishi, 1980). Further experiments have been done with another strain of this species (MB273), which contains plasmid DNA of two sizes, 6.8 Md (pSK1) and 14 Md (Okanishi, 1980). An aureothricin nonproducing variant, 18a, lacked 6.8 Md plasmid DNA but still produced aureothricin on agar media supplemented with oleate, palmitate, or L-cystine. From this finding, it was argued that pSK1 might have an indirect effect on antibiotic production by an involvement in cell membrane function. The significance of later findings by Toyama et al. (1981) that the 6.8-Md plasmid DNA of strain MB273 actually consists of two molecular species (pSK1 and pSK2) with different restriction endonuclease cleavage maps remains to be seen.

2. Streptomycin in *S. griseus*. Zhuang et al. (1980) found that a streptomycin nonproducing variant of *Streptomyces griseus,* lacking a plasmid, could be transformed with plasmid DNA by using as a selective marker a natural lincomycin resistance apparently determined by the plasmid. Transformation was not achieved by the use of protoplasts and PEG but instead made use of mycelia grown to early log phase and treated with DNA in the presence of 10^{-6}M Ca^{2+}. Under these conditions, a transformation frequency of about 10^{-5} could be achieved for the lincomycin resistance marker. The transformants failed to regain aerial mycelium production, which the recipient strain lacked, but many of them (66%) did regain the ability to produce streptomycin, leading to a conclusion that the plasmid was involved in some way with streptomycin biosynthesis.

3. Amrumycin in *S. coelicolor* A3(2). Troost et al. (1979) made the surprising claim that the SCP2 plasmid of *S. coelicolor* A3(2) controls synthesis of at least three antibacterial substances, two described as bacteriocins (streptocins A and B) and the other as a small molecular weight substance (amrumycin). Evidence for the existence of three separate substances rested on antimicrobial spectra, times of onset of inhibition with age of the producing culture, and dialyzability, rather than on chemical criteria. Studies in my laboratory have failed to find reproducible evidence for antimicrobial activity unambiguously associated with the presence of SCP2; inhibition reactions of certain *S. coelicolor* A3(2) strains by others was sometimes observed, but use of several different pairs of isogenic strains, with and without SCP2, showed that this was unrelated to SCP2 status (Bibb and Hopwood, 1981). Moreover, some of the cultures described as SCP1$^-$ by Troost et al. (1979), and kindly supplied to us for testing, showed fertility properties and methylenomycin production characteristic of SCP1-containing strains (J. A. Ewing, personal communication). At least one had the Dag$^-$ phenotype characteristic of NF strains (Hodgson and Chater, 1981); thus the fertility properties attributed to SCP2 variants by Troost et al. (1979) might well have been due to SCP1. The existence of amrumycin and the two streptocins, and particularly their relationship to SCP2, must therefore be regarded as unproven at the present time.

VI. PRODUCTION OF NEW ANTIBIOTICS BY RECOMBINATION

Two different strategies have been discussed for the use of genetic recombination in the discovery of new antibiotics, the "silent gene" and the "hybrid antibiotic" approaches. In the first, the idea is to reveal an antibiotic activity that is within the genetic capacity of a single strain to produce, but which is not normally detected because of overefficient repression; recombination with a second strain lacking the same regulatory genes may have the desired result. In the second, the object is to generate a recombinant strain possessing a new combination of structural genes, and therefore of biosynthetic enzymes, able to cooperate to produce a novel chemical structure. Although the two situations may overlap in practice, and indeed it may be hard to prove which has led to detection of a new antibiotic, it can be argued that random recombination of rather closely related strains by protoplast fusion may be the best approach to unmasking "silent genes," while gene cloning between more distant relatives may be best for generating "hybrid" antibiotics (Hopwood 1981c, d). An additional way of destabilizing the genetic regulation of a strain may be randomly to clone DNA segments, in multiple copies, back into the same strain (Chater et al., 1982).

It is too early for the strategies of protoplast fusion and gene cloning to have been evaluated in relation to antibiotic discovery. However, there are at least three examples of recombination by the natural conjugational route leading to discovery of antibiotics.

The first concerns matings between mutants of *Streptomyces hygroscopicus* blocked in production of the macrolide turimycin and mutants of *Streptomyces violaceus* blocked in the synthesis of the anthracycline violamycin (Schlegel and Fleck, 1980). These crosses were attempted following positive cosynthesis tests in which the *S. hygroscopicus* mutant acted as secretor to the *S. violaceus* mutant, producing a new activity. In the interspecific crosses, between a methionine auxotroph of *S. hygroscopicus* carrying streptomycin resistance and a streptomycin-sensitive prototroph of *S. violaceus,* streptomycin resistant prototrophs arose at a frequency of 5×10^{-5}. Out of 90 recombinants producing some biologically active material, three yielded a red-pigmented (presumably anthracycline) antibiotic, and one of these was studied further as the producer of a new anthracycline called iremycin. Iremycin was subsequently shown (Ihn et al., 1980) to have a γ-rhodomycinone aglycone linked at C-10 to a L-rhodosamine sugar moiety; it is probably identical to γ-rhodomycin I described earlier. The intriguing question arises as to the source(s) of the genetic information leading to iremycin production. It seems likely that information for the biosynthesis of the aglycone comes from the *S. violaceus* parent, even though this particular anthracyclinone is not produced by it. The sugar, L-rhodosamine, occurs in the violamycins produced by *S. violaceus*. Thus one is tempted to postulate that iremycin arose by the unmasking of one or more "silent genes" in this species. However, this would not explain the interspecific cosynthesis that led to the discovery of iremycin.

The second example involved a cross between two aminoglycoside producers (Mazières et al., 1981). An auxotrophic mutant of a strain of *S. rimosus* forma *paromomycinus,* producing mainly paromomycin with a trace of neomycin, was crossed with an auxotroph of *Streptomyces kanamyceticus* producing kanamycin. Prototrophic recombinants arose at low frequency ($<10^{-6}$), and several of them yielded large quantities of neomycin. This example has every appearance of an altered regulation of antibiotic production brought about through recombination.

The same conclusion applies to the third, striking, example in which two derivatives of the same strain of *N. mediterranea* were crossed (Schupp et al., 1981). One (T104) carried three auxotrophic markers and streptomycin resistance but was a wild type for the rifamycin pathway, yielding the end product, rifamycin B. The other strain (T191) was a blocked mutant (*ans-11*) used in the mapping studies cited earlier; it could not convert rifamycin

W to rifamycin S, the immediate precursor of rifamycin B. Recombinants inheriting at least one of the three prototropic alleles of T191 and streptomycin resistance from T104 were screened for production of new rifamycins; two out of 290 were producers, and one of these, prototroph R-21, was studied in detail (Traxler et al., 1981). It was found to produce seven novel rifamycins, of which 3-hydroxyrifamycin S and 3,31-dihydroxyrifamycin S, with hydroxyl groups at position 3 of the chromophore, were particularly interesting. The wealth of new biosynthetic capabilities displayed by recombinant R-21 is a remarkable example of changed regulation brought about by recombination, even between two strains differing only in a number of laboratory-induced mutations. The exact number of mutational steps separating the two parents of R-21 is uncertain, even though the strain pedigree was published, since one parent was itself a recombinant.

These three examples, performed without benefit of the greatly increased power of recombination becoming available through the use of protoplasts in various modes, augur well for the future exploitation of recombination in antibiotic discovery.

REFERENCES

Adams, J. N., and Brownell, G. H. (1976) in *The Biology of the Nocardiae* (Goodfellow, M., Brownell, G. H., and Serrano, J. A., eds.) pp. 285–309, Academic Press, London.

Aguilar, A., and Hopwood, D. A. (1982) *J. Gen. Microbiol. 128,* 1893–1901.

Baltz, R. H. (1978) *J. Gen. Microbiol. 107,* 93–102.

Baltz, R. H. (1980) *Dev. Ind. Microbiol. 21,* 43–54.

Baltz, R. H., and Seno, E. T. (1981) *Antimicrob. Agents Chemother. 20,* 214–225.

Baltz, R. H., Seno, E. T., Stonesifer, J., Matsushima, P., and Wild, G. M. (1981) in *Microbiology—1981* pp. 371–375, American Society for Microbiology, Washington, D.C.

Baumann, R., Hütter, R., and Hopwood, D. A. (1974) *J. Gen. Microbiol. 81,* 463–474.

Bibb, M. J., and Hopwood, D. A. (1981) *J. Gen. Microbiol. 126,* 427–442.

Bibb, M. J., Freeman, R. F., and Hopwood, D. A. (1977) *Mol. Gen. Genet. 154,* 155–166.

Bibb, M. J., Ward, J. M., and Hopwood, D. A. (1978) *Nature, London 274,* 398–400.

Bibb, M. J., Schottel, J. L., and Cohen, S. N. (1980a) *Nature, London 284,* 526–531.

Bibb, M. J., Ward, J. M., and Hopwood, D. A. (1980b) *Dev. Ind. Microbiol. 21,* 55–64.

Bibb, M. J., Ward, J. M., Kieser, T., Cohen, S. N., and Hopwood, D. A. (1981) *Mol. Gen. Genet. 184,* 230–240.

Borisoglebskaya, A. N., Perebityuk, A. N., and Boronin, A. M. (1979) *Antibiotiki 12,* 883–888.

Brownell, G. H., Enquist, L. W., and Denniston-Thompson, K. (1981) in *Actino-*

mycetes (Schaal, K. P., and Pulverer, G., eds.) pp. 363–375, *Zbl. Bakt. Suppl. 11,* Gustav Fischer, Stuttgart, West Germany.

Brufani, M., and Dobler, M. (1968) *Helv. Chim. Acta 51,* 1269–1275.

Chater, K. F., and Hopwood, D. A. (1982) in *The Biology of the Actinomycetes* (Goodfellow, M., Mordarski, M., and Williams, S. T., eds.), Academic Press, London. (in press)

Chater, K. F., Bruton, C. J., Suarez, J. E., and Springer, W. (1981a) in *Microbiology—1981,* pp. 380–383, American Society for Microbiology, Washington, D.C.

Chater, K. F., Bruton, C. J., Springer, W., and Suarez, J. E. (1981b) *Gene 15,* 249–256.

Chater, K. F., Hopwood, D. A., Kieser, T., and Thompson, C. J. (1982) *Curr. Topics Microbiol. Immunol. 96,* 69–95.

Friend, E. J., and Hopwood, D. A. (1971) *J. Gen. Microbiol. 68,* 187–197.

Friend, E. J., Warren, M., and Hopwood, D. A. (1978) *J. Gen. Microbiol. 106,* 201–206.

Ghisalba, O., Auden, J. A. L., Schupp, T., and Nuësch, J. (1982) in *Antibiotic Production by Fermentation Biotechnology* (Vandamme, E. J., ed.), Marcel Dekker, New York. (in press)

Gorst-Allman, C. P., Rudd, B. A. M., Chang, C., and Floss, H. G. (1981) *J. Org. Chem. 46,* 455–456.

Haneishi, T., Terahara, A., Arai, M., Hata, T., and Tamura, C. (1974) *J. Antibiot. 27,* 393–399.

Hodgson, D. A., and Chater, K. F. (1981) *J. Gen. Microbiol. 124,* 339–348.

Hopwood, D. A. (1959) *Ann. N.Y. Acad. Sci. 81,* 887–898.

Hopwood, D. A. (1967) *Bacteriol. Rev. 31,* 373–403.

Hopwood, D. A. (1972) in *Methods in Microbiology* (Norris, J. R., and Ribbons, D. W., eds.) Vol. 7B, pp. 29–158, Academic Press, London.

Hopwood, D. A. (1978) *Ann. Rev. Microbiol. 32,* 373–392.

Hopwood, D. A. (1979) *J. Nat. Products 42,* 596–602.

Hopwood, D. A. (1981a) *Ann. Rev. Microbiol. 35,* 237–272.

Hopwood, D. A. (1981b) in *Actinomycetes* (Schaal, K. P., and Pulverer, G., eds.) pp. 523–531, *Zbl. Bakt. Suppl. 11,* Gustav Fischer, Stuttgart, West Germany.

Hopwood, D. A. (1981c) in *Beta-lactam Antibiotics* (Salton, M. R., and Shockman, G. D., eds.) pp. 585–598, Academic Press, New York.

Hopwood, D. A. (1981d) in *The Future of Antibiotherapy and Antibiotic Research* (Ninet, L., Borst, P. E., Bouanchaud, D. H., and Florent, J., eds.) pp. 407–416, Academic Press, London.

Hopwood, D. A., and Chater, K. F. (1974) in *Handbook of Genetics* (King, R. C., ed.) Vol. 1, pp. 237–255, Plenum, New York.

Hopwood, D. A., and Chater, K. F. (1980) *Phil. Trans. Roy. Soc. B 290,* 313–328.

Hopwood, D. A., and Chater, K. F. (1982) in *Genetic Engineering* (Setlow, J. K., and Hollaender, A., eds.) Vol. IV, Plenum, New York. (in press)

Hopwood, D. A., and Merrick, M. J. (1977) *Bacteriol. Rev. 41,* 595–635.

Hopwood, D. A., and Wright, H. M. (1972) *J. Gen. Microbiol. 71,* 383–398.

Hopwood, D. A., and Wright, H. M. (1976) *J. Gen. Microbiol. 95,* 107–120.

Hopwood, D. A., and Wright, H. M. (1978) *Mol. Gen. Genet. 162,* 307–317.

Hopwood, D. A., and Wright, H. M. (1981) *J. Gen. Microbiol. 126,* 21–27.

Hopwood, D. A., Chater, K. F., Dowding, J. E., and Vivian, A. (1973) *Bacteriol. Rev. 37,* 371–405.

Hopwood, D. A., Wright, H. M., Bibb, M. J., and Cohen, S. N. (1977) *Nature, London 268,* 171–174.

Hopwood, D. A., Bibb, M. J., Ward, J. M., and Westpheling, J. (1979) in *Plasmids of Medical, Environmental and Commercial Importance* (Timmis, K. N., and Pühler, A., eds.) pp. 245–258, Elsevier/North Holland, Amsterdam.

Hopwood, D. A., Thompson, C. J., Kieser, T., Ward, J. M., and Wright, H. M. (1981) in *Microbiology—1981,* pp. 376–379, American Society for Microbiology, Washington, D.C.

Hornemann, U., and Hopwood, D. A. (1981) in *Antibiotics Vol. IV, Biosynthesis* (Corcoran, J. W., ed.) pp. 123–131, Springer, Berlin.

Ihn, W., Schlegel, B., Fleck, W. F., and Sedmera, P. (1980) *J. Antibiot. 33,* 1457–1461.

Ikeda, H., Tanaka, H., and Omura, S. (1982) *J. Antibiot. 35,* 497–516.

Kieser, T., Hopwood, D. A., Wright, H. M., and Thompson, C. J. (1982) *Mol. Gen. Genet. 185,* 223–238.

Kirby, R., and Hopwood, D. A. (1977) *J. Gen. Microbiol. 98,* 239–252.

Komatsu, K., Leboul, J., Harford, S., and Davies, J. (1981) in *Microbiology—1981,* pp. 384–387, American Society for Microbiology, Washington, D.C.

Kruegel, H., Fiedler, G., and Noack, D. (1980) *Mol. Gen. Genet. 177,* 297–300.

Lederberg, J. (1952) *Physiol. Rev. 32,* 403–430.

Makins, J. F., and Holt, G. (1981) *Nature, London 293,* 671–673.

Makins, J. F., and Holt, G. (1982) *J. Chem. Tech. and Biotechnol. 32,* 347–353.

Mazières, N., Peyre, M., and Penasse, L. (1981) *J. Antibiot. 34,* 544–550.

Novick, R., Sanchez-Rivas, C., Gruss, A., and Edelman, I. (1980) *Plasmid 3,* 348–358.

Ochi, K., and Katz, E. (1978) *J. Antibiot. 31,* 1143–1148.

Okanishi, M. (1979) in *Genetics of Industrial Microorganisms* (Sebek, O. K., and Laskin, A. I., eds.) pp. 134–140, American Society for Microbiology, Washington, D.C.

Okanishi, M. (1980) in *Molecular Breeding and Genetics of Applied Microorganisms* (Sakaguchi, K., and Okanishi, M., eds.) pp. 29–46, Academic Press, New York.

Okanishi, M., Manome, T., and Umezawa, H. (1980) *J. Antibiot. 33,* 88–91.

Ōmura, S., Ikeda, H., and Kitao, C. (1979) *J. Antibiot. 32,* 1058–1060.

Pirali, G., Somma, S., Lancini, G. C., and Sala, F. (1974) *Biochim. Biophys. Acta 366,* 310–318.

Randazzo, R., Sciandrello, G., Carere, A., Bignami, M., Velcich, A., and Sermonti, G. (1976) *Mut. Res. 36,* 291–302.

Rhodes, P. M., Winskill, N., Friend, E. J., and Warren, M. (1981) *J. Gen. Microbiol. 124,* 329–338.

Rodicio, M. R., and Chater, K. F. (1982) *J. Bacteriol. 151,* 1078–1085.

Roelants, P., Konvalinkova, V., Mergeay, M., Lurquin, P. K., and Ledoux, L. (1976) *Biochim. Biophys. Acta 442,* 117–122.

Rudd, B. A. M., and Hopwood, D. A. (1979) *J. Gen. Microbiol. 114,* 35–43.

Rudd, B. A. M., and Hopwood, D. A. (1980) *J. Gen. Microbiol. 119,* 333–340.

Schlegel, B., and Fleck, W. F. (1980) *Z. Allg. Mikrobiol. 20,* 527–530.

Schrempf, H., and Goebel, W. (1977) *J. Bacteriol. 131,* 251–258.

Schrempf, H., and Goebel, W. (1979) in *Plasmids of Medical, Environmental and Commercial Importance* (Timmis, K. N., and Pühler, A., eds.) pp. 259–268, Elsevier/North Holland, Amsterdam.

Schrempf, H., Bujard, H., Hopwood, D. A., and Goebel, W. (1975) *J. Bacteriol. 121,* 416–421.
Schupp, T. (1976) in *Second International Symposium on the Genetics of Industrial Microorganisms* (Macdonald, K. D., ed.) pp. 531–533, Academic Press, London.
Schupp, T., and Nuësch, J. (1979) *FEMS Microbiol. Lett. 6,* 23–27.
Schupp, T., Hütter, R., and Hopwood, D. A. (1975) *J. Bacteriol. 121,* 128–136.
Schupp, T., Traxler, P., and Auden, J. A. L. (1981) *J. Antibiot. 34,* 965–970.
Seno, E. T., and Baltz, R. H. (1981) *Antimicrob. Agents Chemother. 20,* 370–377.
Southern, E. M. (1975) *J. Mol. Biol. 98,* 503–517.
Stuttard, C. (1979) *J. Gen. Microbiol. 110,* 479–482.
Stuttard, C. (1981) in *Actinomycetes* (Schaal, K. P., and Pulverer, G., eds.) pp. 533–537, *Zbl. Bakt. Suppl. 11,* Gustav Fischer, Stuttgart, West Germany.
Suarez, J. E., and Chater, K. F. (1980a) *J. Bacteriol. 142,* 8–14.
Suarez, J. E., and Chater, K. F. (1980b) *Nature, London 286,* 527–529.
Thompson, C. J., Ward, J. M., and Hopwood, D. A. (1980) *Nature, London 286,* 525–527.
Thompson, C. J., Ward, J. M., and Hopwood, D. A. (1982a) *J. Bacteriol. 151,* 668–677.
Thompson, C. J., Ward, J. M., Kieser, T., and Hopwood, D. A. (1982b) *J. Bacteriol.* (in press)
Toyama, H., Okanishi, M., and Umezawa, H. (1981) *Plasmid 5,* 306–312.
Traxler, P., Schupp, T., Fuhrer, H., and Richter, W. J. (1981) *J. Antibiot. 34,* 971–979.
Troost, T. R., Danilenko, V. N., and Lomovskaya, N. D. (1979) *J. Bacteriol. 140,* 359–368.
Westpheling, J. (1980) Ph.D. thesis, University of East Anglia, Norwich.
Willetts, N. (1980) *Ann. Rev. Genet. 14,* 41–76.
Yagisawa, M., Huang, T-S. R., and Davies, J. E. (1978) *J. Antibiot. 31,* 809–813.
Zeeck, A., and Christiansen, P. (1969) *Liebigs Ann. Chem. 724,* 172.
Zhuang, Z., Tan, H., Xu, Y., Zhu, Y., and Xue, Y. (1980) *Acta Genet. Sinica 7,* 291–298.

CHAPTER

2

Fungal Genetics and Antibiotic Production

Kenneth D. Macdonald

I. INTRODUCTION

About four decades have passed since antibiotics began to be produced industrially and, from the beginning, the genetic technique of mutagenic treatment followed by selection has been employed to raise antibiotic yield. In the early days, breeding methods based on the exchange of genetic information between strains or its transfer from one to another were simply not feasible because of the lack of genetic knowledge. Later, when it was theoretically possible to consider breeding for increased yield via systems of genetic exchange, the titers of antibiotics had been raised sufficiently that the commercial breeder was not under pressure to exploit these new breeding systems. The relatively unsophisticated method of mutation and selection continued to be used and is still used almost exclusively in industry.

Nevertheless, the possibilities for industrial exploitation of genetic exchange processes in fungi have been studied, and some successes have been reported (Elander, 1967; Ball, 1973a, b; Calam et al., 1976; Hamlyn and Ball, 1979). It would seem clear that, where sexual or parasexual breeding systems are available, continued efforts will be made to employ these in conjunction with mutation and selection to improve industrial strains of fungi. What use will be made of genetic engineering methods remains to be seen.

Fungi elaborate a number of antibiotics, the most important medically being the penicillins, the cephalosporins, and griseofulvin, which are manufactured industrially by *Penicillium chrysogenum, Cephalosporium acremonium,* and *Penicillium patulum. Cephalosporium acremonium* has been renamed *Acremonium chrysogenum,* and *P. patulum* has been renamed *Penicillium griseofulvum* (Pitt, 1979), but the former names of each of these two fungi will be retained in this paper. Most published work on the genetics of antibiotic production in fungi has been concerned with penicillin production in *P. chrysogenum,* but studies on the genetics of this antibiotic have also been carried out in the fungus *Aspergillus nidulans,* which was used as a model because its formal genetics is so well known. Therefore the genetics of antibiotic production in fungi will be discussed mainly in relation to *P. chrysogenum* and *A. nidulans,* although reference will also be made to studies with *C. acremonium* (Treichler et al., 1978; Hamlyn and Ball, 1979). From the preliminary work of Calam et al. (1973), it appears likely that *P. patulum* possesses a parasexual mechanism, although difficulties were experienced in isolating heterozygous diploids and recombinants.

This chapter will begin with a discussion of the methods of genetic analyses available in molds elaborating antibiotics, and then work on the genetics of antibiotic production in the various fungi will be reviewed.

II. METHODS OF GENETIC ANALYSIS

A. Genetic Analysis Via the Sexual Cycle

In this section, a brief outline of the life cycle of *A. nidulans* (Eidam, Winter) will be given with a summary of methods that have been used to estimate linkage following meiotic processes of recombination.

This fungus is an ascomycete (order: Plectascineae, family: Aspergillaceae); like most filamentous fungi that reproduce sexually, it has a vegetative cycle that occurs side by side with a sexual cycle. The asexual spore (conidium), which contains a single haploid nucleus, germinates under suitable conditions to produce colorless septate hyphae with multinucleate cells, some of which form foot cells that differentiate to produce multinucleate conidiophores. These end in globose multinucleate vesicles from the surface of which protrude uninucleate sterigmata, which in turn give rise to two or more secondary sterigmata. The latter remain uninucleate but bud off chains of uninucleate conidia by repeated nuclear division; the youngest conidium is therefore at the base of the chain.

The sexual spores (ascospores) are concentrated in specialized spherical structures termed cleistothecia. A cleistothecium contains up to 10,000 spherically shaped asci, and each ascus holds eight ascospores not arranged in a linear order. It appears likely that just two haploid nuclei multiplying by conjugate division give rise to all the ascus initials in one cleistothecium. The ascus primordium nucleus is diploid as the result of fusion of two haploid nuclei that were dividing conjugately. Following meiosis, the four products divide mitotically. Eight ascospores are produced that are binucleate, so presumably a second mitotic division takes place before their maturation. *Aspergillus nidulans* is homothallic; thus cleistothecia can arise from the growth of a single spore, unlike heterothallic fungi where + and − strains exist and have to come together to produce sexual fruiting bodies (Pontecorvo et al., 1953).

Detailed genetic maps of *A. nidulans* are now available (Clutterbuck, 1974), so when new mutations arise, they can be positioned relative to others on one of the eight chromosomes that *A. nidulans* is known to have. The first step is usually to assign the mutation to a particular chromosome using parasexual haploidization analysis, which will be discussed in detail later. Next, the mutation can be located in relation to others on the chromosome to which it has been assigned following a sexual cross. When sexual crosses are set up between different haploid parents, a preliminary step is the establishment of a heterokaryon, which is a mycelium containing both types of genetically distinct nuclei. Methods of synthesizing heterokaryons will be described later. Following growth in a heterokaryon and assuming random distribution of nuclei, it would be expected that the ratio of selfed to hybrid cleistothecia would be one to one, but in analyzing such a cross, a random sample of ascospores can be isolated from hybrid cleistothecia by selecting those showing recombination for a pair of genetic markers introduced separately. Another method is to select the ascospores for genetic analysis from hybrid cleistothecia. These can be recognized if suitable complementary conidial color mutations are present in each parent. A sample of ascospores streaked on agar from a selfed cleistothecium will show one or other parental color while that from a hybrid cleistothecium will show both parental colors as well as the wild-type green color.

The measurement of linkage is illustrated by the example given in Table 2.1.

B. Genetic Analysis Via the Parasexual Cycle

The parasexual cycle was first demonstrated in *A. nidulans* (Pontecorvo and Roper, 1953) but was later shown to occur in a number of fungi, including *P. chrysogenum* (Pontecorvo and Sermonti, 1953). The latter fungus has an asexual life cycle similar to that of *A. nidulans*, but it lacks a sexual cycle. The importance of detecting a parasexual cycle in *P. chrysogenum* was that it opened the way for breeding by methods of genetic exchange in this asexual industrial organism where strain improvement had been confined previously to the use of mutation and selection.

TABLE 2.1 Illustration of the measurement of linkage between two recessive mutations, *a* and *b*, introduced separately on homologous chromosomes in a sexual cross. Cross between haploid parents of genotype *aB* and *Ab*.

Genotypes of ascospore progeny		*Number of ascospore progeny*
Parental	*aB*	44
Recombinant	*ab*	2
Recombinant	*AB*	2
Parental	*Ab*	44

$$\text{Percentage recombination} = \frac{\text{numbers of } ab + AB}{\text{numbers of } aB + Ab} \times 100$$

$$= 4\%$$

There is 4% recombination between *a* and *b* or a recombinant fraction of 0.04. In map distances, *a* and *b* are 4 map units or 4 centimorgans apart.

1. Steps in the Parasexual Cycle. The steps in the parasexual cycle are as follows.

1. Synthesis of a heterokaryon where haploid nuclei of two parent strains share the same cytoplasm.
2. Fusion of haploid nuclei, one of each parental type, to yield a heterozygous diploid. This happens rarely, and the precise mechanism is unknown; but haploid nuclei lying near to each other probably coalesce during mitosis. Formation of a heterozygous diploid corresponds to formation of a zygote in the sexual cycle. However, heterozygous diploids are not formed in specialized fruiting bodies; they multiply side by side with haploid nuclei in the heterokaryon, and diploid nuclei can be formed.
3. Heterozygous diploids break down by one of two separate processes.
   a. *Mitotic crossing-over.* An exchange of genetic material can occur between homologous chromosomes during mitosis, producing diploids that are homozygous for loci distal to the crossover region, that is, further from the centromere.
   b. *Nondisjunction.* During mitosis of a heterozygous diploid ($2n$), one centromere may fail to divide and give rise to nuclear products of $2n + 1$ and $2n - 1$ constitution. Neither is stable. The former reverts to the diploid state ($2n$), either remaining heterozygous for all chromosomes or becoming homozygous for one chromosome. The latter type ($2n - 1$) reverts to haploidy (n) by a random loss of chromosomes. Since it also appears to be a matter of chance which chromosome is involved in the primary nondisjunction event, then if a population of haploids that arose from different nondis-

junctional events is examined, free recombination of whole chromosomes is evident, with the stipulation of course that each haploid recombinant must carry one or the other of each pair of homologous chromsomes from the heterozygous diploid.

The two methods of genetic analysis available with the aid of the parasexual process are haploidization analysis and analysis based on mitotic crossing-over.

2. Haploidization Analysis. If two different recessive genetic markers are introduced into a heterozygous diploid, one from each haploid parent, analysis of haploid progeny from the heterozygous diploid will lead to the two possibilities shown in Fig. 2.1.

In the first case, where the genetic markers are on homologous chromosomes, only parental-type haploid segregants will be isolated. In the second case, where the genetic markers are on different chromosomes, there will be two parental types, but in addition, two recombinant types will be isolated. In this way genetic markers can be assigned to haploidization groups.

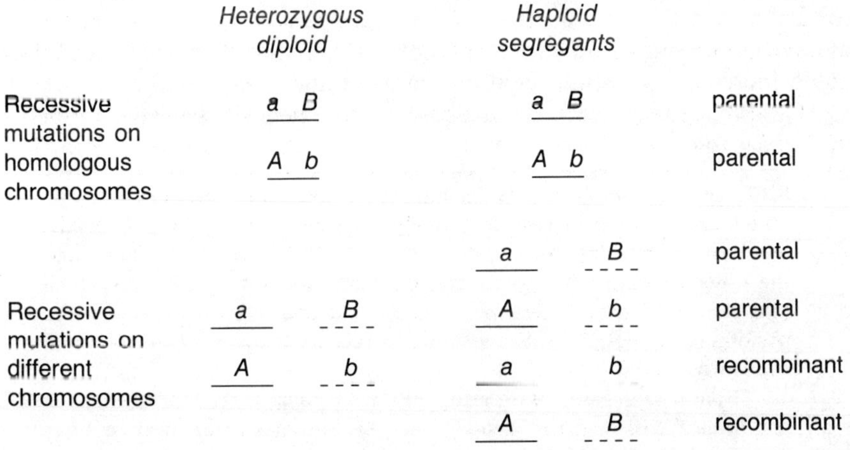

FIGURE 2.1 Segregation of two different recessive mutations (*a* and *b*) introduced into a heterozygous diploid, one from each parent. In the first case, the mutations are on homologous chromosomes; in the second case, they are on different chromosomes.

3. Mitotic Crossing-over. In an organism with a sexual cycle like *A. nidulans,* strains with genetic markers on the same chromosome can be sexually crossed, and mapping data can be gathered as mentioned briefly in the previous section. However, in an asexual mold the next step would be to locate genetic markers on the same chromosome relative to each other by analyzing the progeny from diploids that have gone through mitotic crossing-over events as shown in Fig. 2.2.

Further details of methods of genetic analyses in fungi can be found in Pontecorvo et al. (1953), Clutterbuck (1974), Macdonald and Holt (1976), and Fincham et al. (1979).

III. GENETICS OF PENICILLIN PRODUCTION

Considerable work has gone into investigations on the biochemistry of penicillin production (Arnstein and Morris, 1960; Lemke and Brandon, 1976; Abraham, 1977; Demain, 1982), but much less effort has been devoted to the genetics of penicillin production. Knowledge of the genetics of penicillin biosynthesis should be gained by examining mutants impaired in antibiotic yield, which will include those altered in genes governing different steps in penicillin biosynthesis. Examination of mutants with penicillin yields above their parents should lead to an understanding of the genetic mechanisms involved in overproduction. Possibly, enzymes in the biosynthetic pathway might be altered, or the alterations might be indirect. For example, blockages in reactions that draw off penicillin intermediates could divert metabolism toward increased penicillin production, permeability changes could liberate more penicillin into the environment, or the prevention of penicillin degradation could increase its overall yield (Macdonald and Holt, 1976).

Developments in our understanding of the genetics of penicillin production in *P. chrysogenum* are to a large extent geared to progress in knowledge

---

FIGURE 2.2 Illustration of a method for locating genetic markers following mitotic crossing-over. A heterozygous diploid is synthesized between haploids, one bearing three recessive genetic markers (*a*, *b*, and *x*) as shown and the other bearing wild-type alleles of these mutations. If it is assumed that the positions of the genetic markers *a* and *b* are known and the position of *x* is not, then the position of *x* relative to *a* and *b* can be located as shown.

Diploid segregants arise after mitotic crossing-over, but whether or not homozygosity occurs depends on the centromere orientation after division. Only 50% of the daughter nuclei can become homozygous, since if the centromeres split 1 with 4 and 2 with 3, and not as shown in the illustration, then the daughter nuclei remain heterozygous. In the illustration, homozygous diploids of phenotypic constitution *ABX* (and *abx*), *BX* (and *bx*), and *X* (and *x*) can be isolated, thus locating *x* distal to *a* and *b*. If homozygous diploids of phenotypic constitution *AXB* (and *axb*), *XB* (and *xb*), and *B* (and *b*) arose, then this would position *x* between *a* and *b*. If homozygous diploids of phenotypic constitution *XAB* (and *xab*), *AB* (and *ab*), and *B* (and *b*) arose, then *x* would be located between the centromere and *a*.

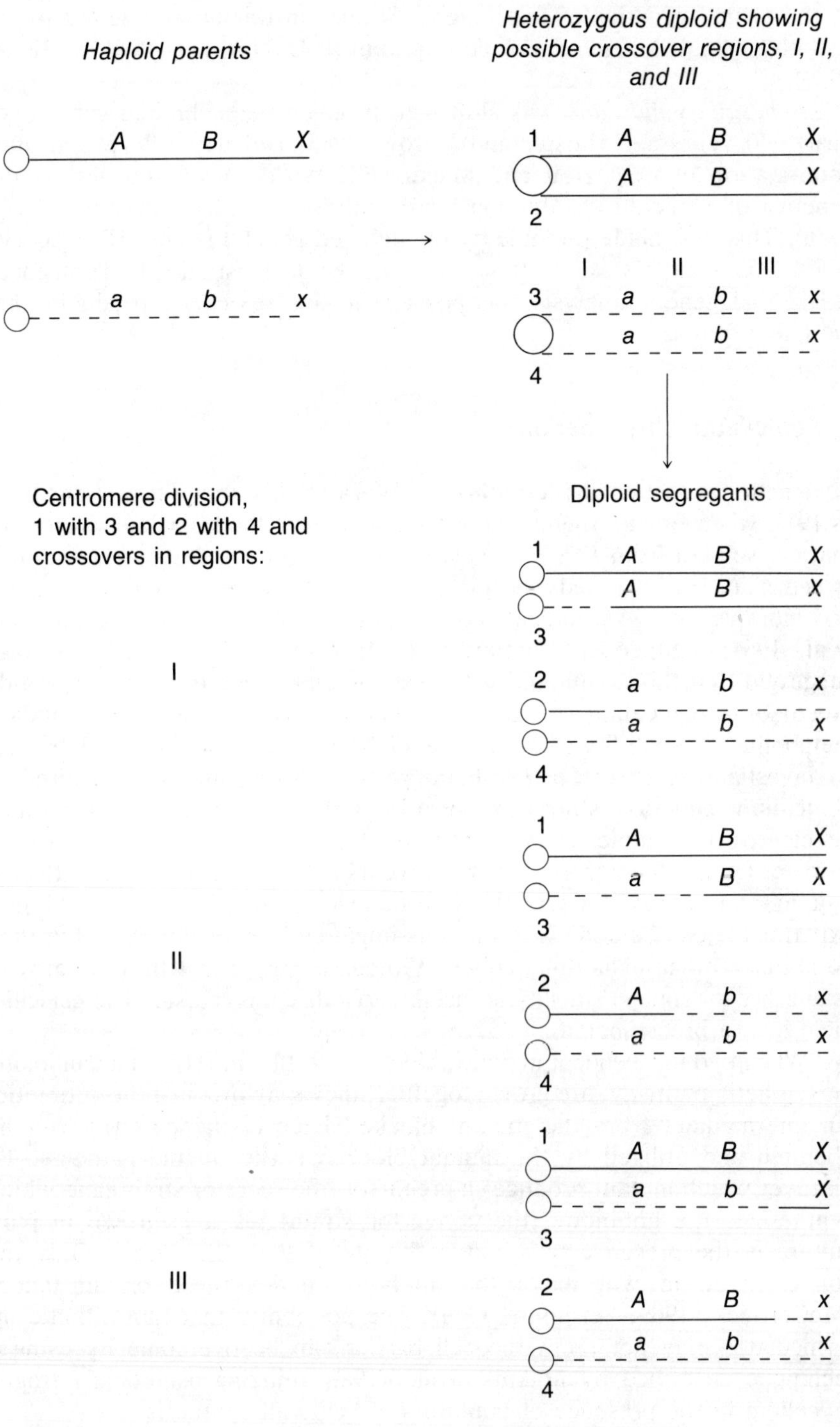
Haploid parents
Heterozygous diploid showing possible crossover regions, I, II, and III
A B X
a b x
1 A B X
A B X
2
I II III
3 a b x
a b x
4
Centromere division, 1 with 3 and 2 with 4 and crossovers in regions:
Diploid segregants
I
1 A B X
A B X
3
2 a b x
a b x
4
II
1 A B X
a B X
3
2 A b x
a b x
4
III
1 A B X
a b X
3
2 A B x
a b x
4

of the formal genetics of this fungus. Studies in this area have led to the tentative establishment of six linkage groups (I. D. Normansell, 1978; Blake, 1980).

Aspergillus nidulans was shown to produce penicillin-like substances nearly 40 years ago (Foster and Karow, 1945; Dulaney, 1947), and after confirmation of this (Holt and Macdonald, 1968a), investigations on the genetics of penicillin production were undertaken (Macdonald and Holt, 1976). This was made possible by the detailed genetic studies (Pontecorvo et al., 1953; Clutterbuck, 1974) that followed the demonstration by Pontecorvo (1948) that genetic analysis was possible in this sexually reproducing homothallic fungus.

A. *Penicillium chrysogenum*

1. Genetics of Penicillin Biosynthesis. Work in this area dates as far back as 1947, when Bonner found that about a quarter of the lysineless auxotrophs that he isolated from *P. chrysogenum* did not produce detectable amounts of penicillin. He suggested that lysine and penicillin had a common precursor, and more recent work indicates that this is α-aminoadipic acid (Somerson et al., 1961; Goulden and Chattaway, 1969; Nash et al., 1973). Later studies suggested that this compound with cysteine and valine formed a tripeptide precursor of penicillin (Arnstein et al., 1959). Since penicillin is a secondary metabolite, the inability of a mutant to elaborate it does not lead to lethality, so investigation of the biosynthetic pathway using mutants impaired in penicillin production is not as straightforward as that involving auxotrophs deficient, for example, in the synthesis of a particular amino acid. In the latter case, the feeding of possible precursors and the ability of different mutants to use these for growth can establish a synthetic sequence. Similar experiments can be done with mutants impaired in penicillin yield but they do not give an all-or-nothing effect. With auxotrophs, growth or no growth establishes the mutant's ability or inability to utilize a precursor. The penicillin yield has to be estimated.

When two independent mutants, blocked at different steps in an antibiotic biosynthetic pathway, are grown together, they may produce the antibiotic. An intermediate from the mutant blocked later in the pathway can be liberated and utilized by the mutant blocked earlier in the pathway. To discover which mutant produces a precursor (the secretor strain) and which synthesizes the antibiotic (the convertor strain), each is grown in pure culture in the presence of a culture filtrate of the other strain. Only the convertor strain will make the antibiotic under these circumstances (McCormick, 1969; Sermonti, 1969). The possibility that biosynthetic intermediates may not cross the cell barrier can be overcome by using a technique described by Makins et al., (1980) utilizing osmotically fragile mycelium in the presence of inhibitors of cell wall synthesis.

Genetic complementation tests can reveal whether impaired mutants are at the same or at different loci. In *P. chrysogenum,* a heterozygous diploid is synthesized between different mutants; if its penicillin yield remains low or undetectable like that of the mutants, then the mutations are on the same complementation group, while if the diploid's penicillin titer approaches that of the mutant's parents, then the mutations are in different complementation groups. Caglioti and Sermonti (1956) and Sermonti (1956) analyzed a number of mutants deficient in penicillin production; complementation analysis indicated the presence of two genetic loci. Nash et al. (1974) examined several impaired mutants and showed differences in their ability to make penicillin pathway intermediates, many of which remained uncharacterized.

Recent studies on the genetics of penicillin biosynthesis by P. J. Normansell et al. (1979) involved examining 12 mutants of *P. chrysogenum* producing 10% or less penicillin than their parents. Testing a series of heterozygous diploids between pairs of these mutants established the existence of five complementation groups designated V, W, X, Y, and Z. Seven mutants belonged to complementation group Y. Parasexual haploidization analysis indicated that mutations from complementation groups W, Y, and Z were on the same chromosome and that mutations of groups V and X were on separate chromosomes from each other and the chromosome bearing mutations in groups W, Y, and Z. While mutants at groups X, Y, and Z were incapable of synthesizing the Arnstein tripeptide δ-(α-aminoadipyl) cysteinylvaline (Arnstein and Morris, 1969), mutants in groups V and W elaborated this compound. A mutant from group X had no acyltransferase activity, a mutant from group V possessed just a trace of this activity, and a member of group W had almost full acyltransferase activity. The possible sites of some of these genetic blocks are illustrated in Fig. 2.3.

Like P. J. Normansell et al. (1979), Sermonti (1956) also found that most mutants impaired in penicillin production belonged to one complementation group. He suggested a number of possible explanations, including a high rate of mutation in one of several loci controlling penicillin biosynthesis, and a single locus controlling production of a multifunctional enzyme concerned with condensation of α-aminoadipic acid, cysteine, and valine (Sermonti, 1959b).

In this connection, R. H. Baltz (cited as a personal communication in Hopwood, 1981) has suggested that many missense mutations may go undetected in genes coding for single enzymes. However, in genes coding for polypeptides, which form spatially complicated enzyme complexes, not only nonsense but also missense mutations could abolish enzyme function because of the three-dimensional integrity required.

An enzyme concerned in the final stage of tripeptide synthesis has been isolated: δ-(α-aminoadipyl)cysteinylvaline synthetase catalyzes synthesis of the tripeptide from δ-(L-α-aminoadipyl)-L-cysteine and valine but apparently not from L-α-aminoadipic acid and L-cysteinyl-L-valine (Loder and Abraham,

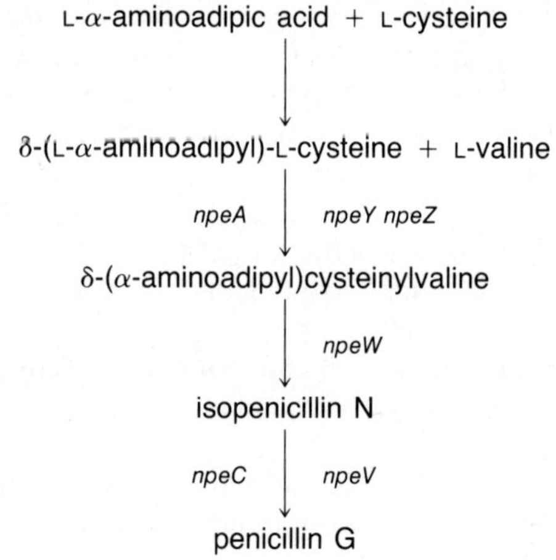

FIGURE 2.3 Steps in the biosynthesis of penicillin and the possible sites of genetic blocks. Mutations in *Aspergillus nidulans: npeA* and *npeC*. Mutations in *Penicillium chrysogenum: npeV, npeW, npeY,* and *npeZ*.

1971b; Fawcett and Abraham, 1975). A number of other enzymes concerned with the biosynthesis of penicillin have been recognized, and further work correlating mutations with enzyme activities will develop knowledge of the genetic regulation of this antibiotic's synthesis in *P. chrysogenum*.

2. Genetics of Penicillin Overproduction. Mutations that lead to increased antibiotic production may alter the activities of penicillin biosynthetic enzymes but, as was discussed earlier, could also alter reactions not directly concerned with the penicillin biosynthetic pathway. Unraveling processes in the genetic regulation of penicillin overproduction is therefore not likely to prove easy, but studies in this area are worth pursuing because of the potential benefits.

As was already mentioned, since the earliest days of the industrial production of penicillin, efforts have been made to improve antibiotic yield. Environmental alterations have certainly raised titers, but the most dramatic advances in productivity have been due to induced mutational changes. Since *P. chrysogenum* strain NRRL 1951 was first isolated and used in deep culture nearly 40 years ago (Raper and Alexander, 1945), mutational changes have probably increased penicillin yield several hundredfold. The benefits of the mutation and selection approach have therefore been very great indeed. Nevertheless, as mutant succeeds mutant in a strain selection program, it appears to become increasingly difficult to isolate better-yielding strains. This is not unexpected, considering that fewer and fewer sites will be available on the genome at which beneficial mutations can occur. Also, more subtle changes could take place during a long program of strain selection, and mutations deleterious to penicillin productivity could be acquired unwittingly in strains selected for high yield. These cryptic mutations could reduce the penicillin titer to below its potential level.

When parasexuality was shown to occur in *P. chrysogenum* after its discovery in *A. nidulans,* it led to the hope that undesirable characteristics, such as those considered above, could be bred out; it led as well, of course, to the hope that hybridization could combine the useful traits of two strains into one.

Early work on the genetics of penicillin production threw up a number of problems. For example, one of the commoner genetic labeling mutations, auxotrophy, often led to reduced penicillin yields, and this was normally a pleiotropic effect of the auxotrophic mutation (Macdonald et al., 1963a). Therefore care had to be taken for genetic studies of penicillin production to select auxotrophs that retained their parents' titers, to avoid the extra confusion of labeling mutations affecting penicillin yield. It was also found difficult to synthesize heterokaryons between auxotrophs. In *A. nidulans,* two auxotrophs with complementary growth factor requirements were grown on complete medium and then inoculated together on minimal medium. Enough growth factors from the complete medium were presumably carried over with the auxotrophs to permit the small amount of growth by each auxotroph needed to allow anastomoses to form so that outgrowth of heterokaryotic mycelium was possible. This method was not successful in *P. chrysogenum,* apparently because growth was too slow and lysis occurred before heterokaryons could arise. Techniques for heterokaryon production in *P. chrysogenum* were devised by Sermonti (1954) and Macdonald et al. (1963b). In the former case, the parent auxotrophs were plated together on a medium supplying limited amounts of the growth factors needed by the auxotrophic parents. Occasionally, tufts of mycelium grew between colonies of different genetic constitution; this proved to be heterokaryotic. Macdonald et al. (1963b) grew the parent auxotrophs together on complete medium and then fragmented the mixed mycelium and plated on minimal medium. It was presumed that during growth on complete medium, some heterokaryotic mycelium would have arisen and that heterokaryotic fragments would grow on minimal medium, which is what was found.

By using Jinks' (1954) heterokaryon test, it was shown that penicillin production was mainly under nuclear control rather than controlled by elements in the cytoplasm (Sermonti, 1959a; Macdonald et al., 1963b), so the way was opened for hybridization studies to begin. Heterozygous diploids were relatively easy to synthesize even between strains of distant lineage (Macdonald et al., 1963c). The low penicillin yields of heterozygous diploids made between highly developed strains and their ancestral strain indicated the recessive nature of positive penicillin mutations. Heterozygous diploids between mutants of divergent lineage were also synthesized; each was selected after many steps of mutation and selection. Most of the progeny turned out to have the genome of one or other haploid parent (Macdonald,

1968a), which led to the proposal that during their derivation in different sequential mutation and selection programs, the mutants used to make the heterozygous diploid had accumulated different chromosomal rearrangements that prevented isolation of recombinant progeny (Macdonald, 1968a). It was suggested that strains used in parasexual breeding programs should be derived with mutagens not causing gross chromosomal damage and that crosses between sisters or cousins rather than between widely divergent strains would have a better expectation of a useful outcome (Macdonald, 1968a).

Practical successes in exploiting parasexual processes in *P. chrysogenum* were reported by several workers (Alikhanian and Borisova, 1956; Alikhanian, 1962; Elander, 1967; Ball, 1973a, b; Elander et al., 1973; Calam et al., 1976). A haploid strain used for industrial production of penicillin was labeled with suitable genetic markers, and a homozygous diploid was synthesized (Elander, 1967; Elander et al., 1973). The diploid elaborated more penicillin than the original haploid. A possible explanation was that the haploid bore a number of recessive genes of high mutability, which led to reduced antibiotic yield; these could accumulate during growth and lead to reduced penicillin titer. If these mutations happened at random, their effects could be concealed to some extent in the diploid, thus raising its penicillin titer relative to the original haploid. Certainly an examination of morphological variation indicated that this was much lower in the diploid than in the haploid (Elander et al., 1973).

Calam et al. (1976) set out to improve the sporulation of their current production strain. Good sporulation is important in preparing seed stages in an industrial plant. They isolated a number of heterozygous diploids made between their current production strain and a sister strain of an ancestor of the production strain. This sister strain had a lower penicillin titer than the production strain but a sporulation rate ten times as great. The heterozygous diploids had penicillin yields a little lower than the lower-yielding haploid parent, but they sporulated as well as the better-sporing parent, indicating the recessive nature of mutations responsible for reduced sporulation. Some of the heterozygous diploids were treated with nitrogen mustard, and strains with better penicillin yields than their parent diploids were isolated. These strains, in turn, were subjected to a further round of nitrogen mustard treatment when a strain was isolated that had a conidial size suggesting that it was diploid. The putative diploid yielded at least as much penicillin as the haploid production strain, and its sporulation rate was five to ten times better. Therefore, it was adopted for the industrial production of penicillin. Mutagens can cause mitotic crossing-over and nondisjunction, so it is possible that the second generation diploid selected was a diploid homozygous for beneficial penicillin and sporulation genes.

Ball (1973b) has pointed out that if mutagens are used to induce segregation, it might be useful to treat first with an agent that increases mitotic crossing-over and then follow this by treatment with a haploidization agent such as *p*-fluorophenylalanine. This could lead to the isolation of useful and stable segregants. The problem with diploids is that they can continue to segregate; and if, for example, a diploid were still heterozygous for useful dominant genes such as those controlling good sporulation, it might conceivably lose these on further segregation. Methods have been proposed for preserving the heterozygous diploid state under plant fermentation conditions. In one of these, each haploid parent would be treated with a mutagen known to cause chromosomal rearrangements and strains with recessive mutations restricting growth would be isolated for diploid synthesis (Macdonald, 1964). If a diploid were heterozygous for different chromosomal rearrangements, it would yield only haploid progeny with parental genomes that would be at a selective disadvantage because of their poor growth properties. Such a heterozygous diploid would therefore retain its stability in an industrial fermentation. Another method proposed the mutagenic induction of nonallelic recessive lethal mutations in a diploid (Azevedo and Roper, 1967).

Alikhanian and his associates found, not unexpectedly, as discussed earlier, that it became increasingly difficult to isolate better penicillin yielders following mutagenic treatment of highly developed haploid strains (Alikhanian and Borisova, 1956; Alikhanian, 1962). A heterozygous diploid was synthesized between two such strains, and the diploid and both parental haploids were treated with ultraviolet light. A higher percentage of isolates with increased penicillin titers were discovered among the progeny from the diploid than among the progeny from either haploid parent. Some of the isolates from the diploid could have retained beneficial mutations and lost deleterious mutations by recombination processes induced by ultraviolet light.

As was discussed earlier, the lack of recombination following parasexual crosses between haploid strains of divergent lineage indicated that attention would have to be paid to the type of mutagen employed in strain selection programs and that crosses between distantly connected strains should be avoided (Macdonald, 1968a). Using closely related strains, Ball (1973a) found that obstacles to recombination vanished, and he isolated haploid recombinant progeny that bore positive penicillin mutations from both parents and had penicillin titers greater than either.

It can be seen from this review of the genetics of penicillin overproduction in *P. chrysogenum* that the work that has been done has been almost exclusively concerned with practical issues. It is hoped that more fundamental studies on the formal genetics of *P. chrysogenum* and on the genetics of penicillin biosynthesis (I. D. Normansell, 1978; Blake, 1980) will lead to similar studies on the genetics of penicillin overproduction in this organism. Developments in this area will certainly aid the commercial breeder to plan more rational hybridization programs leading to the isolation of improved strains.

B. Aspergillus nidulans

1. Genetics of Penicillin Biosynthesis. A number of mutants impaired in penicillin production have been isolated in *A. nidulans* (Edwards et al., 1974), and diploid complementation tests have indicated the presence of four complementation groups, A, B, C, and D (Holt et al., 1976). As in *P. chrysogenum,* most mutants belonged to one complementation group; 20 out of 28 were assigned to group A. This gene, given the symbol *npeA* (non-penicillin-producing), was located on chromosome VI (Holt et al., 1976), and strains bearing mutants at this locus were found to have their ability to make penicillin restored in the presence of Arnstein's tripeptide, δ-(α-aminoadipyl)cysteinylvaline (J. F. Makins, personal communication). Therefore this locus appears to be analogous with the *npeY* locus in *P. chrysogenum.* In both *A. nidulans* and *P. chrysogenum,* the majority of mutations occur at loci that control formation of the tripeptide. Wild-type isolates of *A. nidulans* can be arranged into different heterokaryon compatibility classes, and there is a tendency for those in the same heterokaryon group to have similar characteristics, including the same level of penicillin productivity. Wild-type isolates not making detectable amounts of penicillin were confined to two heterokaryon-compatible groups and all were impaired at the *npeA* locus, which emphasizes its importance in penicillin biosynthesis (Cole et al., 1976).

The other three loci, *npeB, npeC,* and *npeD* were positioned on the genetic map of *A. nidulans* as shown in Fig. 2.4 (J. F. Makins, G. Holt, and K. D. Macdonald, personal communication).

Makins et al. (1981) have investigated the possibility of intergeneric cosynthesis of penicillin between impaired mutants of *A. nidulans* and *P. chrysogenum.* They used a technique described by Makins et al. (1980). Mycelium, treated with an enzyme or enzyme complex capable of lysing the cell wall, was grown in a medium in which the osmotic pressure was sufficiently above that of the cell contents to prevent protoplast formation; inhibitors of cell wall synthesis were also added to the medium, and it was presumed that regions of discontinuity in the cell wall, resulting from the action of the lytic enzymes, were maintained in the presence of these inhibitors. The impermeability of fungal cell walls to intermediates of penicillin biosynthesis has been implicated as the cause of failure to observe penicillin production when pairs of mutants were grown together (Bonner, 1947; Nash et al., 1974; Macdonald and Holt, 1976). It was proposed by Makins et al. (1980) that diffusion of intermediates from one mycelium to another might take place under conditions where gaps occurred in the cell wall, yet not in normal cultures with intact mycelium where the intermediates were incapable of crossing the cell wall barrier.

Intergeneric cosynthesis was demonstrated between mutants of *A. nidulans* and *P. chrysogenum* by Makins et al. (1981), and their studies tended

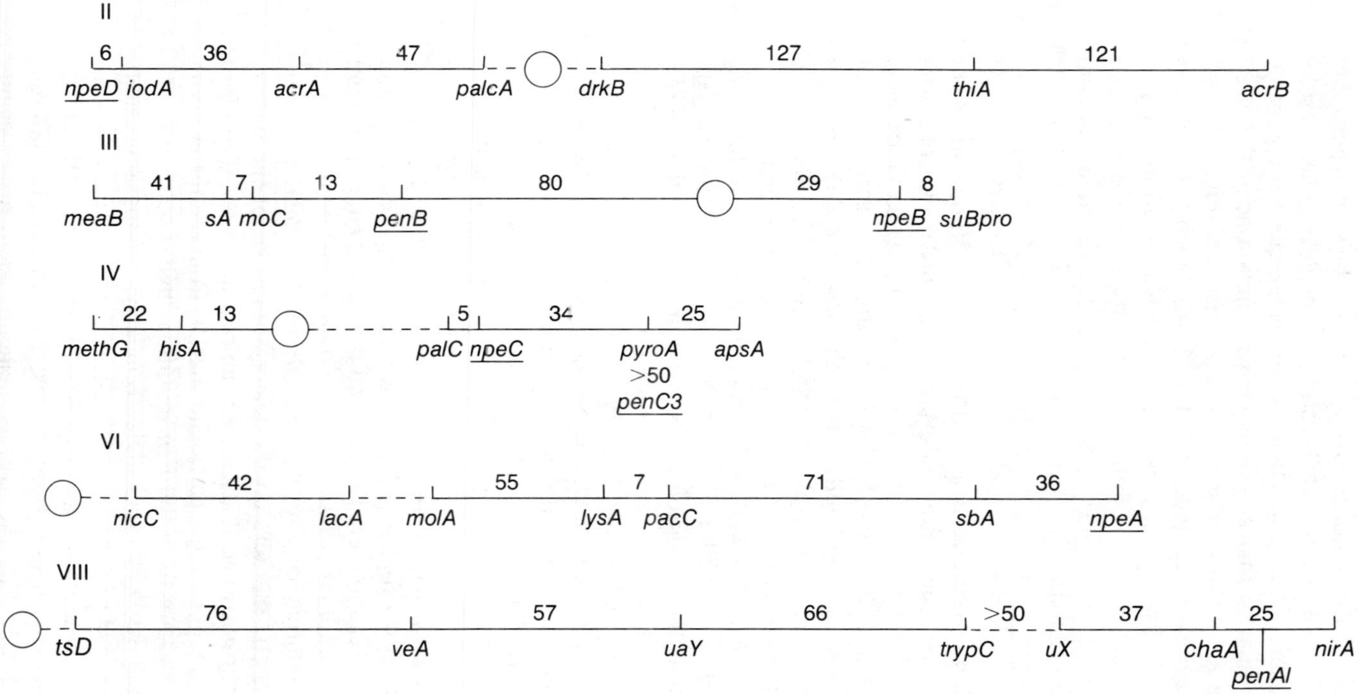

FIGURE 2.4 **Genetic map of *Aspergillus nidulans* showing the position of mutations concerned with biosynthesis and overproduction of penicillin. With these mutations, which are underlined, a few other genetic markers are shown to enable the penicillin mutations to be located in relation to other loci.**

Explanations of symbols: 0, centromere; *npeA*, *npeB*, *npeC*, *npeD*, impairing penicillin titer; *penA1*, *penB2*, *penC3*, increasing penicillin titer; *acrA*, *acrB*, acriflavine resistant; *apsA*, anucleate primary sterigmata; *chaA*, chartreuse-colored conidia; *drkB*, dark-colored conidia; *iodA*, iodoacetate-resistant; *hisA*, histidine-requiring; *lacA*, lactose-nonutilizing; *lysA*, lysine-requiring; *meaB*, methylammonium-resistant; *methG*, methionine-requiring; *moC*, morphologically abnormal; *molA*, molybdate-resistant; *nicC*, nicotinic acid–requiring; *nirA*, nitrite-nonutilizing; *pacC*, acid phosphatase–deficient; *palC*, alkaline phosphatase–deficient; *palcA*, acid and alkaline phosphatase–deficient; *pyroA*, pyridoxin-requiring; *sA*, sulphate-nonutilizing; *sbA*, sorbitol nonutilizing; *suBpro*, supressor of proline requirement; *thiA*, thiazole-requiring; *trypC*, tryptophan-requiring; *tsD*, temperature-sensitive; *uX*, urea-nonutilizing; *uaY*, uric acid–nonutilizing; *veA*, velvet morphology.

to confirm that mutations at *npeA* or *npeY* both blocked the step in penicillin synthesis concerned with synthesis of Arnstein's tripeptide. Their results also suggested that strains carrying mutations at the *npeC* locus were probably impaired at a step similar to that in strains bearing mutations at the *npeV* locus and lacked the ability to produce acyltransferase (see Fig. 2.3).

As was pointed out by Macdonald and Holt (1976), biochemical studies with mutants impaired in penicillin biosynthesis are hampered by the low titers in this mold, which makes it difficult to identify intermediates. The technique of intergeneric cosynthesis that has been described should help to circumvent this obstacle.

2. Genetics of Penicillin Overproduction. After confirmation that strains of *A. nidulans* elaborated penicillin (Holt and Macdonald, 1968a), this organism was chosen for study of the formal genetics of penicillin production because its genetics was so well understood. Following treatment of derivatives of the Glasgow strain NRRL 194 with ultraviolet light (254 nm), three mutants with raised penicillin yields were isolated (Ditchburn et al., 1976). These mutants were given the symbol *pen* (for penicillin producers), and all three were shown to be under nuclear rather than cytoplasmic control following heterokaryon tests. The penicillin yield of the wild-type strain was approximately 5 u/ml and the yields of strains bearing the mutations were: *penA1*, 20 u/ml; *penB2*, 12 u/ml; and *penC3*, 20 u/ml. When heterozygous diploids were made between strains bearing these mutations and strains bearing their wild-type alleles, *penA1* proved to be a recessive mutation, *penB2* was a dominant mutation, and *penC3* was a semidominant mutation.

These mutations were then positioned on specific chromosomes following parasexual haploidization. A heterozygous diploid was synthesized between a strain bearing a *pen* marker and a tester strain genetically labeled on all the eight chromosomes that *A. nidulans* is known to have. The diploid was then treated with *p*-fluorophenylalanine to induce haploidization, and a number of haploid segregants were isolated and characterized for penicillin yield and the genetic markers they bore. Because there is free recombination of whole chromosomes during haploidization, the *pen* marker would recombine freely with all the genetic markers on the tester strain except the one on the homolog of the chromosome that carried the *pen* mutation. Therefore among haploid segregants bearing a particular marker from the tester strain on a different linkage group from the *pen* mutation, half would carry the *pen* mutation and half its wild-type allele. Similarly, among segregants bearing the wild-type allele of the genetic marker from the tester strain, half would carry the *pen* mutation and half its wild-type allele. When the *pen* mutation and a genetic marker from the tester strain were on different linkage groups, it followed that the average penicillin yield of segregants with a particular marker from the tester strain would be similar to the average yield of segregants with the wild-type allele of the same marker. If, however, the genetic marker from the tester strain was on the same linkage group as the *pen* marker, then segregants bearing the genetic marker from the tester strain would all carry the wild-type allele of the *pen* mutation, and their average penicillin yield would be below that of segregants bearing the wild-type allele of the marker from the tester strain, which would all carry the mutant *pen* marker. In this way, *penA1, penB2,* and *penC3* were positioned on chromosomes VIII, III, and IV, respectively (Macdonald et al., 1972).

To locate these mutations in relation to others on the three chromosomes, sexual crosses were made between strains bearing *penA1, penB2,* or *penC3* and other strains with suitable genetic labels on chromosomes VIII, III, or IV. Further details of the procedures employed will be found in Ditchburn et al. (1976) and Macdonald and Holt (1976). The three *pen* mutations were positioned as shown in Fig. 2.4.

Since all three mutations were on different chromosomes, it was considered worthwhile to search for haploid recombinants bearing two or more different *pen* mutations. Parasexual crosses were made between suitably labeled strains, and haploid recombinants with the following assortments of markers were isolated: *penA1 penB2, penA1 penC3,* and *penB2 penC3* (Ditchburn et al., 1976; Macdonald and Holt, 1976). None were isolated bearing all three *pen* mutations, possibly because an insufficient number of segregants were examined. Haploid segregants of type *penA1 penB2* had penicillin titers similar to those of haploids carrying only *penA1;* haploid segregants of type *penA1 penC3* had penicillin titers like those of haploids bearing only *penA1;* and haploid segregants of type *penB2 penC3* had titers similar to those of haploids with *penC3* only. The conclusions drawn were that *penA1* was epistatic to *penB2* and *penC3* and that *penC3* was epistatic to *penB2*.

In earlier studies, a cross was made between a natural isolate of *A. nidulans* with a penicillin yield of about 20 u/ml and a strain bearing *penA1* also with a yield of 20 u/ml. Additive effects on penicillin titer were demonstrated, and some recombinants with titers approaching 40 u/ml were isolated (Holt and Macdonald, 1968b). The possibility of using biometric techniques to examine the genetics of penicillin production in *A. nidulans* was explored (Merrick 1975a, b, 1976; Merrick and Caten, 1975). Relying only on natural variability, Merrick was able, after alternating hybridization and selection procedures, to isolate strains that had about three times the penicillin yield of the strains with which he had begun (Merrick, 1976). More recently, Simpson and Caten (1979) initiated a selection program with two strains isogenic except for genetic markers induced for crossing purposes. Both strains were subjected to six cycles of mutagenic treatment and selection, using in one case near-ultraviolet light (365 nm) in conjunction with 8-methoxysporalen for mutagenic purposes and in the other, ethyl methanesulphonate. In both lines, penicillin yields of the end strains were increased by about 300% over the original parents. Backcrosses to the ancestral strains suggested that the mutations increasing penicillin yield that had accumulated in one line were recessive, while those in the other appeared semidominant, possibly suggesting a balance between accumulation of recessive and dominant mutants. A heterozygous diploid made between the two high-yielding strains had a penicillin yield lower than either, suggesting that at least some of the mutations increasing penicillin titer in both strains were different (Simpson and Caten, 1980).

Sexual crosses were then set up between the two end strains of both six-cycle mutation and selection programs. Segregants with penicillin yields similar to the ancestral strains were isolated, but complementary segregants with higher penicillin titers than either parent were not found (Simpson and Caten, 1981). The inability to isolate high-yielding recombinants could be blamed on unfavorable reactions between the induced mutants in both selection lines. Simpson and Caten (1981) suggested that, after recurrent mutagenesis in their experiments, balanced polygenic systems could be built up that might restrict the gains expected following recombination. Whether this would be widespread with different organisms or different mutagens would require further experimentation to decide. Another possibility is that an insufficient number of segregants were isolated from these crosses to permit the detection of high-yielding recombinants. There is evidence in *P. chrysogenum* that high yielders are at a selective disadvantage (Macdonald, 1968b, 1972).

When a homozygous diploid is synthesized between suitably labeled haploids, each bearing the same mutations increasing penicillin yield, the evidence suggests that normally it does not yield more penicillin than a haploid carrying those mutations, although there can be exceptions (Elander, 1967; Elander et al., 1973). While *pen* mutations are duplicated in a homozygous diploid, deleterious mutations that prevent full expression of the former may also be duplicated. Therefore it would be useful to examine the effect of duplicating only chromosomal material bearing *pen* mutations. Upshall et al. (1979) have devised methods of producing stable disomic strains, that is, strains having a haploid number of chromosomes with just one present in duplicate. They identified mutants that failed to grow at 42°C but did grow at 37°C after prolonged incubation, when they gave rise to colonies disomic for a specific chromosome. These remained stable during culture at 37°C, only reverting to haploidy when transferred to 30°C. Methods of producing stable disomic strains in *A. nidulans* with duplications of *pen* genes are at present under investigation (O'Donnell, A. Upshall, and K. D. Macdonald, unpublished).

IV. GENETICS OF CEPHALOSPORIN PRODUCTION

A. Genetics of Cephalosporin Biosynthesis

Studies of blocked mutants have helped to explain the biosynthesis of cephalosporin C. *Cephalosporium acremonium* produces Arnstein's tripeptide (Loder and Abraham, 1971a), and this is believed to be a precursor of cephalosporin C as it is for penicillin in *P. chrysogenum*. After ring closure, the tripeptide yields isopenicillin N and then penicillin N (Fawcett et al., 1976; Demain, 1982). Expansion of the five-membered ring of penicillin N gives the six-membered ring of cephalosporin (Demain, 1982; Felix et al., 1981). The final steps of synthesis proceed from deacetoxycephalosporin C to deacetylcephalosporin C to cephalosporin C (see Chapter 4).

Because of its importance in the biosynthesis of cephalosporin, the sulfur metabolism of *C. acremonium* has been investigated in some detail using mutants blocked at various steps (Treichler et al., 1979). While in *P. chrysogenum* the sulfur of penicillin is acquired from inorganic sulfur via the sulfate reduction pathway, the sulfur of cephalosporin normally comes from methionine by reverse transsulfuration (Caltrider and Niss, 1966; Nüesch et al., 1973), although mutants of *C. acremonium* exist that can utilize inorganic sulfur for cephalosporin synthesis (Niss and Nash, 1973). Cysteine was thought to be a precursor of β-lactam antibiotics; but when it was supplied in radioactive form, there was little incorporation (Döbeli and Nüesch, 1980). It has been shown that production of cephalosporin C is dependent on the presence of γ cystathionase, which suggests that a precursor of cysteine derived from cystathione rather than cysteine itself forms part of the cephalosporin molecule (Treichler et al., 1978). For maximum expression of this enzyme, the mold is grown on a medium rich in methionine, which suppresses sulfate-assimilating enzymes. However, if a block is introduced preventing the synthesis of cysteine from inorganic sulfur, the organism must synthesize cysteine via the reverse transsulfuration pathway even when methionine is absent. Thus cephalosporin is made in the absence of methionine and can be synthesized from inorganic sulfur (Treichler et al., 1979).

B. Genetics of Cephalosporin Overproduction

Not much is known about the parasexual genetics of *C. acremonium;* of a number of studies reported, the most comprehensive has been that of Hamlyn and Ball (1979). They used genetically labeled strains that belonged to one of three groups A, B, or C. Strains from an individual group normally differed only by the few mutations necessary for genetic labeling. Those of group C were derived from a strain having a relatively low titer of cephalosporin, and groups A and B were divergent lines obtained from group C after a number of mutation and selection steps for increased yield of cephalosporin C. Crosses of types A × A, C × C, A × B, and A × C were set up, that is, two sets of sister strain crosses, divergent strain crosses, and a cross of derivatives of a highly developed strain with derivatives of an ancestral strain. The methods of Nüesch et al. (1973) were employed initially, but the results of over 40 crosses suggested that fusion between nuclei of different strains did not happen easily. Unfortunately, hyphal cells of *C. acremonium* are mainly uninucleate, so the ability to isolate heterokaryons and heterozygous diploids is restricted. Hamlyn and Ball (1979) decided to examine protoplast fusion as a means of enabling nuclei from different strains to come together in a common cytoplasm (Peberdy, 1979). This proved a very successful method of initiating crosses; heterozygotes were obtained from sister strain crosses and divergent strain crosses and these yielded recombinants spontaneously.

A strain from group A could synthesize cephalosporin C from inorganic sulfur and yielded three times the amount of cephalosporin made by a strain from group B. The former was asporous and grew slowly, while the latter sporulated and grew at about four times the rate of the strain from group A. Crosses initiated by protoplast fusion were set up between these two strains, and from one a prototroph was selected that segregated progeny with different growth rates and varied morphologies. About 600 segregants were yield-tested, and one gave 40% more cephalosporin C than the prototrophic parent of the group A genetically labeled parent and also grew faster and sporulated better than this prototroph.

The reults suggested that in *C. acremonium,* heterozygous diploids are short-lived and that a high frequency of segregation occurs, possibly in the main at an intrachromosomal level, since most segregants appear to be haploid. Whatever the details of the genetic processes turn out to be, it seems clear that *C. acremonium* has a mechanism of genetic recombination that can be utilized to breed strains of commercial value.

REFERENCES

Abraham, E. P. (1977) *J. Antibiot. 33,* 1–26.

Alikhanian, S. I. (1962) *Adv. Appl. Microbiol. 4,* 1–50.

Alikhanian, S. I., and Borisova, L. N. (1956) *Bull. Acad. Sci. USSR Biol. Ser. No. 2,* 74–90.

Arnstein, H. R. V., and Morris, D. (1960) *Biochem. J. 76,* 357–361.

Arnstein, H. R. V., Morris, D., and Toms, E. J. (1959) *Biochim. Biophys. Acta 35,* 561–562.

Azevedo, J. L., and Roper, J. A. (1967) *J. Gen. Microbiol. 49,* 149–155.

Ball, C. (1973a) in *First International Symposium on the Genetics of Industrial Microorganisms* (Vaněk, Z., Hošťálek, Z., and Cudlín, J., eds.) pp. 227–237, Academia, Prague.

Ball, C. (1973b) in *Progress in Industrial Microbiology* (Hockenhull, D. J. D., ed.) Vol. XII, pp. 47–72, Churchill Livingstone, Edinburgh.

Blake, C. E. M. (1980) "Recombination in *Penicillium chrysogenum,*" 230 pp., Ph.D. thesis, Polytechnic of Central London, London.

Bonner, D. (1947) *Arch. Biochem. 13,* 1–9.

Caglioti, M. T., and Sermonti, G. (1956) *J. Gen. Microbiol. 14,* 38–46.

Calam, C. T., Daglish, L. B., and Gaitskell, W. S. (1973) in *First International Symposium on the Genetics of Industrial Microorganisms* (Vaněk, Z., Hošťálek, Z., and Cudlín, J., eds.) pp. 265–282, Academia, Prague.

Calam, C. T., Daglish, L. B., and McCann, E. P. (1976) in *Second International*

Symposium on the Genetics of Industrial Microorganisms (Macdonald, K. D., ed.) pp. 273–287, Academic Press, London.
Caltrider, P. G., and Niss, H. F. (1966) *Appl. Microbiol. 14,* 746–753.
Clutterbuck, A. J. (1974) *Handbook of Genetics* Vol. I *Bacteria, Bacteriophages and Fungi* (King, R. C., ed.) pp. 447–510, Plenum Press, London.
Cole, D. S., Holt, G., and Macdonald, K. D. (1976) *J. Gen. Microbiol. 96,* 423–426.
Demain, A. L. (1982) in *Antibiotics Containing the β-Lactam Structure* (Demain, A. L., and Solomon, N. A., eds.) Springer, Berlin.
Ditchburn, P., Holt, G., and Macdonald, K. D. (1976) in *Second International Symposium on the Genetics of Industrial Microorganisms* (Macdonald, K. D., ed.) pp. 213–227, Academic Press, London.
Döbeli, H., and Nüesch, J. (1980) *Antimicrob. Agents Chemother. 18,* 111–117.
Dulaney, E. L. (1947) *Mycologia, 39,* 582–586.
Edwards, G. F. St. L., Holt, G., and Macdonald, K. D. (1974) *J. Gen. Microbiol. 84,* 420–422.
Elander, R. P. (1967) *Abh. Deut. Akad. Wiss. Berlin 4,* 403–423.
Elander, R. P., Espenshade, M. A., Pathak, S. G., and Pan, C. H. (1973) in *First International Symposium on the Genetics of Industrial Microorganisms* (Vaněk, Z., Hošťálek, Z., and Cudlín, J., eds.) pp. 239–253, Academia, Prague.
Fawcett, P. A., and Abraham, E. P. (1975) in *Methods in Enzymology* Vol. 43, *Antibiotics* (Hash, J. H., ed.) pp. 471–473, Academic Press, London.
Fawcett, P. A., Usher, J. J., and Abraham, E. P. (1976) in *Second International Symposium on the Genetics of Industrial Microorganisms* (Macdonald, K. D., ed.) pp. 129–138, Academic Press, London.
Felix, H. R., Peter, H. H., and Treichler, H. J. (1981) *J. Antibiot. 34,* 567–575.
Fincham, J. R. S., Day, P. R., and Radford, A. (1979) *Fungal Genetics* 4th ed., 636 pp., Blackwell Scientific Publications, Oxford.
Foster, J. W., and Karow, E. O. (1945) *J. Bacteriol. 49,* 19–29.
Goulden, S. A., and Chattaway, F. W. (1969) *Biochem. J. 110,* 55–56.
Hamlyn, P. F., and Ball, C. (1979) in *Third International Symposium on the Genetics of Industrial Microorganisms* (Sebek, O. K., and Laskin, A. I., eds.) pp. 185–191, American Society for Microbiology, Washington, D.C.
Holt, G., and Macdonald, K. D. (1968a) *Antonie van Leeuwenhoek J. Microbiol. Serol. 34,* 409–416.
Holt, G., and Macdonald, K. D. (1968b) *Nature, London 219,* 636–637.
Holt, G., Edwards, G. F. St. L., and Macdonald, K. D. (1976) in *Second International Symposium on the Genetics of Industrial Microorganisms* (Macdonald, K. D., ed.) pp. 199–211, Academic Press, London.
Hopwood, D. A. (1981) in *31st Symposium of the Society of General Microbiology* (Glover, S. W., and Hopwood, D. A., eds.) pp. 187–218, Cambridge University Press, Cambridge.
Jinks, J. L. (1954) *Nature, London 174,* 409–410.
Lemke, P. A., and Brandon, D. R. (1976) in *Cephalosporins and Penicillins: Chemistry and Biology* (Flynn, E. H., ed.) pp. 370–437, Academic Press, London.
Loder, P. B., and Abraham, E. P. (1971a) *Biochem. J. 123,* 471–476.
Loder, P. B., and Abraham, E. P. (1971b) *Biochem. J. 123,* 477–486.
Macdonald, K. D. (1964) *Nature, London 204,* 404–405.
Macdonald, K. D. (1968a) *Mut. Res. 5,* 302–305.

Macdonald, K. D. (1968b) *Nature, London 218,* 371–372.
Macdonald, K. D. (1972) *Appl. Microbiol. 23,* 990–993.
Macdonald, K. D., and Holt, G. (1976) *Sci. Prog. Oxf. 63,* 547–573.
Macdonald, K. D., Hutchinson, J. M., and Gillett, W. A. (1963a) *J. Gen. Microbiol. 33,* 365–374.
Macdonald, K. D., Hutchinson, J. M., and Gillett, W. A. (1963b) *J. Gen. Microbiol. 33,* 375–383.
Macdonald, K. D., Hutchinson, J. M., and Gillett, W. A. (1963c) *J. Gen. Microbiol. 33,* 385–394.
Macdonald, K. D., Holt, G., and Ditchburn, P. (1972) in *Proceedings of the Fourth International Fermentation Symposium* (Terui, G., ed.) pp. 251–257, Society of Fermentation Technology, Japan.
Makins, J. F., Holt, G., and Macdonald, K. D. (1980) *J. Gen Microbiol. 119,* 397–404.
Makins, J. F., Allsop, A., and Holt, G. (1981) *J. Gen. Microbiol. 122,* 339–343.
McCormick, J. R. D. (1969) in *Genetics and Breeding of Streptomyces* (Sermonti, G., and Alačevíc, M., eds.) pp. 163–176, Yugoslav Academy of Sciences and Arts, Zagreb.
Merrick, M. J. (1975a) *J. Gen. Microbiol. 91,* 278–286.
Merrick, M. J. (1975b) *J. Gen. Microbiol. 91,* 287–294.
Merrick, M. J. (1976) in *Second International Symposium on the Genetics of Industrial Microorganisms* (Macdonald, K. D., ed.) pp. 229–242, Academic Press, London.
Merrick, M. J., and Caten, C. E. (1975) *J. Gen. Microbiol. 86,* 283–293.
Nash, C. H., de la Higuera, N., Neuss, N., and Lemke, P. A. (1974) *Dev. Ind. Microbiol. 15,* 114–132.
Niss, H. F., and Nash, C. H. (1973) *Antimicrob. Agents Chemother. 4,* 474–478.
Normansell, I. D. (1978) "The Genetics of *Penicillium chrysogenum* with Particular Reference to Penicillin Biosynthesis," 235 pp., Ph.D. thesis, Polytechnic of Central London, London.
Normansell, P. J., Normansell, I. D., and Holt, G. (1979) *J. Gen. Microbiol. 112,* 113–126.
Nüesch, J., Treichler, H. J., and Liersch, M. (1973) in *First International Symposium on the Genetics of Industrial Microorganisms* (Vaněk, Z., Hošťálek, Z., and Cudlín, J., eds.) pp. 309–334, Academia, Prague.
Peberdy, J. F. (1979) in *Third International Symposium on the Genetics of Industrial Microorganisms* (Sebek, O. K., and Laskin, A. I., eds.), pp. 192–196, American Society for Microbiology, Washington, D.C.
Pitt, J. I. (1979) *The Genus Penicillium and Its Teleomorphic States, Eupenicillium and Talaromyces,* 634 pp., Academic Press, London.
Pontecorvo, G. (1948) *Proc. 8th Int. Cong. Genet. Hereditas Supp.* 642–643.
Pontecorvo, G., and Roper, J. A. (1953) *Advan. Genet. 5,* 218–233.
Pontecorvo, G., and Sermonti, G. (1953) *Nature, London 172,* 126–127.
Pontecorvo, G., Roper, J. A., Hemmons, L. M., Macdonald, K. D., and Bufton, A. W. J. (1953) *Advan. Genet. 5,* 141–238.
Raper, K. B., and Alexander, D. F. (1945) *J. Elisha Mitchel Scientific Soc. 61,* 74–113.
Sermonti, G. (1954) *R. C. 1st Sup Sanit. Roma 17,* 213–239.
Sermonti, G. (1956) *J. Gen. Microbiol. 15,* 599–608.
Sermonti, G. (1959a) *Sel. Sci. Papers 1st Sup. Sanit. Roma 2,* 407–415.

Sermonti, G. (1959b) *Ann. N.Y. Acad. Sci. 81,* 950–973.

Sermonti, G. (1969) *Genetics of Antibiotic-Producing Microorganisms,* 389 pp., Wiley-Interscience, London.

Simpson, I. N., and Caten, C. E. (1979) *J. Gen. Microbiol. 113,* 209–217.

Simpson, I. N., and Caten, C. E. (1980) *J. Gen. Microbiol. 121,* 5–16.

Simpson, I. N., and Caten, C. E. (1981) *J. Gen. Microbiol. 126,* 311–319.

Somerson, N. L., Demain, A. L., and Nonheimer, T. D. (1961) *Arch. Biochem. Biophys. 93,* 238–241.

Treichler, H. J., Liersch, M., and Nüesch, J. (1978) in *Antibiotics and Other Secondary Metabolites* (Hütter, R., Leisinger, T., Nüesch, J., and Wehrli, W., eds.) pp. 177–199, Academic Press, London.

Treichler, H. J., Liersch, M., Nüesch, J., and Döbeli, H. (1979) in *Third International Symposium on the Genetics of Industrial Microorganisms* (Sebek, O. K., and Laskin, A. I., eds.) pp. 97–104, American Society for Microbiology, Washington, D.C.

Upshall, A., Giddings, B., Teow, S. C., and Mortimore, I. D. (1979) in *Third International Symposium on the Genetics of Industrial Microorganisms* (Sebek, O. K., and Laskin, A. I., eds.) pp. 197–204, American Society for Microbiology, Washington, D.C.

CHAPTER

3

Metabolic Control of Secondary Biosynthetic Pathways

Arnold L. Demain
Yair Aharonowitz
Juan-Francisco Martín

I. INTRODUCTION

Primary metabolism involves an interconnected series of enzyme-mediated catabolic, amphibolic, and anabolic pathways, which provide biosynthetic intermediates and energy and convert precursors into essential macromolecules, such as deoxyribonucleic acid (DNA), ribonucleic acid (RNA), protein, lipids, and polysaccharides. Primary metabolism is essentially identical for all living beings. Its metabolic reactions are finely balanced, and metabolic intermediates other than those necessary for cell growth rarely accumulate. In addition to this general type of metabolism, certain taxonomic groups are capable of synthesizing special metabolites. Sometimes the same general enzymes are used, but more often, special synthetases produced by specific cells under specific nutritional conditions are needed; this is secondary metabolism.

Secondary metabolites have also been termed "idiolites" (J. B. Walker, 1974) because they are formed during the idiophase (production phase) of batch cultures. These special metabolites often possess bizarre chemical structures and are not essential for growth of the producing organism, although they probably have survival functions in nature. Many secondary metabolites possess some kind of biological activity that is expressed outside the producing organism. When such activity involves interference with the growth processes of organisms other than the producer, the secondary metabolite is termed an antibiotic. Secondary metabolites are produced only by some species of a genus and by some strains of a species, usually as families of closely related components. One *Micromonospora* strain is reported to produce 48 aminocyclitol antibiotics (Berdy, 1974).

Because of the low specificity of some enzymes involved in secondary metabolism, the proportion of each component in a mixture is influenced by genetic and environmental factors. In contrast, primary metabolism has narrow specificity because errors in the biosynthesis of essential cell components are usually lethal. Errors in secondary metabolism are usually of no consequence to producing cells, especially since the modified metabolite often retains biological activity.

An important characteristic of secondary metabolism is that idiolites are usually produced only at low specific growth rates. Some pathways are also affected by regulatory mechanisms, such as induction, catabolite regulation, and end product regulation. In other words, secondary metabolism seems to be governed by (1) overall regulatory controls that operate as functions of growth rate and (2) specific regulatory effects on individual pathways.

II. TROPHOPHASE AND IDIOPHASE

In batch cultures containing nutritionally rich media, high levels of antibiotics are usually produced only after most of the cellular growth has occurred (Weinberg, 1970). This is clearly observed in antibiotic-producing unicellular bacterial cultures with well-defined exponential growth phases (Banerjee and Bose, 1974; Bernlohr and Novelli, 1963; Fujikawa et al., 1968; Hurst, 1966; Kurahashi et al., 1969; Kurylo-Borowska, 1967; Paulus, 1967; Sarkar and Paulus, 1972). The growth phase is called the "trophophase," while the production phase is termed the "idiophase" (Bu'Lock et al., 1975).

The separation between trophophase and idiophase is not as clear-cut in fermentations with filamentous microorganisms (actinomycetes and molds). Here, dry cell weight often continues to increase significantly during the idiophase, although at a lower rate than in the trophophase. However, dry weight is a poor criterion of true growth. Cellular mass consists not only of structural material required for cell replication (cell walls, membranes, cytoplasmic organelles, ribosomes, nuclei, etc.) but also of assimilatory reserve materials, such as polyols, lipids, polyphosphates, and nonstructural carbohydrates. These reserve materials may account for 50–60% of the dry weight at the end of a fermentation. Other reasons for the overlapping of trophophase and idiophase include the use of inocula already engaged in secondary metabolism and the presence in the fermentation stage of pellets constituting a very heterogeneous cell population (GrootWassink and Gaucher, 1980). Probably the best parameter with which to measure true replicatory growth is the increase in DNA (Martin and McDaniel, 1975). Then cell growth can often be clearly dissociated from antibiotic production. Other changes that can indicate the end of replicatory growth are a drop in respiratory activity (Martin and McDaniel, 1975) and a decrease in the rate of RNA synthesis (G. H. Jones, 1975; G. H. Jones and Weissbach, 1970; Liras et al., 1977; Mertz and Doolin, 1973).

In many antibiotic fermentations (for example, chloramphenicol, colistin, penicillin, and bacitracin), typical trophophase–idiophase dynamics occur in complex media capable of supporting rapid growth, but the two phases overlap in defined media supporting slow growth (Frøyshov, 1977; Haavik, 1974a, b; Hanlon and Hodges, 1981; Hendlin, 1949; Ito et al., 1969; Lurie et al., 1975; Malik and Vining, 1970; Pirt and Righelato, 1967; Snoke and Cornell, 1965). The factor that controls the onset of antibiotic biosynthesis is probably a deficiency of one or more growth-limiting nutritional components. This arrests growth and initiates idiolite biosynthesis by mechanisms that are just beginning to be understood. In a defined medium supporting only slow growth, some nutritional factor may be growth-limiting from the start of cultivation, thus favoring antibiotic production while growth is still occurring.

We emphasize that the timing of product formation should not be used to define a secondary metabolite. As was discussed above, trophophase and idiophase ideally occur at two separate times in batch cultures, but they often overlap. A secondary metabolite is secondary only because it is not necessary for vegetative growth of the producing culture.

Microorganisms appear to be programmed to produce antibiotics only when the specific growth rate decreases below a certain level (Bu'Lock, 1974). This phenomenon probably evolved in nature in response to competitive pressures. We believe that one role of antibiotics in nature is to inhibit or kill other microorganisms (Katz and Demain, 1977) and that another is to stimulate differentiation processes (Piret and Demain, 1981). The occurrence of antibiotic production only after growth nears completion is useful to the producing strains, since during growth these species are sensitive to the antibiotic that they produce. Sensitivity is usually restricted to the growth phase before antibiotic production begins; only later does the producing strain become resistant (e.g., Hitchcock and Katz, 1978). Antibiotic-producing species avoid suicide (Demain, 1974) by mechanisms that include (1) modification (and thus detoxification) of the antibiotic by enzymes formed by the producer, (2) alteration of the antibiotic's target in the producing cell, and (3) decreased inward permeability to the antibiotic after it has been excreted.

III. REPRESSION OF ANTIBIOTIC SYNTHESIS

Many investigators have monitored the activity of antibiotic synthetases and observed late enzyme formation (Table 3.1). Such control of gene expression might occur by interference with transcription or translation. Several mechanisms appear to be involved. In one model, a small molecule acts as a co-repressor or inhibitor, repressing the formation or inhibiting the action of antibiotic synthetases. The co-repressor or inhibitor must be depleted or neutralized before antibiotic synthesis can occur. This model fits some of the experimental data on carbon catabolite regulation, nitrogen metabolite regulation, and phosphate control. In a second regulatory model, an inducer or activator must be synthesized by the producing culture or added to it in order to initiate biosynthesis.

A. Carbon Catabolite Repression

Glucose, usually an excellent carbon source for growth, interferes with the biosynthesis of many antibiotics (Table 3.2). During studies on fermentation medium development, polysaccharides or oligosaccharides are often found to be better carbon sources for antibiotic production. In a medium containing glucose plus a more slowly utilized carbon source, glucose is usually used first, and no antibiotic is produced. After glucose is depleted, the second carbon source may then be used for antibiotic biosynthesis. Depending on the microorganism, the carbon regulatory effect may be exerted by rapidly used carbon sources other than glucose. For example, citrate is favored over glucose by the novobiocin producer *Streptomyces niveus,* resulting in diauxic growth. Novobiocin production is suppressed during citrate utilization but not during the (secondary) phase when glucose is utilized (Kominek, 1972). Thus we should not think of this phenomenon as a "glucose effect."

Carbon catabolite repression affects the formation of specific enzymes involved in the biosynthesis of secondary metabolites, for example, phenoxazinone synthase in the actinomycin fermentation (Gallo and Katz, 1972) and the ring expansion enzyme of cephalosporin biosynthesis (Demain, 1982; Zanca et al., 1982). Carbon catabolite repression of inducible catabolic enzymes in enteric bacteria involves cyclic adenosine 3′,5′-monophosphate (cAMP) as a positive effector (Pastan and Adhya, 1976). Uptake of glucose inhibits the activity of adenylate cyclase, thereby decreasing the intracellular level of cAMP. When present, cAMP interacts with a receptor protein; the cAMP receptor protein–cAMP complex binds at the promoter site of target operons and activates gene transcription.

TABLE 3.1 Antibiotic Synthetases Repressed During Growth

Enzyme	*Process*	*Organism*	*Reference*
Amidinotransferase	Streptomycin	*Streptomyces griseus*	J. B. Walker and Hnilica (1964)
Streptidine kinase	Streptomycin	*S. griseus*	J. B. Walker and Hnilica (1964)
Mannosidostreptomycinase	Streptomycin	*S. griseus*	Inamine et al. (1969)
Thymidine diphosphate glucose oxidoreductase	Tylosin	*Streptomyces rimosus*	Matern et al. (1973)
Thymidine diphosphate 4-keto-deoxyglucose methylase	Tylosin	*S. rimosus*	Matern et al. (1973)
Guanosine triphosphate 8-formylhydrolase	Toyocamycin	*S. rimosus*	Elstner and Suhadolnik (1971, 1975)
	Sangivamycin	*S. rimosus*	Elstner and Suhadolnik (1971, 1975)
Gramicidin S synthetases	Gramicidin S	*Bacillus brevis*	Kurahashi et al. (1969); Matteo et al. (1975), Tomino et al. (1967)
Tyrocidine synthetases	Tyrocidine	*B. brevis*	Fujikawa et al. (1968)
Bacitracin synthetases	Bacitracin	*Bacillus licheniformis*	Weinberg and Tonnis (1967)
Phenoxazinone synthase	Actinomycin	*Streptomyces antibioticus*	Gallo and Katz (1972)
Neomycin phosphatase	Neomycin	*Streptomyces fradiae*	Majumdar and Majumdar (1971, 1972)
O-demethylpuromycin methyltransferase	Puromycin	*Streptomyces alboniger*	Sankaran and Pogell (1973)
N-acetylkanamycin amidohydrolase	Kanamycin	*Streptomyces kanamyceticus*	Satoh et al. (1976)
Cephalosporin C acetylhydrolase	Cephalosporin C	*Cephalosporium acremonium*	Hinnen and Nuesch (1976)
PABA synthetase	Candicidin	*S. griseus*	Martin et al. (1981)

TABLE 3.2 Carbon Catabolite Regulation of Antibiotic Biosynthesis

Antibiotic	*Interfering Carbon Source*	*Noninterfering Carbon Source*	*Reference*
Penicillin	Glucose	Lactose	Soltero and Johnson (1953)
Actinomycin	Glucose	Galactose	Gallo and Katz (1972)
Streptomycin	Glucose	Mannan, slowly fed glucose	Inamine et al. (1969)
Siomycin	Glucose	Maltose	Kimura (1967)
Indolmycin	Glucose	Fructose	Hurley and Bialek (1974)
Bacitracin	Glucose	Citrate	Haavik (1974a, b)
Cephalosporin C	Glucose	Sucrose	Demain (1963)
Chloramphenicol	Glucose	Glycerol	C. G. Smith and Hinman (1963)
Violacein	Glucose	Maltose	DeMoss (1967)
Prodigiosin	Glucose	Galactose	Ramsey et al. (1973)
Mitomycin	Glucose	Low glucose	Kirsch (1967)
Neomycin	Glucose	Maltose	Majumdar and Majumdar (1971)
Kanamycin	Glucose	Galactose	Basak and Majumdar (1973)
Enniatin	Glucose	Lactose	Audhya and Russell (1975)
Puromycin	Glucose	Glycerol	Sankaran and Pogell (1975)
Novobiocin	Citrate	Glucose	Kominek (1972)
Candicidin	Glucose	Slowly fed glucose	Martin and McDaniel (1974)
Candihexin	Glucose	Slowly fed glucose	Martin and McDaniel (1974)
Butirosin	Glucose	Glycerol	Howells et al. (1972)
Cephamycin	Glycerol	Asparagine, starch	Aharonowitz and Demain (1978)

It is not known whether cAMP is involved in carbon catabolite regulation of antibiotic synthesis. cAMP or its dibutyryl derivative does not reverse glucose repression of penicillin biosynthesis in *Penicillium chrysogenum* (G. Revilla and J. F. Martin, submitted for publication), of phenoxazinone synthase in *Streptomyces antibioticus* (E. Katz, personal communication), or of cephalosporin C acetylhydrolase in *Cephalosporium acremonium* (Hinnen and Nuesch, 1976). On the other hand, it is claimed (Satoh et al., 1976) that cAMP relieves glucose repression of N-acetylkanamycin amidohydrolase in *Streptomyces kanamyceticus*. cAMP and its binding protein (and cyclic guanosine 3′,5′-monophosphate) have been found in *Streptomyces hygroscopicus*, the producer of turimycin (a 16-member macrolide antibiotic), but it is not clear that these compounds play any role in regulating turimycin biosynthesis (Gersch et al., 1978). Although the cellular cAMP content drops at the onset of turimycin formation and added cAMP stimulates growth and interferes with turimycin synthesis, phosphate regulation might be responsible. The cAMP content in *Streptomyces griseus* drops to 10% of its trophophase peak level before streptomycin formation (Ragan and Vining, 1978); a similar situation occurs in the chloramphenicol-producing *Streptomyces venezuelae* (Chatterjee and Vining, 1982). These observations suggest that a high cAMP level does not relieve repression of antibiotic formation, in contrast to its effect on inducible enzymes in enteric bacteria. Indeed, although glucose represses the inducible β-galactosidase in *S. venezuelae*, cAMP fails to rise when glucose is exhausted and induction to lactose utilization begins (Chatterjee and Vining, 1982).

B. Nitrogen Metabolite Repression

The regulatory mechanisms that control the use of nitrogen sources in bacteria, yeasts, and molds are now partially understood (Dubois et al., 1974; Magasanik et al., 1974). Ammonia (or some readily used nitrogen source) represses enzymes involved in the use of other nitrogen sources; these enzymes include nitrite reductase, nitrate reductase, nicotinamide adenine dinucleotide (NAD) dependent glutamate dehydrogenase, arginase, ornithine transaminase, extracellular protease, acetamidase, threonine dehydratase, allantoinase, and those enzymes dealing with purine degradation, transport of urea and glutamate, and histidine usage.

That nitrogen regulation plays a role in antibiotic biosynthesis was emphasized by the discovery (Aharonowitz and Demain, 1979) that NH_4^+ interferes with cephalosporin formation in *Streptomyces clavuligerus*. Addition of ammonium to *S. clavuligerus* cultures grown on asparagine, glutamine, and other amino acids as a nitrogen source suppressed antibiotic production (Aharonowitz and Demain, 1979). The ammonium effect might be explained by inhibition of antibiotic synthetases or, alternatively, by an effect at some other step(s) in the synthesis of antibiotics. To distinguish between these alternatives, excess ammonium was added at different times during growth, and antibiotic production was monitored. The ammonium effect expressed itself early in the growth phase. Furthermore, no inhibition was found when ammonium was added after the onset of antibiotic production. It appears from this observation that inhibition of antibiotic synthesis by ammonium was not due to inhibition of the activity of antibiotic synthetases. Rather, an earlier process was affected, involving either the biosynthesis of antibiotic synthetases or other metabolic steps required to enhance production. As a result of this work, isolated reports in the literature (Dulaney, 1948; Martin and McDaniel, 1974; R. L. Smith et al., 1962) could be tied together into a working hypothesis suggesting that high levels of NH_4^+ or certain amino acids repress antibiotic biosynthesis during growth (Aharonowitz, 1979). Included in the list of processes regulated by nitrogen sources are rifamycin (Jiao et al., 1979), chloramphenicol (Westlake et al., 1968), actinomycin (Katz, 1980), leucomycin (Ōmura et al., 1980), and aflatoxin (Kachholz and A. L. Demain, unpublished data).

Several recent publications describe nitrogen metabolism of *Streptomyces noursei* in relation to the biosynthesis of nourseothricin, a streptothricin-type antibiotic (Gräfe et al., 1974a; 1977). *o*-Aminobenzoic (anthranilic) acid stimulates nourseothricin biosynthesis and has a marked effect on the regulation of enzymes of nitrogen metabolism. It represses glutamine synthetase and alanine dehydrogenase while stimulating the formation of nicotinamide adenine dinucleotide phosphate (NADP)-dependent glutamate dehydrogenase and the antibiotic (Gräfe et al., 1977). The activity of the NAD-specific alanine dehydrogenase of *S. hygroscopicus,* the producer of macrolide antibiotic JA 6599, also appears to be correlated inversely with antibiotic production (Gräfe et al., 1974b).

The relationship between the biosynthesis and activity of glutamine synthetase (an enzyme that occupies a key position in nitrogen metabolism) and antibiotic production has recently been studied in several laboratories. With *S. clavuligerus,* the highest specific cephamycin production was obtained under those growth conditions that led to high glutamine synthetase activity (Aharonowitz, 1979). When cells were grown on media containing ammonium in excess, both glutamine synthetase activity and specific antibiotic production decreased. Similar observations were reported for thienamycin production by *Streptomyces cattleya* (Wax and Snyder, 1980) and for chloramphenicol production by *S. venezuelae* (Vining and Chatterjee, 1982). However, the inference that a specific nitrogen signal acts, directly or indirectly through the nitrogen assimilatory pathway, to regulate antibiotic gene expression, remains speculative owing to the lack of studies on mutants possessing altered nitrogen metabolism.

C. Phosphate Regulation

Phosphate is the crucial growth-limiting nutrient in many antibiotic fermentations; in cultures of *Streptomyces aureofaciens,* it is depleted before the onset of tetracycline production, and it is virtually exhausted during growth of *S. griseus* (a candicidin producer) before the onset of candicidin synthesis (Liras et al., 1977). Extracellular phosphate remains at very low levels during the entire idiophase of the candicidin fermentation. If increasing concentrations of phosphate are included in the medium, phosphate depletion is delayed, as is the onset of antibiotic synthesis. An initial phosphate concentration of 10 mM saturates the growth requirement; cell mass increases throughout the fermentation, and no antibiotic synthesis occurs. Thus candicidin synthesis is initiated by phosphate depletion. A similar phenomenon occurs in vancomycin fermentations (Mertz and Doolin, 1973). Phosphate also inhibits candicidin synthesis once it has started (Liu et al., 1975; Martin and Demain, 1976).

Synthesis of β-lactam antibiotics by both actinomycetes and fungi is influenced by the phosphate content of the medium. In *S. clavuligerus* cultures, phosphate levels higher than 25 mM reduce antibiotic production (Aharonowitz and Demain, 1977). With *C. acremonium* grown in a defined medium, the optimum phosphate concentration for cephalosporin synthesis was 2.75 mM; higher levels reduced synthesis by more than 30%. However, this negative effect could be overcome by controlling the glucose supply in fed-batch and chemostat experiments (Kuenzi, 1980). Phosphate regulates the synthesis of many antibiotics. These belong to diverse biosynthetic groups, such as peptides, polyene macrolides, tetracyclines, and biosynthetically complex compounds, and many control mechanisms may exist. However, the different biosynthetic pathways could also be controlled by a common regulatory effector (Martin, 1976).

In the last few years, studies on candicidin biosynthesis by *S. griseus* (Martin, 1978a, b) have confronted the important question: Is intracellular orthophosphate the ultimate effector, or does it merely regulate the level of some other intracellular effector that controls expression of antibiotic biosynthesis? In investigating the possible role of nucleotides, Martin and Demain (1977a) showed by uptake studies that the inhibition of candicidin biosynthesis in *S. griseus* by exogenous ribonucleotides (including cyclic nucleotides) of adenine, guanine, cytosine, and uracil, but not by ribonucleosides or their bases, was the result of nucleotide cleavage. The effect was due to the released phosphate (Martin and Demain, 1977b). Mediation of the phosphate effect by cyclic nucleotides might be suspected from changes in intracellular cAMP concentration observed during the growth of a streptomycin-producing strain of *S. griseus* when different levels of inorganic phosphate were included in the medium. However, cAMP levels could not be correlated with differences in streptomycin production (Terry and Springham, 1981). There is stronger evidence that adenosine triphosphate (ATP) is involved. Its concentration doubles or triples within 5 min of adding phosphate to candicidin-producing cells (Martin and Demain, 1976). The increase occurs before the inhibition of antibiotic synthesis, which is detected 15 min after phosphate addition. This suggests that ATP may be the intracellular effector that controls antibiotic synthesis. Involvement of ATP is supported by evidence (Janglova et al., 1969) that ATP levels are lower in improved tetracycline-producing strains than in their poor-producing ancestral strains. Also, Fynn and Davison (1976) have pointed out the possible involvement of ATP in regulating tyrothricin biosynthesis. Thus it is possible, though not proven, that a high ATP level interferes with antibiotic synthesis. This is consistent with data from time course studies on ATP concentration during antibiotic fermentations, which show a rapid decrease in intracellular ATP just before the onset of antibiotic synthesis (Liras et al., 1977).

Some progress is being made in identifying enzymes of secondary metabolism repressed by phosphate. These include anhydrotetracycline oxygenase (Behal et al., 1979), three enzymes of tylosin biosynthesis (Madre and Pape, (1980), and PABA synthetase of the candicidin pathway (Martin et al., 1981).

Phosphatase regulation by phosphate also appears to be involved in antibiotic biosynthesis. In the pathways by which antibiotics such as streptomycin (Miller and Walker, 1970), viomycin (Pass and Raczynska-Bojanowska, 1968), and neomycin (Majumdar and Majumdar, 1970) are made, some intermediates are phosphorylated, whereas the end products are not. In general, the phosphatases that cleave phosphorylated intermediates are regulated via feedback inhibition or repression by inorganic phosphate. Streptomycin biosynthesis, which is markedly inhibited by phosphate, includes at least three phosphate-cleaving steps in the formation of the streptidine moiety (Demain and Inamine, 1970). Miller and Walker (1969) reported that a phosphorylated streptomycin accumulates in cultures of *S. griseus* growing in excess inorganic phosphate. This compound contains a phosphate ester at carbon-6 of the streptidine moiety and is biologically inactive (Miller and Walker, 1970; Nomi and Nimi, 1969; Nomi et al., 1968). Further experiments suggest that the phosphate group is introduced during biosynthesis of the streptidine moiety (J. B. Walker and Walker, 1967). The dihydrostreptose moiety is transferred enzymatically from deoxythymidine diphosphate-dihydrostreptose to streptidine 6-phosphate to form O-α-L-dihydrostreptose (1→4)-streptidine-6-phosphate (Kniep and Grisebach, 1976). Streptomycin phosphate is probably an obligatory intermediate in streptomycin biosynthesis (Nomi and Nimi, 1969). Streptomycin phosphatase, an idiophase enzyme present only in streptomycin producers, is inhibited but not repressed by phosphate (M. S. Walker and Walker, 1971).

Phosphatase control of the biosynthesis of neomycin B in *Streptomyces fradiae* involves an alkaline phosphatase that is inhibited and repressed by inorganic phosphate (Majumdar and Majumdar, 1971). This enzyme is synthesized late in the fermentation, and there appears to be a direct relationship between enzyme activity and neomycin formation. That the alkaline phosphatase activity arises from de novo protein synthesis was demonstrated by inhibiting its formation with chloramphenicol (Bandyopadhyay and Majumdar, 1974).

IV. INDUCTION

Enzyme induction is involved in controlling the biosynthesis of antibiotics, although it is sometimes difficult to determine whether a stimulatory response is the result of true induction or merely due to increased availability of precursors.

An intriguing induction phenomenon is the stimulation of streptomycin biosynthesis by A-factor (2(*S*)-isocapryloyl-3(*R*)-hydroxymethyl-γ-butyrolactone) (Kleiner et al., 1976). It has been reported that 1 μg of pure A-factor added at the time of inoculation to a mutant of *S. griseus* blocked in streptomycin biosynthesis induces the production of 1 g of streptomycin; that is, the induction coefficient is 10^6 (Khokhlov and Tovarova, 1979). A-factor restored the ability to produce streptomycin in 114 of 119 mutant strains that could not synthesize the streptidine moiety. A-factor is formed by all wild-type and industrial streptomycin-producing strains of *S. griseus* that have been investigated. Mutants unable to synthesize A-factor cannot form streptomycin. A-factor is also produced by *Streptomyces bikiniensis*, but not by other species, such as *Streptomyces albus* and *Streptomyces mashuensis*, that synthesize streptomycin (Khokhlov et al., 1973; Khokhlov and Tovarova, 1979). The stimulatory effect is intense when the compound is added at the time of inoculation but absent when the compound is added after 48 h. A short treatment period of 3–4 min, followed by washing, is sufficient to induce streptomycin formation.

A-factor is also involved in *S. griseus* differentiation. Antibiotic-producing strains differentiate normally, releasing aerial spores in solid media or in submerged cultures. Mutants are not able to differentiate unless supplemented with A-factor. The supplement also causes the formation of well-developed intracellular membranes and tubular structures that are similar to those in the producing strains.

The molecular mechanism by which A-factor induces streptomycin biosynthesis is not known. The activity of glucose-6-phosphate-nicotinamide adenine dinucleotide phosphate(NADP)-oxidoreductase (glucose-6-phosphate dehydrogenase; E.C. 1.1.149) is reported to be high in nonproducing mutants, whereas it is practically undetectable in high-producing strains (Khokhlov and Tovarova, 1979). Addition of A-factor to nonproducing strains immediately decreases the activity of this enzyme. However, A-factor does not affect glucose-6-phosphate dehydrogenase in vitro. This suggests that it induces the formation of another compound, which in turn inhibits the enzyme. Khokhlov and Tovarova (1979) suggest that the second compound might be adenosine diphosphoribose, a breakdown product of NADP by the action of NADP-glycohydrolase. They further propose that when the pentose phosphate pathway is blocked at the glucose-6-phosphate dehydrogenase stage, glucose is preferentially transformed into streptomycin instead of undergoing oxidation.

Another induction-like phenomenon is the stimulation of cephalosporin C biosynthesis by methionine in *C. acremonium*. Methionine appears to be both a sulfur precursor and an inducer of cephalosporin biosynthesis. Norleucine, the nonsulfur analog of methionine, is also an inducer (Demain et al., 1963). Studies with sulfur amino acid auxotrophs of *C. acremonium* supported earlier evidence that methionine is a regulatory effector and emphasized the importance of endogenous methionine in fermentations in which the amino acid was not added externally. In a double mutant (strain 11-8) of *C. acremonium* blocked in the paths from sulfate to cysteine and from cysteine to methionine, norleucine was able to replace excess methionine in stimulating cephalosporin production (Drew and Demain, 1975a, b, c). This is particularly significant because norleucine contains no sulfur and thus cannot be a sulfur precursor.

Despite general agreement between the data of Drew and Demain (1975a, b) and those of Treichler et al. (1979), the two groups have tentatively reached somewhat different conclusions. Drew and Demain (1975c) favor the hypothesis that methionine acts as an inducer of one or more β-lactam synthetases; Treichler et al. (1979) believe that cystathionine-γ-lyase is crucial in the methionine effect, this cystathionine-cleaving enzyme somehow transferring the cysteine moiety into the cephalosporin biosynthetic pathway.

Methionine induces swelling and fragmentation of the mycelium simultaneously with induction of cephalosporin biosynthesis. It is not clear, however, whether the antibiotic is required for morphological differentiation or whether cephalosporin production is a consequence of morphological differentiation.

In the patulin pathway of *Penicillium urticae*, at least some of the pathway enzymes are coordinately induced by intermediates such as 6-methylsalicylic acid and *m*-hydroxybenzyl alcohol (Gaucher et al., 1981).

The stimulatory effects of 5,5-diethylbarbiturate (barbital) on the biosynthesis of rifamycin and anthracycline antibiotics may also be induction phenomena. Barbital is not a precursor of rifamycins, since, whether ring or side chain labeled, it is neither incorporated into rifamycin nor metabolized to malonate by *Nocardia mediterranea* (Kluepfel et al., 1965; Lancini and White, 1973). It exerts its effect only when added in the first 48 h of a fermentation (Margalith and Pagani, 1961). Studies with washed mycelium of *N. mediterranea* led to the conclusion that additions during the growth phase increased the activity of enzymes transforming rifamycin SV into rifamycin B (Lancini and White, 1973). Production of the anthracycline galirubin by *Streptomyces galileus* is also stimulated by barbital if the addition is made before 36 h of cultivation; the effect is more intense under conditions of limited aeration. Barbital may affect specific enzymes that hydroxylate aklavinone (Kralovcova et al., 1977); its regulatory effect is not a general one, since barbital does not stimulate production of daunomycinone glycosides by *Streptomyces coeruleorubidus* (Blumauerova et al., 1977).

Benzylthiocyanate stimulates tetracycline biosynthesis if added to the culture during the first 12 h of a fermentation, before the tetracene nucleus is formed (Hostalek, 1964a, b, c). This suggests a regulatory effect on primary metabolism and/or induction of tetracycline-synthesizing enzymes. Other possible examples of induction are the increase in staphylomycin (synonym: virginiamycin) biosynthesis by a substance produced by *Streptomyces virginiae* (Yanagimoto and Terui, 1971a, b) and the stimulation of fosfomycin biosynthesis by methionine (Rogers and Birnbaum, 1974). The substance promoting staphylomycin biosynthesis is formed just before the antibiotic appears; most of the inducer is excreted into the broth.

V. GROWTH RATE

The molecular mechanism of catabolite regulation may be related to growth rate control of antibiotic biosynthesis (Bu'Lock, 1974; Martin and Demain, 1978). It seems that a certain low specific growth rate must be maintained in an organism to exploit its full capacity for antibiotic biosynthesis. Ryu and Hospodka (1980) determined the specific growth rates of *P. chrysogenum* that are optimal for high specific penicillin production. By correlating these with the appropriate specific uptake rates of nutrients such as carbon, nitrogen, phosphate, and oxygen, they designed an improved process for penicillin production based on optimal feed rates. Slow addition of glucose to obtain a slow growth rate eliminates glucose interference with penicillin biosynthesis (Soltero and Johnson, 1954). Similarly, slow feeding of glucose stimulates formation of the polyene macrolide antibiotics, candidin and candihexin (Martin and McDaniel, 1974).

In some fermentations, growth rate alone appears to regulate secondary metabolism; for example, the production of gramicidin S (Matteo et al., 1976), bacitracin (Hanlon and Hodges, 1981), patulin (Gaucher et al., 1981), and cephamycin (Lilley et al., 1981) occurs at a low growth rate irrespective of the type of nutrient limitation employed to produce the low growth rate. However, in some cases, a specific phosphate deficiency appears to be needed in addition to the low growth rate. One such case involves thienamycin production by *Streptomyces cattleya* (Lilley et al., 1981).

VI. CONTROL IN BRANCHED PATHWAYS

Some antibiotics have branched pathways of biosynthesis that give rise to a primary metabolite on the one hand and to a secondary metabolite on the other. The primary end product may then feedback inhibit the common part of the pathway and thus inhibit antibiotic biosynthesis. Lysine inhibits penicillin production by *P. chrysogenum* (Masurekar and Demain, 1972) and cephalosporin production by *Paecilomyces persicinus* P-10 (D'Amato and Pisano, 1976). In *P. chrysogenum* strain Wis 54-1255, more than 90% of the lysine in the broth must be exhausted before penicillin biosynthesis begins (Luengo et al., 1979). The molecular mechanism of the lysine regulatory effect has been elucidated in recent years. It is now well established that the penicillin biosynthetic pathway branches from the lysine pathway at the α-aminoadipic (or the adenylaminoadipic) acid (AAA) stage. Demain and Masurekar (1974) suggested that lysine acts at the level of homocitrate synthase, the first enzyme in the lysine biosynthetic pathway. This suggestion has been confirmed by reversal studies (Friedrich and Demain, 1977) and by in vitro studies in which cell-free extracts of low- and high-producing *P. chrysogenum* strains were used (Luengo et al., 1979).

The interrelationship between lysine metabolism and AAA is very different in actinomycetes. In two cephamycin-producing streptomycetes, AAA was shown to be a catabolic product of lysine rather than an intermediate in lysine biosynthesis (Kirkpatrick et al., 1973; Whitney et al., 1972). The enzyme catalyzing the first reaction in the conversion of L-lysine to AAA has been isolated (Kern et al., 1980).

Mendelovitz and Aharonowitz (1982) have examined whether regulatory effects exerted at the early stages of lysine biosynthesis affect antibiotic synthesis in the β-lactam producer, *S. clavuligerus*. When lysine, AAA, and diaminopimelic acid (DAP) were added to chemically defined media, production increased about twofold. This suggested that the release of enzymes involved in lysine biosynthesis from feedback effects might increase the intracellular pool of metabolic intermediates and catabolic products, causing an increase in antibiotic synthesis. Aspartokinase (AK), the first enzyme in the pathway of lysine biosynthesis, is subject to concerted feedback inhibition by lysine and threonine. Dihydrodipicolinate (DDP)-synthetase, the first specific enzyme of the lysine biosynthetic branch, was not inhibited by lysine but was partially inhibited by concentrations of AAA and D,L-DAP. Also, biosynthesis of DDP-synthetase was slightly decreased upon addition of DAP; no other amino acids had any effect. These results suggested that AK activity controls carbon flow to lysine. Mutants resistant to the lysine analog S-(2-aminoethyl)cysteine (AEC) included many in which AK was deregulated, being no longer inhibited by the concerted effect of lysine and threonine. Among them, a large number produced amounts of antibiotics ranging from 1.5 to 8 times that found in wild-type fermentations. Two that were examined showed a dramatic elevation in the concentration of DAP in their intracellular amino acid pools (S. Mendelovitz and Y. Aharonowitz, submitted for publication).

The increase in DAP, rather than lysine or AAA, is probably due to inhibition of DAP-decarboxylase by lysine, as was proposed earlier by Kirkpatrick et al. (1973). Such a regulation of DAP-decarboxylase is important for the cell in order to ensure efficient cell wall formation. This suggests that DAP-decarboxylase activity might be rate-limiting in the formation of the cephamycins. The work demonstrates how simple genetic and biochemical methods may be used in strain development programs. The isolation of lysine analog–resistant mutants in the fungal and actinomycete systems had two different rationales and led to opposite results. *Penicillium chrysogenum* mutants which were resistant to lysine-antimetabolites and which over-produced lysine underproduced penicillin (Masurekar and Demain, 1974). On the other hand, lysine analog–resistant mutants of *S. clavuligerus* accumulated DAP and overproduced cephamycins (S. Mendelovitz and Y. Aharonowitz, unpublished data).

Another example of branched pathway effects is the suppression of candicidin biosynthesis by aromatic amino acids (Liu et al., 1972; Martin et al., 1979a). The target for regulation is *p*-aminobenzoate synthetase and the main mechanism is enzyme repression. Aromatic amino acids have a cumulative effect and also inhibit DAHP synthetase, the first enzyme of aromatic amino acid biosynthesis (G. Naharro and J. F. Martin, unpublished data).

VII. TERMINATION OF THE IDIOPHASE

After antibiotic synthetases are formed, the rate of antibiotic biosynthesis is linear for a while. In some cultures, the phase of active antibiotic synthesis is rather short, lasting 4–20 h. In others, it is much longer than the trophophase, and it may be prolonged for days by continuously or intermittently providing nonrepressive or noninhibitory levels of carbon source (Liu et al., 1975; Martin and McDaniel, 1974; Pirt, 1976). Invariably, however, the rate of antibiotic formation decreases with time. Cessation may not be due to loss of viability of the producing cells since lysis is avoided as long as the carbon source is not depleted. There are two principal reasons why antibiotic biosynthesis stops: (1) irreversible decay of one or more enzymes of the pathway, the loss occurring at the primary or secondary metabolic level, and (2) feedback effects of the accumulated antibiotic.

The activity of peptide antibiotic synthetases decreases rapidly a few hours after the onset of antibiotic production. In vivo stabilization of such enzymes would prolong the period of product formation. Tyrocidine synthetase activity disappears from the soluble portion of *Bacillus brevis* cells and appears to associate with the forespore membrane at stage IV of the sporulation process (Lee et al., 1975). The membrane-bound synthetase activity associated with the forespore disappears a few hours later. Similarly, soluble bacitracin synthetase is present for only a short period during the growth of *Bacillus licheniformis* and then becomes predominantly cell-bound (Frøyshov, 1977). This may represent natural immobilization, which protects the enzyme complex against intracellular proteolytic activities appearing at the end of the growth phase (Schaeffer, 1969). Inactivation of gramicidin S synthetase in frozen and thawed cells or in cell-free extracts requires oxygen and is independent of protease action and of energy metabolism (Friebel and Demain, 1977a, b). The L-ornithine- and D-phenylalanine-activating activities of gramicidin S synthetase are also lost during aeration of the cells. This oxygen-dependent inactivation may involve oxidation and inactivation of some of the sulfhydryl groups that exist in gramicidin S synthetase (Laland and Zimmer, 1973). It is interesting to note that removal of oxygen in vivo by sparging with nitrogen prevents inactivation of gramicidin S synthetase (Friebel and Demain, 1977a, b). Oxygen is required for growth, but control of dissolved oxygen, to produce a high degree of oxygen transfer during exponential growth followed by a moderate degree of oxygen transfer during the stationary phase, might improve antibiotic production.

It has become evident that several antibiotics interfere with their own biosynthesis. In a few cases, they are known to inhibit the activity of their synthetases. End product regulation of mycophenolic acid biosynthesis in *Penicillium stoloniferum* involves inhibition of S-adenosylmethionine: demethylmycophenolic acid O-methyltransferase, the enzyme that catalyzes the final step in the synthesis of mycophenolic acid (Muth and Nash, 1975). Similarly, puromycin inhibits S-adenosylmethionine:O-demethylpuromycin O-methyltransferase, the last enzyme of its biosynthetic pathway (Sankaran and Pogell, 1975). Elymoclavine, the terminal alkaloid produced by a *Claviceps* strain, inhibits the first specific enzyme in its own biosynthesis (Cheng and Floss, 1980).

Feedback repression of arylamine synthetase, the first enzyme unique to the biosynthesis of chloramphenicol, limits production of the antibiotic (A. Jones and Vining, 1976). The enzyme, which has been found only in chloramphenicol-producing cultures, is fully repressed when the chloramphenicol concentration in the medium reaches 100 mg/l. This concentration affects neither cell growth nor the formation of other enzymes of the aromatic amino acid pathway, such as chorismate mutase, prephenate dehydratase, and anthranilate synthetase (A. Jones and Westlake, 1974). Staphylomycin appears to repress its own biosynthesis by inhibiting formation of the staphylomycin inducer (Yanagimoto and Terui, 1971a, b). Kanamycin also appears to act by repressing an acetyltransferase involved in its biosynthesis (Satoh et al., 1975). Other examples of antibiotics that limit their own synthesis include aurodox (Liu et al., 1977), cycloheximide (Kominek, 1975a, b), bacitracin (Frøyshov et al., 1980), indolmycin (Speedie et al., 1975), tylosin (Seno and Baltz, 1981), ristomycin (Egorov et al., 1971), fungicidin (Spizek et al., 1965), and candihexin (Martin and McDaniel, 1974). The interpretation of data (Gordee and Day, 1972) on feedback control by penicillin is controversial (Revilla et al., 1978; Nestaas and Demain, 1981).

VIII. STRAIN IMPROVEMENT AND CONTROL MECHANISMS

Strain development programs in industry have markedly increased the productivity of many antibiotic-producing cultures. Most such programs have been carried out by random testing of survivors of mutagenesis treatments (Demain, 1973). These strains usually exhibit weaker regulation of antibiotic production. For example, a high-producing mutant of *P. chrysogenum* was found to be less sensitive to valine inhibition of acetohydroxy acid synthetase than its ancestral strain (Goulden and Chattaway, 1969). Acetohydroxy acid synthetase is the first enzyme in the pathway for valine biosynthesis; valine is a precursor of penicillin. Most of the mutants obtained in these strain development programs are probably altered in regulation. The concept that control mechanisms can be bypassed by mutations is now being explored, and while no one has yet discovered a method to select directly for high production of secondary metabolites, attempts toward more rational screening procedures have become common.

Among mutants of *Pseudomonas aureofaciens* that are resistant to tryptophan analogs, some are overproducers of pyrrolnitrin (Elander et al., 1971). The increased production is presumably due to removal of the regulatory mechanism that controls tryptophan synthesis, since tryptophan is a precursor of pyrrolnitrin. A similar case is the trifluoroleucine-resistant mutant of *Streptomyces lipmanii* (Godfrey, 1973) that excretes up to sixfold more cephamycin than its parent strain. The mutant is probably deregulated in the pathway for branched amino acid biosynthesis that supplies valine for cephamycin production.

Suppression of an auxotrophic mutation by a second mutation sometimes yields strains that are feedback-resistant in the primary metabolic pathway. The procedure may also be of importance in secondary metabolism. The suppression of a *cys* auxotrophic mutation in *S. lipmanii* led to improved production of cephamycin (Godfrey, 1973). Suppression of a *met* auxotrophic mutation in *Streptomyces viridifaciens* (Dulaney and Dulaney, 1967) and of an *ilv* mutation in *Streptomyces antibioticus* (Polsinelli et al., 1965) improved production of chlortetracycline and actinomycin, respectively. Another strategy for eliminating mechanisms controlling antibiotic biosynthesis is to suppress antibiotic-nonproducing mutants back to production. Dulaney and Dulaney (1967) reported that a significant proportion of such revertants of *S. viridifaciens* are overproducers of chlortetracycline. Unowsky and Hoppe (1978) obtained a 20% increase in aurodox production by such a manipulation.

Although the above-mentioned screening methods yield superior producers, they are still rather indirect. Recently, there has been an attempt to screen directly for mutants that are deregulated in a control mechanism of major importance to antibiotic formation. Martin et al. (1979b) devised a method to visualize candicidin production on agar and selected mutants of *S. griseus* that were altered in such a way that phosphate no longer inhibited candicidin biosynthesis. Some of these mutants produced better than the wild type in both phosphate-supplemented and nonsupplemented media. These were not all phosphate permeability mutants; some appeared to be deregulated in phosphate control of candicidin synthesis. Phosphate-deregulated mutants may have considerable industrial significance. Production of antibiotics that are sensitive to phosphate must now be carried out in low-phosphate media. Use of such mutants in antibiotic fermentations should make it possible to use complex media with high phosphate contents.

Knowledge of the biochemical genetics of antibiotic biosynthetic pathways has advanced slowly in comparison with the rapid progress in basic molecular biology and genetics. However, developments in the latter have provided a sound basis for increased understanding of gene expression and control of antibiotic biosynthesis. Of special interest will be the genetic engineering of antibiotic-producing strains, including the manipulation of plasmids containing structural or regulatory genes coding for antibiotic production (Hopwood and Merrick, 1977).

IX. SUMMARY

Unlike primary metabolites, which are produced or required by all microorganisms, secondary metabolites are restricted in distribution and are not required for growth in culture, although they probably have survival value in nature. Secondary metabolites are structurally diverse and unusual, generally are produced in mixtures with other members of the same chemical family, and usually are formed at low specific growth rates. In batch cultures, processes leading to the production of antibiotics are often sequential; the cultures exhibit a distinct growth phase (trophophase) followed by a production phase (idiophase). However, trophophase and idiophase may overlap. Timing depends on the nutritional environment presented to the culture.

Antibiotic production in nature may allow producing organisms to compete effectively with other forms of life and to differentiate. Delay in antibiotic production until after trophophase is useful to a producing organism, since during growth the organism is sensitive to its own antibiotic. Resistance develops during idiophase. Resistance mechanisms include enzymatic modification of the antibiotic, alteration of the target, and decreased uptake of excreted antibiotic.

Structural genes coding for antibiotic synthetases are usually chromosomal. However, in one case, the enzymes are coded by plasmid genes. Specific mechanisms regulating the onset of antibiotic synthesis include carbon catabolite repression, nitrogen metabolite repression, phosphate regulation, and induction. Mechanisms that control the initiation of antibiotic synthesis include repression and induction of antibiotic synthetases. Cessation of antibiotic biosynthesis occurs via decay of antibiotic synthetases as well as feedback inhibition and repression of these enzymes.

Where branched pathways lead to a primary metabolite and an antibiotic, the primary metabolite may interfere with antibiotic formation by inhibiting an early step of the common pathway, thus preventing accumulation of the antibiotic precursor.

Routine strain improvement programs have successfully yielded high-producing mutants of antibiotic producers. Attempts at intentional deregulation of antibiotic producers have begun only recently but have been successful. Undoubtedly, the antibiotic industry will increase its use of such rational screening techniques and feature more genetic engineering of antibiotic-producing strains.

REFERENCES

Aharonowitz, Y. (1979) in *Genetics of Industrial Microorganisms* (Sebek, O., and Laskin, A. I., eds.) pp. 210–217, American Society for Microbiology, Washington, D.C.

Aharonowitz, Y., and Demain, A. L. (1977) *Arch. Microbiol. 115,* 169–173.

Aharonowitz, Y., and Demain, A. L. (1978) *Antimicrob. Agents Chemother. 14,* 159–164.

Aharonowitz, Y., and Demain, A. L. (1979) *Can. J. Microbiol. 25,* 61–67.

Audhya, T. K., and Russell, D. W. (1975) *J. Gen. Microbiol. 86,* 327–332.

Bandyopadhyay, S. K., and Majumdar, S. K. (1974) *Antimicrob. Agents Chemother. 5,* 431–434.

Banerjee, A. B., and Bose, S. K. (1974) *J. Bacteriol. 87,* 1397–1406.

Basak, K., and Majumdar, S. K. (1973) *Antimicrob. Agents Chemother. 4,* 6–10.

Behal, V., Hostalek, Z., and Vanek, Z. (1979) *Biotechnol. Lett. 1,* 177–182.

Berdy, J. (1974) *Adv. Appl. Microbiol. 18,* 309–406.

Bernlohr, R. W., and Novelli, G. D. (1963) *Arch. Biochem. Biophys. 103,* 94–104.

Blumauerova, M., Mateju, J., Stajner, K., and Vanek, Z. (1977) *Folia Microbiol. 22,* 275–285

Bu'Lock, J. D. (1974) in *Industrial Aspects of Biochemistry* (Spencer, B., ed.) Vol. I, pp. 335–346, North Holland, Amsterdam.

Bu'Lock, J. D., Hamilton, D., Hulme, M. A., Powell, A. J., Smalley, H. M., Shepherd, D., and Smith, G. N. (1975) *Can. J. Microbiol. 11,* 765–778.

Chatterjee, S., and Vining, L. C. (1982) *Can. J. Microbiol. 28,* 311–317.

Cheng, L. J., and Floss, H. G. (1980) *J. Natural Prod. 43,* 329–339.

D'Amato, R. F., and Pisano, M. A. (1976) *Antonie van Leeuwenhoek J. Microbiol. Serol. 42,* 299–308.

Demain, A. L. (1963) *Clin. Med. 70,* 2045–2051.

Demain, A. L. (1973) *Adv. Appl. Microbiol. 16,* 177–202.

Demain, A. L. (1974) *Ann. N.Y. Acad. Sci. 235,* 601–612.

Demain, A. L. (1982) Abstracts of the Fourth International Symposium on Genetics of Industrial Microorganisms, p. 13, Kyoto, Japan.

Demain, A. L., and Inamine, E. (1970) *Bacteriol. Rev. 34,* 1–19.

Demain, A. L., and Masurekar, P. (1974) *J. Gen. Microbiol. 82,* 143–151.

Demain, A. L., Newkirk, J. F., and Hendlin, D. (1963) *J. Bacteriol. 85,* 339–344.

DeMoss, R. D. (1967) in *Antibiotics* Vol. II, *Biosynthesis* (Gottlieb, D., and Shaw, P. D., eds.) pp. 77–81, Springer, New York.

Drew, S. W., and Demain, A. L. (1975a) *Antimicrob. Agents Chemother. 8,* 5–10.

Drew, S. W., and Demain, A. L. (1975b) *Eur. J. Appl. Microbiol. 2,* 121–128.

Drew, S. W., and Demain, A. L. (1975c) *J. Antibiot. 28,* 889–895.

Dubois, E., Grenson, M., and Wiame, J. M. (1974) *Eur. J. Biochem. 48,* 603–616.

Dulaney, E. L. (1948) *J. Bacteriol. 56,* 305–313.

Dulaney, E. L., and Dulaney, D. D. (1967) *Trans. N.Y. Acad. Sci. 29,* 782–789.
Egorov, N. S., Toropova, E. G., and Suchkova, L. A. (1971) *Microbiologiya 40,* 475–480.
Elander, R. P., Mabe, J. A., Hamill, R. L., and Gorman, M. (1971) *Folia Microbiol. 16,* 156–165.
Elstner, E. F., and Suhadolnik, R. J. (1971) *J. Biol. Chem. 246,* 6973–6987.
Elstner, E. F., and Suhadolnik, R. J. (1975) *Methods Enzymol. 43,* 515–520.
Friebel, T. E., and Demain, A. L. (1977a) *J. Bacteriol. 130,* 1010–1016.
Friebel, T. E., and Demain, A. L. (1977b) *FEMS Microbiol. Lett. 1,* 215–218.
Friedrich, C. G., and Demain, A. L. (1977) *J. Antibiot. 30,* 760–761.
Frøyshov, O. (1977) *FEBS Lett. 81,* 315–318.
Frøyshov, O., Mathiesen, A., and Haavik, H. I. (1980) *J. Gen. Microbiol. 117,* 163–167.
Fujikawa, K., Suzuki, T., and Kurahashi, K. (1968) *Biochim. Biophys. Acta 161,* 232–246.
Fynn, G. H., and Davison, J. A. (1976) *J. Gen. Microbiol. 94,* 68–74.
Gallo, M., and Katz, E. (1972) *J. Bacteriol. 109,* 659–667.
Gaucher, G. M., Lam, K. S., GrootWassink, J. W. D., Neway, J., and Deo, Y. M. (1981) *Dev. Ind. Microbiol. 22,* 219–232.
Gersch, D., Romer, W., Bocker, H., and Thrum, H. (1978) *FEMS Microbiol. Lett. 3,* 39–41.
Godfrey, O. (1973) *Antimicrob. Agents Chemother. 4,* 73–79.
Gordee, E. Z., and Day. L. E. (1972) *Antimicrob. Agents Chemother. 1,* 315–322.
Goulden, S. A., and Chattaway, F. W. (1969) *J. Gen. Microbiol. 59,* 111–118.
Gräfe, U., Bocker, H., Reinhardt, G., and Thrum, H. (1974a) *Z. Allg. Mikrobiol. 14,* 659–673.
Gräfe, U., Bocker, H., Reinhardt, G., Tkocz, H., and Thrum, H. (1974b) *Z. Allg. Mikrobiol. 14,* 181–192.
Gräfe, U., Bocker, H., and Thrum, H. (1977) *Z. Allg. Mikrobiol. 17,* 201–209.
GrootWassink, J. W. D., and Gaucher, G. M. (1980) *J. Bacteriol. 141,* 443–455.
Haavik, H. I. (1974a) *J. Gen. Microbiol. 81,* 383–390.
Haavik, H. I. (1974b) *J. Gen. Microbiol. 84,* 321–326.
Hanlon, G. W., and Hodges, N. A. (1981) *J. Bacteriol. 147,* 427–431.
Hendlin, D. (1949) *Arch. Biochem. Biophys. 24,* 435–446.
Hinnen, A., and Nuesch, J. (1976) *Antimicrob. Agents Chemother. 9,* 824–830.
Hitchcock, M. J. M., and Katz, E. (1978) *Antimicrob. Agents Chemother. 13,* 104–114.
Hopwood, D. A., and Merrick, M. J. (1977) *Bacteriol. Rev. 41,* 595–635.
Hostalek, Z. (1964a) *Folia Microbiol. 9,* 78–88.
Hostalek, Z. (1964b) *Folia Microbiol. 9,* 89–95.
Hostalek, Z. (1964c) *Folia Microbiol. 9,* 96–102.
Howells, J. D., Anderson, L. E., Coffey, G. I., Senos, G. D., Underhill, M. A., Vogler, D. L., and Erlich, J. (1972) *Antimicrob. Agents Chemother. 2,* 79–83.
Hurley, L. H., and Bialek, D. (1974) *J. Antibiot. 27,* 49–56.
Hurst, A. (1966) *J. Gen. Microbiol. 45,* 503–513.
Inamine, E., Lago, B. D., and Demain, A. L. (1969) in *Fermentation Advances* (Perlman, D., ed.) pp. 199–221, Academic Press, New York.
Ito, M., Aida, J., and Uemura, T. (1969) *Agr. Biol. Chem. 33,* 262–269.
Janglova, Z., Suchy, J., and Vanek, Z. (1969) *Folia Microbiol. 14,* 208–210.

Jiao, R.-S., Chen, Y.-M., Wu, M.-G., and Gu, W.-L. (1979) *Acta Physiol. Sinica 5,* 395–402.

Jones, A., and Vining, L. C. (1976) *Can. J. Microbiol. 22,* 237–244.

Jones, A., and Westlake, D. W. S. (1974) *Can. J. Microbiol. 20,* 1599–1611.

Jones, G. H. (1975) *Biochem. Biophys. Res. Comm. 63,* 469–475.

Jones, G. H., and Weissbach, H. (1970) *Arch. Biochem. Biophys. 137,* 558–573.

Katz, E. (1980) in *Biochemical and Medical Aspects of Tryptophan Metabolism* (Hayaishi, O., Ishimura, Y., and Kido, R., eds.) pp. 159–177, Elsevier/North-Holland Biomedical Press, Amsterdam.

Katz, E., and Demain, A. L. (1977) *Bacteriol. Rev. 41,* 449–474.

Kern, B. A., Hendlin, D., and Inamine, E. (1980) *Antimicrob. Agents Chemother. 17,* 679–685.

Khokhlov, A. S., and Tovarova, I. I. (1979) in *Regulation of Secondary Product and Plant Hormone Metabolism* (Luckner, M., and Schreiber, K., eds.) pp. 133–145, Pergamon Press, New York.

Khokhlov, A. S., Anisova, L. N., Tovarova, I. I., Kleiner, E. M., Kovalenki, I. V., Krasilnikova, O. I., Kornitskaja, E. Y. and Pliner, S. A. (1973) *Z. Allg. Mikrobiol. 13,* 647–665.

Kimura, A. (1967) *Agr. Biol. Chem. 31,* 845–852.

Kirkpatrick, J. R., Doolin, J. L., and Godfrey, O. W. (1973) *Antimicrob. Agents Chemother. 4,* 542–550.

Kirsch, E. J. (1967) in *Antibiotics* Vol. II, *Biosynthesis* (Gottleib, D., and Shaw, P. D., eds.) pp. 66–76, Springer, New York.

Kleiner, E. M., Pliner, S. A., Soifer, V. S., Onoprienko, V. V., Balashova, T. A., Rozynov, B. V., and Khokhlov, A. S. (1976) *Bioorg. Khim. 2,* 1142–1147.

Kluepfel, D., Lancini, G. C., and Startori, G. (1965) *Appl. Microbiol. 13,* 600–614.

Kniep, B., and Grisebach, H. (1976) *FEBS Lett. 65,* 44–46.

Kominek, L. A. (1972) *Antimicrob. Agents Chemother. 1,* 123–134.

Kominek, L. A. (1975a) *Antimicrob. Agents Chemother. 7,* 856–860.

Kominek, L. A. (1975b) *Antimicrob. Agents Chemother. 7,* 861–863.

Kralovcova, E., Blumauerova, M., and Vanek, Z. (1977) *Folia Microbiol. 22,* 182–188.

Kuenzi, M. T. (1980) *Arch. Microbiol. 128,* 78–83.

Kurahashi, K., Yamada, M., Mori, K., Fujikawa, K., Kambe, M., Imae, E., Takahashi, H., and Sakamoto, Y. (1969) *Cold Spring Harbor Symp. Quant. Biol. 34,* 815–826.

Kurylo-Borowska, Z. (1967) in *Antibiotics* Vol. II, *Biosynthesis* (Gottlieb, D., and Shaw, P. D., eds.) pp. 342–352, Springer, New York.

Laland, S. G., and Zimmer, T. L. (1973) *Essays Biochem. 9,* 31–57.

Lancini, G. C., and White, R. C., (1973) *Process Biochem. 8,* 14–16.

Lee, S. G., Littau, V., and Lipmann, F. (1975) *J. Cell Biol. 66,* 238–242.

Lilley, G., Clark, A. E., and Lawrence, G. C. (1981) *J. Chem. Tech. Biotechnol. 31,* 127–134.

Liras, P., Villanueva, J. R., and Martin, J. F. (1977) *J. Gen. Microbiol. 102,* 269–277.

Liu, C. M., McDaniel, L. E., and Schaffner, C. P. (1972) *J. Antibiot. 25,* 116–121.

Liu, C. M., McDaniel, L. E., and Schaffner, C. P. (1975) *Antimicrob. Agents Chemother. 7,* 196–202.

Liu, C. M., Hermann, T., and Miller, P. A. (1977) *J. Antibiot. 30,* 244–251.

Luengo, J. M., Revilla, G., Villaneuva, J. R., and Martin, J. F. (1979) *J. Gen. Microbiol. 115,* 207–211.
Lurie, L. M., Verkhotseva, T. P., and Levitov, M. M. (1975) *Antibiotiki 20,* 291–295.
Madre, N., and Pape, H. (1980) *Hoppe-Seyler's Z. Physiol. Chem. 361,* 299.
Magasanik, B., Prival, M., Brenchley, J., Tyler, B., Deleo, A., Streicher, S., Bender, R., and Paris, C. (1974) *Curr. Top. Cell Regul. 8,* 118–138.
Majumdar, M. K., and Majumdar, S. K. (1970) *Biochem. J. 120,* 271–278.
Majumdar, M. K., and Majumdar, S. K. (1971) *Folia Microbiol. 16,* 285–292.
Majumdar, M. K., and Majumdar, S. K. (1972) *Biochem. J. 122,* 397–404.
Malik, V. S., and Vining, L. C. (1970) *Can. J. Microbiol. 16,* 173–179.
Margalith, P., and Pagani, H. (1961) *Appl. Microbiol. 9,* 325–336.
Martin, J. F. (1976) in *Microbiology—1976* (Schlessinger, D., ed.) pp. 548–552, American Society for Microbiology, Washington, D.C.
Martin, J. F. (1978a) *Adv. Biochem. Eng. 6,* 105–127.
Martin, J. F. (1978b) in *Antibiotics and Other Secondary Metabolites: Biosynthesis and Production* (Hütter, R., Leisinger, T., Neusch, J., and Wehrli, W., eds.) pp. 19–37, Academic Press, New York.
Martin, J. F., and Demain, A. L. *Biochem. Biophys. Res. Comm. 71,* 1103–1109.
Martin, J. F., and Demain, A. L. (1977a) *Can. J. Microbiol. 23,* 1334–1339.
Martin, J. F., and Demain, A. L. (1977b) *J. Bacteriol. 132,* 590–595.
Martin, J. F., and Demain, A. L. (1978) in *The Filamentous Fungi* (Smith, J. E., and Berry, D. R., eds.) Vol. III, pp. 426–450, Edward Arnold, London.
Martin, J. F., and McDaniel, L. E. (1974) *Dev. Ind. Microbiol. 15,* 324–337.
Martin, J. F., and McDaniel, L. E. (1975) *Biotechnol. Bioeng. 17,* 925–938.
Martin, J. F., Gil, J. A., Naharro, G., Liras, P., and Villanueva, J. R. (1979a) in *Genetics of Industrial Microorganisms* (Sebek, O. K., and Laskin, A. I., eds.) pp. 205–209, American Society for Microbiology, Washington, D.C.
Martin, J. F., Naharro, G., Liras, P., and Villanueva, J. R. (1979b) *J. Antibiot. 32,* 600–606.
Martin, J. F., Alegre, M. T., Gil, J. A., and Naharro, G. (1981) in *Advances in Biotechnology: Fermentation Products* (Vezina, C., and Singh, K., eds.) Vol. III, pp. 129–134, Pergamon Press, Toronto, Canada.
Masurekar, P., and Demain, A. L. (1972) *Can. J. Microbiol. 18,* 1045–1048.
Masurekar, P. S., and Demain, A. L. (1974) *Antimicrob. Agents Chemother. 6,* 366–368.
Matern, U., Brillinger, G. U., and Pape, H. (1973) *Arch. Microbiol. 88,* 37–48.
Matteo, C. C., Glade, M., Tanaka, A., Piret, J., and Demain, A. L. (1975) *Biotechnol. Bioeng. 17,* 129–142.
Matteo, C. C., Cooney, C. L., and Demain, A. L. (1976) *J. Gen. Microbiol. 96,* 415–422.
Mendelovitz, S., and Aharonowitz, Y. (1982) *Antimicrob. Agents Chemother. 21,* 74–84.
Mertz, F. P., and Doolin, L. E. (1973) *Can. J. Microbiol. 19,* 263–270.
Miller, A. L., and Walker, J. B. (1969) *J. Bacteriol. 99,* 401–405.
Miller, A. L., and Walker, J. B. (1970) *J. Bacteriol. 104,* 8–12.
Muth, W. L., and Nash, C. H. (1975) *Antimicrob. Agents Chemother. 8,* 321–327.
Nestaas, E., and Demain, A. L. (1981) *Eur. J. Appl. Microbiol. Biotechnol. 12,* 170–172.
Nomi, R., and Nimi, O. (1969) *Agr. Biol. Chem. 33,* 1459–1463.

Nomi, R., Nimi, O., and Kado, T. (1968) *Agr. Biol. Chem. 32,* 1256–1260.
Ōmura, S., Tanaka, Y., Kitao, C., Tanaka, H., and Iwai, Y. (1980) *Antimicrob. Agents Chemother. 18,* 691–695.
Pass, L., and Raczyńska-Bojanowska, K. (1968) *Acta Biochim. Pol. 15,* 355–367.
Pastan, L., and Adhya, S. (1976) *Bacteriol. Rev. 40,* 527–551.
Paulus, H. (1967) in *Antibiotics* Vol. II, *Biosynthesis* (Gottlieb, D., and Shaw, P. D., eds.) pp. 254–267, Springer, New York.
Piret, J. M., and Demain, A. L. (1981) in *Sporulation and Germination* (Levinson, H. S., Sonenshein, A. L., and Tipper, D. J., eds.) pp. 243–245, American Society for Microbiology, Washington, D.C.
Pirt, S. J. (1976) *Dev. Ind. Microbiol. 17,* 1–10.
Pirt, S. J., and Righelato, R. C. (1967) *Appl. Microbiol. 15,* 1284–1290.
Polsinelli, M., Albertini, A., Cassani, G., and Ciferri, O. (1965) *J. Gen. Microbiol. 39,* 239–246.
Ragan, C. M., and Vining, L. C. (1978) *Can. J. Microbiol. 24,* 1012–1015.
Ramsey, H. H., Hussain, S. M., and Williams, R. P. (1973) Abstacts of the Annual Meeting of the American Society for Microbiology, p. 180.
Revilla, G., Luengo, J. M., and Martin, J. F. (1978) Abstracts of the First European Congress on Biotechnology, pp. 3/51–3/52, Interlaken, Switzerland.
Rogers, T. O., and Birnbaum, J. (1974) *Antimicrob. Agents Chemother. 5,* 121–132.
Ryu, D. D. Y., and Hospodka, J. (1980) *Biotechnol. Bioeng. 22,* 289–298.
Sankaran, L., and Pogell, B. M. (1973) *Nature New Biol. 245,* 260.
Sankaran, L., and Pogell, B. M. (1975) *Antimicrob. Agents Chemother. 8,* 721–732.
Sarkar, N., and Paulus, H. (1972) *Nature New Biol. 239,* 228–230.
Satoh, A., Ogawa, H., and Satomura, Y. (1975) *Agr. Biol. Chem. 39,* 2331–2336.
Satoh, A., Ogawa, H., and Satomura, Y. (1976) *Agr. Biol. Chem. 40,* 191–196.
Schaffer, P. (1969) *Bacteriol. Rev. 33,* 48–71.
Seno, E. T., and Baltz, R. H. (1981) *Antimicrob. Agents Chemother. 20,* 370–377.
Smith, C. G., and Hinman, J. W. (1963) *Prog. Ind. Microbiol. 4,* 137–163.
Smith, R. L., Bungay, H. R., and Pittenger, R. C. (1962) *Appl. Microbiol. 10,* 293–296.
Snoke, J. E., and Cornell, N. (1965) *J. Bacteriol. 89,* 415–420.
Soltero, F. V., and Johnson, M. J. (1953) *Appl. Microbiol. 1,* 52–57.
Soltero, F. V., and Johnson, M. J. (1954) *Appl. Microbiol. 2,* 41–44.
Speedie, M. K., Hornemann, U., and Floss, H. G. (1975) *J. Biol. Chem. 250,* 7819–7825.
Spizek, J., Malek, I., Suchy, J., Vondracek, M., and Vanek, Z. (1965) *Folia Microbiol. 10,* 263–266.
Terry, J., and Springham, D. G. (1981) *Can. J. Microbiol. 27,* 1044–1047.
Tomino, S., Yamada, M., Itoh, H., and Kurahashi, K. (1967) *Biochemistry 6,* 2552–2560.
Treichler, H. J., Liersch, M., Nuesch, J., and Dobeli, H. (1979) in *Genetics of Industrial Microorganisms* (Sebek, O. K., and Laskin, A. I., eds.) pp. 97–104, American Society for Microbiology, Washington, D.C.
Unowsky, J., and Hoppe, D. C. (1978) *J. Antibiot. 31,* 662–666.
Vining, L. C., and Chatterjee, S. (1982) in *Overproduction of Microbial Products* (Kramphanel, V., Sikyfa, B., and Vanek, Z., eds.) pp. 35–46, Academic Press, London.

Walker, J. B. (1974) *J. Biol. Chem. 249,* 2397–2404.
Walker, J. B., and Hnilica, V. S. (1964) *Biochim. Biophys. Acta 89,* 473–482.
Walker, J. B., and Walker, M. S. (1967) *Biochemistry 6,* 3821–3829.
Walker, M. S., and Walker, J. B. (1971) *J. Biol. Chem. 246,* 7034–7040.
Wax, R., and Snyder, L. (1980) Abstracts of the Annual Meeting of the American Society for Microbiology, p. 187.
Weinberg, E. D. (1970) *Adv. Microb. Physiol. 4,* 1–44.
Weinberg, E. D., and Tonnis, S. M. (1967) *Can. J. Microbiol. 13,* 614–615.
Westlake, D. W. S., Sala, F., McGrath, R., and Vining, L. C. (1968) *Can. J. Microbiol. 14,* 587–593.
Whitney, J. G., Brannon, D. R., and Wickner, K. J. (1972) *Antimicrob. Agents Chemother. 1,* 247–251.
Yanagimoto, M., and Terui, G. (1971a) *J. Ferment. Technol. 49,* 604–610.
Yanagimoto, M., and Terui, G. (1971b) *J. Ferment. Technol. 49,* 611–618.
Zanca, D. M., Villanueva, J. R., and Martin, J. F. (1982) Abstracts of the Fourth International Symposium on Genetics of Industrial Microorganisms, p. 67, Kyoto, Japan.

CHAPTER

4

β-Lactams

Joseph O'Sullivan
Christopher Ball

I. INTRODUCTION

β-Lactam antibiotics can be grouped into at least six families; of these, the penicillins and cephalosporins are the most important clinically, and the most biochemical and genetic data has been accumulated about them. In the last decade, four additional families of β-lactam antibiotics have been discovered. The cephamycins are similar to the cephalosporins, but the clavams, the carbapenems, and the monocyclic β-lactams all represent significant structural departures from the classical penicillins and cephalosporins. In this review, we present an outline of the biochemical, microbiological, and genetic research that has led to our present understanding of β-lactam antibiotics. Because of the breadth of the subject matter, we have been selective in our coverage, but wherever necessary, we refer to recent interviews that consider aspects of the subject in greater detail.

II. BIOCHEMISTRY OF β-LACTAM ANTIBIOTICS

A. Biosynthesis

The main features of the biosynthesis of penicillins and cephalosporins have now been worked out. They are shown in Figs. 4.1 and 4.2.

1. Isopenicillin N. Formation of the dipeptide L-α-aminoadipyl-L-cysteine (AC) is the first step in producing penicillins and cephalosporins. No information is available on its biosynthesis, although this step may be seen as the point of commitment to β-lactam formation and therefore under greater regulatory control than other steps in the pathway. The AC then condenses with L-valine to give the tripeptide δ-(L-α-aminoadipyl)-L-cysteinyl-D-valine (ACV) with epimerization of the L-valine. Tripeptide synthesis from AC and valine was studied by Loder and Abraham (1971b) in cell-free extracts of *Cephalosporium* sp. They found that the enzyme responsible was particulate, in contrast to glutathione synthase, which was soluble. In addition to the ACV tripeptide, *Cephalosporium acremonium* produces two related peptides, one of which contains α-aminoadipic acid, cysteine, valine, and glycine, while the other contains α-aminoadipic acid, cysteine, β-hydroxyvaline, and glycine (Loder and Abraham, 1971a). Eriquez and Pisano (1979) isolated the same or similar peptides from *Paecilomyces* and concluded that the glycine was N-terminal. *Penicillium chrysogenum* produces α-aminoadipylserinylvaline, α-aminoadipylserinylisodehydrovaline, and α-aminoadipylalanylvaline in addition to ACV (Neuss et al., 1980), and the prokaryote *Streptomyces clavuligerus* produces two dipeptides, α-aminoadipylcysteine and γ-glutamylcysteine, as well as ACV (J. O'Sullivan and Abraham, 1981).

The relative abundance of α-aminoadipic acid–containing peptides in penicillin-, cephalosporin-, and cephamycin-producing organisms may reflect broad specificity of the peptide synthases involved or poor regulation of peptide formation. Since a definite biosynthetic role has been ascribed to only one of the peptides, namely ACV, it may be that control of the biosynthetic pathway is exerted predominantly at the point of ACV cyclization to give isopenicillin N. Cyclization of ACV to give a penicillin was first demonstrated by Fawcett et al. (1976) using lysed protoplasts of *C. acremonium*. Subsequently, it was shown in both *C. acremonium* (J. O'Sullivan et al., 1979b, Konomi et al., 1979) and *P. chrysogenum* (Meesschaert et al., 1980) that the product was isopenicillin N and that the cyclization took place without prior hydrolysis of the tripeptide (J. O'Sullivan et al., 1979b). The isopenicillin N synthase has been partially purified (Abraham et al., 1980) and found to require the SH form of the tripeptide, O_2, a reducing agent, and ferrous ion for activity, but 2-oxoglutarate was not required. The enzyme displays a marked specificity for LLD-ACV, although a penicillin product was generated when the valine of ACV was replaced by isoleucine (Abraham et al., 1980, Bahadur et al., 1981).

2. Cephalosporins. The biosynthesis of β-lactams diverges at isopenicillin N; cephalosporins are formed in fungi such as *C. acremonium,* whereas penicillins with nonpolar side chains are formed in fungi such as *P. chrysogenum*. Penicillin N is formed from isopenicillin N by epimerization of

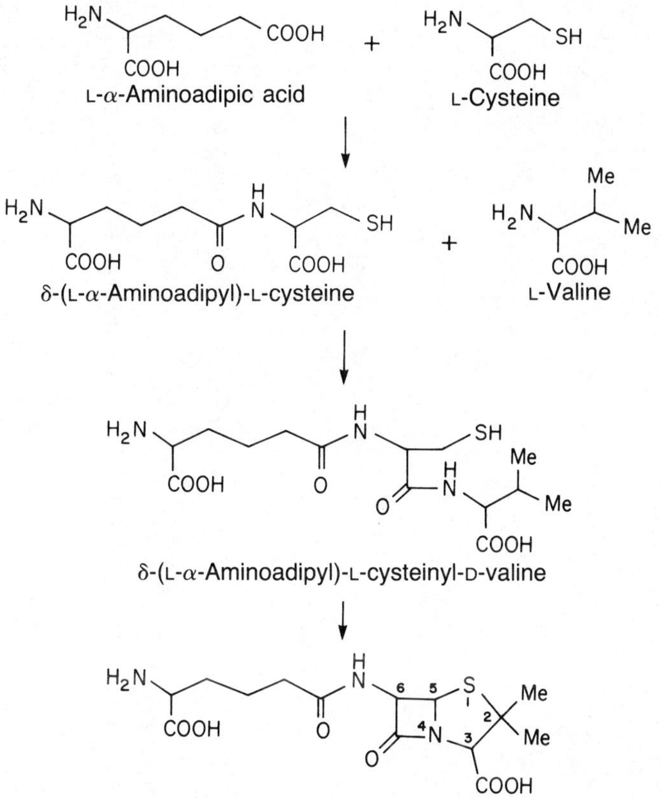

FIGURE 4.1 Biosynthesis of isopenicillin N.

the L-α-aminoadipic side chain (Abraham et al., 1980; Jayatilake et al., 1981). The enzyme responsible is very labile and has not been purified.

Although for many years it seemed likely that cephalosporins were derived from penicillins, it was only when a protoplast lysate was made from a β-lactam nonproducing strain of *C. acremonium* that this could be demonstrated (Kohsaka and Demain, 1976). Detection of the reaction was made possible by the availability of a β-lactam supersensitive strain of *Escherichia coli* (Aoki et al., 1976). Subsequently, using ^{3}H- and ^{14}C-labeled penicillin N in the cell-free system, it was unequivocally shown that ^{3}H-, ^{14}C-labeled desacetoxycephalosporin C was formed (Baldwin et al., 1980). The reaction was studied in more detail by Sawada et al., (1979) and Hook et al., (1979) and shown to be stimulated by ferrous ion and ascorbate but not by 2-oxoglutarate. Later, Felix et al. (1981), using permeabilized cells of *C. acremonium,* indicated a marked stimulation of activity by 2-oxoglutarate. Resolution of these differences and a fuller analysis of the specificity of the enzyme must await its purification. Specificity data would help to

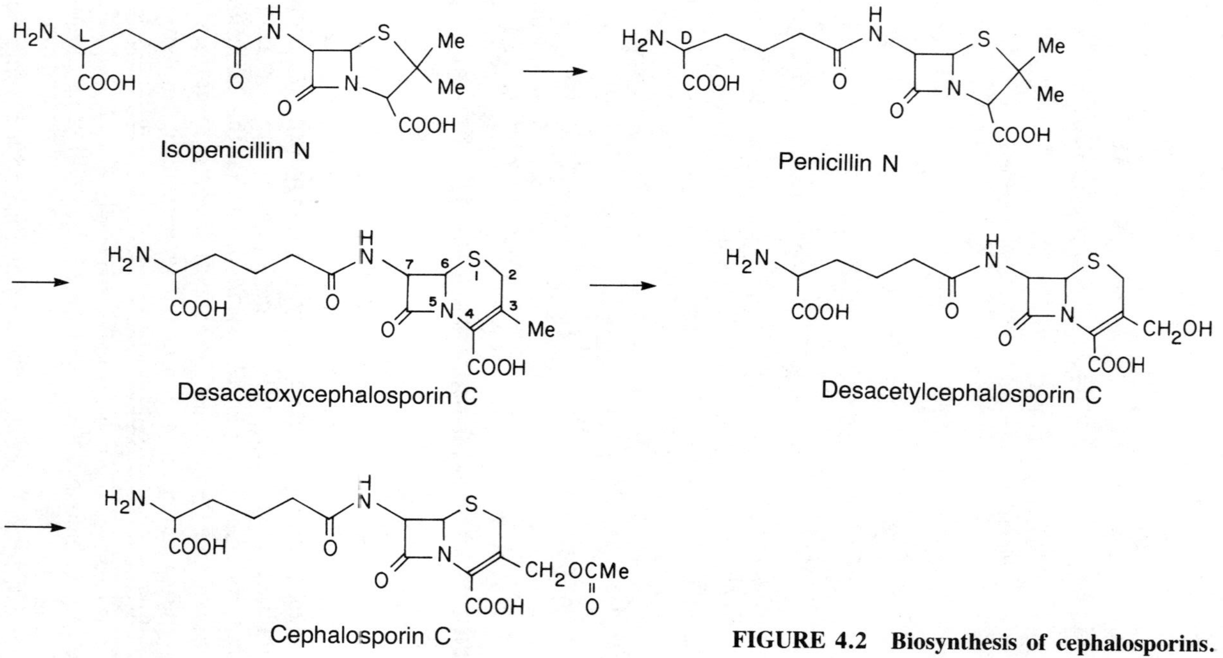

FIGURE 4.2 Biosynthesis of cephalosporins.

explain why such an enzyme exists in *C. acremonium* but not in *P. chrysogenum,* and also why a mutant of *C. acremonium* that converts the putative tripeptide LLD-S-carboxymethylcys-*cys-val* (Troonen et al., 1976) to the corresponding penicillin cannot, apparently, form the corresponding desacetoxycephalosporin.

The conversion of desacetoxycephalosporin C to desacetylcephalosporin C, which has been studied with partially purified enzymes from *C. acremonium* and *S. clavuligerus,* is a dioxygenase-catalyzed reaction by virtue of its requirement for O_2, Fe^{2+}, a reducing agent, and 2-oxoglutarate. (Turner et al., 1978); it has also been established that the oxygen atom at the exocyclic methylene of cephalosporin C is derived from O_2 (Stevens et al., 1975). The final step in forming cephalosporin C is acetylation of desacetylcephalosporin C; this reaction occurs with acetyl-CoA as the acetyl donor in the presence of Mg^{2+} (Fujisawa et al., 1973; Liersch et al., 1976). Support for the sequence desacetoxycephalosporin C → desacetylcephalosporin C → cephalosporin C has been provided by the isolation of mutants which are desacetylcephalosporin C– and cephalosporin C–negative and therefore accumulate desacetoxycephalosporin C, or are cephalosporin C–negative and accumulate desacetylcephalosporin C (Queener et al., 1974; Fujisawa et al., 1975; Liersch et al., 1976).

(I)

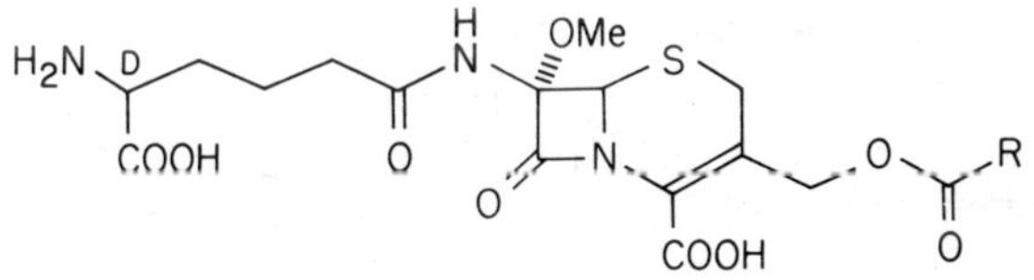

3. Cephamycins. Cephamycins typically have Structure I, where R is —$C(OCH_3){=}CHC_6H_4OH$ in cephamycin A, —$C(OCH_3){=}CHC_6H_4OSO_3H$ in cephamycin B, and —NH_2 in cephamycin C; a desmethoxycephamycin C has also been reported (Higgens and Kastner, 1971; Nagarajan et al., 1971). The cephamycins are produced by streptomycetes and are derived from α-aminoadipic acid, cysteine, and valine as are penicillins and cephalosporins (Whitney et al., 1972). The oxygen of the methoxyl group is derived from molecular oxygen (J. O'Sullivan et al., 1979a), and the methoxy methyl from methionine (Whitney et al., 1972). An enzyme system from *S. clavuligerus* methoxylates cephalosporin C and carbamoyldesacetylcephalosporin C when O_2, Fe^{2+}, 2-oxoglutarate, S-adenosylmethionine, and a reducing agent are present (J. O'Sullivan and Abraham, 1980). In a separate study, an enzyme from *S. clavuligerus* was found to transfer a carbamoyl group from carbamoyl phosphate to desacetyl-7α-methoxycephalosporin C or to desacetylcephalosporin C (Brewer et al., 1980). The specificity of carbamoylation and methoxylation and the distribution of cephamycins in natural isolates suggest the pathway:

desacetoxycephalosporin C → desacetylcephalosporin C →
O-carbamoyldesacetylcephalosporin C → cephamycin C.

4. Penicillins with nonpolar side chains. The biosynthesis of penicillins with nonpolar side chains differs from that of penicillin N, cephalosporins, and cephamycins in that the side chain can be varied by adding monosubstituted acetic acids, such as phenylacetic acid, to the fermenting organism which is, typically, *P. chrysogenum*. When *P. chrysogenum* is grown in the absence of a suitable side chain precursor, 6-aminopenicillanic acid (6-APA) (Structure II) is produced. When a side chain precursor is added, penicillin with the

(II)

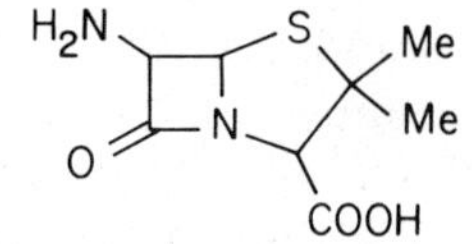

corresponding side chain is excreted into the medium. This suggests the presence of enzymes in *P. chrysogenum* that remove the α-aminoadipyl side chain of isopenicillin N, activate the side chain precursor, and transfer the activated side chain to the penicillin nucleus. Thus

$$\text{isopenicillin N} + \text{phenylacetyl CoA} \rightarrow \text{penicillin G.}$$

The side chain precursor must be transported into the cells, and the penicillin formed must be transported out. Apart from this, the formation of penicillins appears to be a relatively simple enzymatic process. However, many details remain unclear; in particular, it is not known whether 6-APA is a free intermediate in the conversion of isopenicillin N to penicillin G and how many enzymes are involved. Numerous enzymatic activities relating to the terminal stages of penicillin formation have been detected in cell extracts of *P. chrysogenum,* and these have been reviewed in detail by Neuss and Queener (1982). Activation of the side chain precursor has been demonstrated in cell-free enzyme systems, as has acylation of 6-APA and transacylation between benzyl penicillin and other penicillins. Fewer studies have been carried out with isopenicillin N, however, owing largely to the difficulty of acquiring this material. Using a crude extract of *P. chrysogenum,* Loder (1972) showed the formation of ^{14}C-benzylpenicillin from ^{14}C-phenylacetyl CoA and isopenicillin N. This was confirmed and expanded by Fawcett et al. (1975), who showed that 6-APA and isopenicillin N, both labeled with ^{3}H in the 2β-methyl position, were substrates for an acyltransferase in *P. chrysogenum*.

The question of what happens to the α-aminoadipic acid after removal from isopenicillin N was examined by Friedrich and Demain (1978) using ^{14}C-α-aminoadipic acid. Seven to ten moles of penicillin were formed per mole of α-aminoadipic acid. The presence of 6-oxopiperidine-2-carboxylic acid (cyclic α-aminoadipic acid, Structure III) was reported by Trown et al. (1963) in a *Cephalosporium* sp. and was recently detected at a level of 28 mole % of penicillin V in fermentations of a high-titer strain of *P. chrysogenum* (Brundidge et al., 1980). The ratio of III to penicillin V was constant throughout the growth cycle, implying a close relationship between formation of penicillin V and production of III; it is not clear whether production of III is a general feature of the penicillin fermentation or is confined to certain highly mutated strains. Further work in this area, especially with cell-free systems, is likely to throw substantial light on the central enzymatic questions of the *Penicillium* fermentation.

(III) 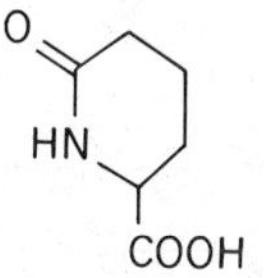

Penicillin acylases, which are of great importance in commercial development of the clinically useful penicillins, are discussed in Section II.B.

The subcellular localization of penicillin in *P. chrysogenum* was examined by Kurylowicz et al. (1969) using enzymatic and immunological techniques. They concluded that benzylpenicillin was present in, and exported by, the Golgi apparatus. Now that several of the enzymes of penicillin (and cephalosporin) biosynthesis can be measured in cell-free systems, it should be possible to design further histochemical tests for electron microscopy and also to analyze subcellular localizations by density-gradient centrifugation. In this way, knowledge of the subcellular distribution of the metabolic processes leading to β-lactam formation, and of the secretion process, may be gained.

5. Clavulanic acid. Using a β-lactamase inhibition screen, Reading and Cole (1977) isolated clavulanic acid (Structure IV, R = H) from *S. clavuligerus*.

(IV) 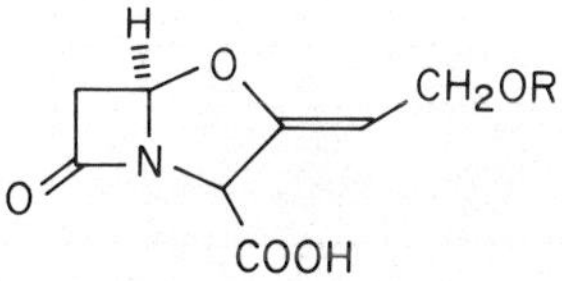

This organism also produces penicillin N, deacetoxycephalosporin C, and cephamycin C. Clavulanic acid consists of a β-lactam ring fused to an oxazolidine ring and therefore differs from classical β-lactams in having an oxygen atom in place of sulfur; there is also no substitution at the C-6 position. The same culture contained a second clavulanate (Structure IV, R = $COCH_2CH_2OH$) with a β-hydroxypropionyl group (Beecham Ltd., 1979) and three related compounds (Structure V, R = COOH, CH_2OH, or CH_2OCHO) lacking the carboxyl group (D. Brown et al., 1979). An additional

(V) 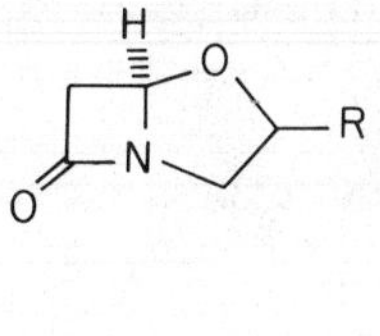

member (Structure VI) of the clavam series was recently isolated from *Streptomyces antibioticus* in a screen for antifungal agents (Wanning et al., 1981).

(VI)

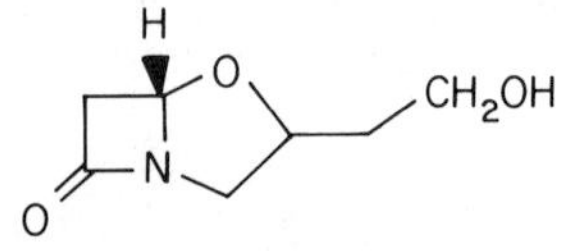

The biosynthesis of clavulanic acid was studied using ^{13}C-NMR (Elson and Oliver, 1978) and ^{14}C labeling (Sterling and Elson, 1979). Glycerol was shown to provide the carbon atoms of the β-lactam ring, and the remainder of the molecule was probably derived from glutamate via 2-oxoglutarate. The coincidence of clavams and classical β-lactams in the same organism raises the question whether a uniform mechanism exists for closing the β-lactam ring in this organism. A single-point-mutant leading to the simultaneous loss of clavams and penams in *S. clavuligerus* has yet to be reported. No evidence for a peptide precursor of clavulanic acid has been reported, and specificity data for the ACV cyclase (Abraham et al., 1980) would seem, at this time, to preclude cyclization of a peptide lacking cysteine.

6. The carbapenems. The carbapenems are a group of naturally occurring β-lactams produced mainly by streptomycetes. A large number have been described to date (Table 4.1). They resemble penicillins in having a five-membered ring fused to the β-lactam but differ in that the sulfur of penicillin is replaced by carbon and there is a double bond between C-2 and C-3.

Thienamycin, the first carbapenem reported from nature, is produced by *Streptomyces cattleya* together with N-acetylthienamycin and N-acetyldehydrothienamycin (Kahan et al., 1979). The epithienamycins (Cassidy et al., 1981) produced by strains of *Streptomyces flavogriseus* (Stapley et al., 1981) appear to be synonymous with olivanic acids (MM series, Table 4.1) (A. G. Brown et al., 1976). The PS series (Okamura et al., 1978; Shibamoto et al., 1980), C-1393-S_2, and C-1393-H_2 (Harada et al., 1980), which appear to be identical to the carpetimycins (Nakayama et al., 1981), and the asparenomycins (Tanaka et al., 1981) extend this group. A ring-saturated carbapenam has also been reported (Sankyo Co., Ltd., 1980). The configuration at C-8 of the side chain and of the hydrogens on the β-lactam ring is also of importance in differentiating the carbapenems and in determining their properties. Recently, a simple carbapenem, SQ 27,860, has been isolated from species of *Serratia* and *Erwinia* (Parker et al., 1982).

The thienamycin nucleus is biosynthetically derived from acetate (β-lactam carbons) and glutamate (five-membered ring) (Albers-Schonberg et al., 1976). Box et al. (1979) studied the interrelationships of the *cis*-olivanic acids and reported that the three sulfated compounds (MM 13902, 17880, 4550) were derived from MM 22380, whereas MM22382 appeared to be the

TABLE 4.1 Structures of Carbapenems

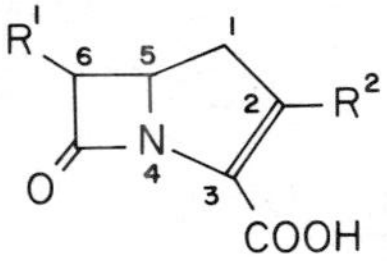

Substance	R^1	R^2	*β-lactam configuration*
Thienamycin	—$CH(CH_3)OH$	—$SCH_2CH_2NH_2$	*trans*
N-acetylthienamycin	—$CH(CH_3)OH$	—$SCH_2CH_2NHCOCH_3$	*trans*
Epithienamycin A/MM22380	—$CH(CH_3)OH$	—$SCH_2CH_2NHCOCH_3$	*cis*
Epithienamycin B/MM22382	—$CH(CH_3)OH$	—$SCH{=}CHNHCOCH_3$	*cis*
Epithienamycin C/MM22381	—$CH(CH_3)OH$	—$SCH_2CH_2NHCOCH_3$	*trans*
Epithienamycin D/MM22383	—$CH(CH_3)OH$	—$SCH{=}CHNHCOCH_3$	*trans*
Epithienamycin E/MM13902	—$CH(CH_3)OSO_3H$	—$SCH{=}CHNHCOCH_3$	*cis*
Epithienamycin F/MM17880	—$CH(CH_3)OSO_3H$	—$SCH_2CH_2NHCOCH_3$	*cis*
MM4550	—CH_2CH_3	—$SCH_2CH_2NHCOCH_3$	*cis*
PS-5	—CH_2CH_3	—$SCH_2CH_2NHCOCH_3$	*trans*
PS-6	—$CH(CH_3)_2$	—$SCH_2CH_2NHCOCH_3$	*trans*
PS-7	—CH_2CH_3	—$SCH{=}CHNHCOCH_3$	*trans*
Carpetimycin A/C-1393-H_2	—$CH(CH_3)O$	—$SCH{=}CHNHCOCH_3$	*cis*
Carpetimycin B/C-1393-S_2	—$CH(CH_3)OSO_3H$	—$SCH{=}CHNHCOCH_3$	*cis*
Asparenomycin	$=C(CH_3)CH_2OH$	—$SCH{=}CHNHCOCH_3$	—
SQ 27,860	—H	—H	—

direct precursor of MM13902 and MM4550. Relationships between these and other carbapenems were further discussed by Shibamoto et al. (1980). PS-6, the carpetimycin/C-1393 compounds, and asparenomycin all have three carbons in the C-6 side chain and may represent a distinct biosynthetic pathway or pathways. As with clavulanic acid, the carbapenems are produced by organisms that, in most reported cases, also produce penicillin and cephalosporin, but so far, no data linking the biosynthetic pathways have been reported.

7. Monocyclic β-lactams. The monocyclic β-lactams are a group of compounds characterized by having a single unfused β-lactam ring. The earliest reported examples were the alkaloids, pachystermine A and B (Kikuchi and Uyeo, 1967), tabtoxin, the agent of wildfire disease in tobacco plants, which is produced by *Pseudomonas tabaci* (Stewart, 1971), and a similar compound, (*S*)-alanyl-3-[α-(*S*)-chloro-3(*S*)-hydroxy-2-oxo-3-azetidinylmethyl]-(*S*)-alanine (Scannell et al., 1975), produced by a streptomycete. Our discussion here will concentrate on two other classes of monocyclic β-lactams, the nocardicins and the monobactams.

(VII)

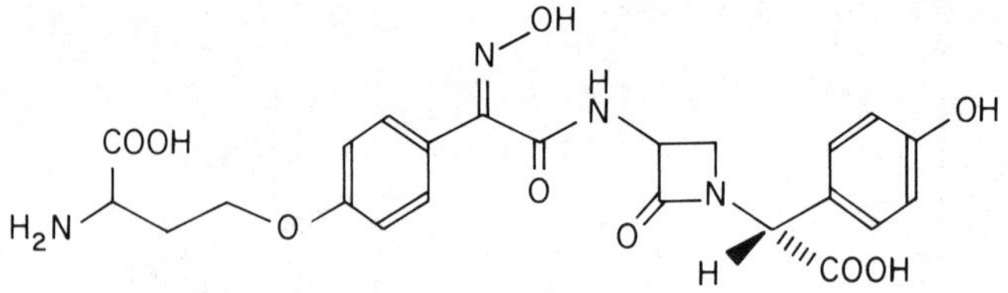

The nocardicins, a group of seven related compounds, were isolated from *Nocardia uniformis* subsp. *tsuyamanensis* fermentation broths (Aoki et al., 1976; Hosoda et al., 1977). Nocardicin A, the prototype of the group, has Structure VII. Its production was stimulated by adding tyrosine, *p*-hydroxypyruvic acid, DL-*p*-hydroxymandelic acid, or L-phenylglycine to fermentations (Hosoda et al., 1977). Radiolabel experiments showed that ^{14}C from tyrosine, serine, homoserine, and glycine was incorporated into nocardicin A. [U-^{14}C]Serine was incorporated to a level of 18%, and when nocardicin was hydrolyzed to 3-aminocardicinic acid (Structure VIII), 84% of the added ^{14}C was recovered. This left little doubt that serine was incorporated into the β-lactam ring (Hosoda et al., 1977).

(VIII)

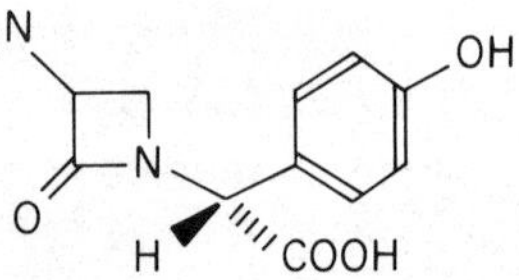

The origin of the ether-linked homoserine moiety was less clear, however, since homoserine did not stimulate nocardicin A production and was incorporated to only 4% (Hosoda et al., 1977). This problem was reexamined by Townsend and Brown (1981), who showed that methionine stimulated production of nocardicin A and was incorporated to a level about 12 times that of homoserine. They also showed that, while [U-^{14}C]tyrosine was well incorporated into the two aromatic segments of nocardicin A, [1-^{14}C]tyrosine was poorly incorporated, suggesting that the true precursor was *p*-hydroxyphenylglycine. A series of experiments using specifically labeled glycine and serine established that serine was incorporated into the β-lactam ring of nocardicin A with retention of ^{3}H at the C-3 position. An S_n2 displacement reaction was postulated as a mechanism for ring closure, and phosphorylated serine was suggested as a likely intermediate (Townsend and Nguyen, 1981). No information on peptide precursors of nocardicin A or on the genetics of the producing organism has yet been reported.

The most recent β-lactam antibiotics to be discovered are monocyclic compounds of bacterial origin with a sulfate group attached to the nitrogen of the ring (Table 4.2). The first reported examples were from *Pseudomonas* sp. and proved to be isomers; they were given the names sulfazecin and isosulfazecin (Imada et al., 1981). Independently, Sykes et al. (1981) reported the isolation of three classes of monocyclic β-lactams from diverse bacterial taxons. This group of compounds was given the name monobactams.

The first to be isolated and the simplest structurally is SQ 26,180, produced by *Chromobacterium violaceum*. It has weak activity against a

TABLE 4.2 Structures of Monobactams and Sulfazecins

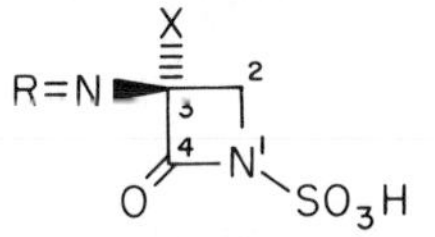

Compound	*X*	*R*
Sulfazecin/SQ 26,445	—OCH_3	γ-D-*glu*-D-*ala*-, H—
Isosulfazecin	—OCH_3	γ-D-*glu*-L-*ala*-, H—
SQ 26,180	—OCH_3	CH_3CO—, H—

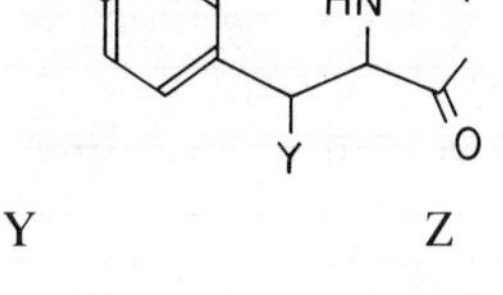

		Y	Z
SQ 26,700	—H	—H	—OH
SQ 26,875	—OCH_3	—H	—OH
SQ 26,823	—OCH_3	—H	—H
SQ 26,970	—OCH_3	—OH	—OSO_3H
SQ 26,812	—OCH_3	—OSO_3H	—OSO_3H

range of bacteria and is stable to β-lactamases; it inhibits *Streptomyces* R61-DD-carboxypeptidase and binds to *E. coli* penicillin-binding proteins (PBP)1a, 4, and 5/6 as well as to PBP4 of *Staphylococcus aureus* (Wells et al., 1982a). The compound SQ 26,445 is the same as sulfazecin. It was reported from numerous strains of *Acetobacter* and *Gluconobacter* (Sykes et al., 1981) as well as from *Pseudomonas acidophila* (Imada et al., 1981; Asai et al., 1981). *Agrobacterium radiobacter* produces a family of five monobactams with weak antibacterial activity (Sykes et al., 1981; Parker and Rathnum, 1982). They inhibit the *Streptomyces* R61-DD-carboxypeptidase and are resistant to β-lactamases, with the exception of SQ 26,700, which lacks a 3α-methoxyl group, and of SQ 26,823, which is susceptible to the *S. aureus* penicillinase (Wells et al., 1982b).

Studies on the biosynthesis of monobactams have shown that the carbon atoms of the β-lactam ring are derived from serine in all three types, represented by SQ 26,180, SQ 26,445, and SQ 26,812. This suggests a uniform ring closure mechanism. The sulfamate sulfur of monobactams could be supplied by a wide variety of organic and inorganic sulfur-containing compounds, presumably via active sulfate. A group of intracellular peptides that have been reported might be intermediates in the biosynthesis of monobactams (J. O'Sullivan et al., 1982).

The methoxy methyl group is derived from methionine (J. O'Sullivan et al., 1982), and by analogy with the cephamycin C methoxyl group; the oxygen presumably is derived from molecular oxygen. Further, since a desmethoxymonobactam has been isolated (SQ 26,700, Table 4.2) it is likely that here, as with cephamycin C, the methoxyl group is added after ring closure.

B. Other Biochemical Reactions of β-Lactams

1. Enzymic Synthesis and Degradation. a. *Acylases and Esterases.* Most of the commercially available penicillins and cephalosporins are semisynthetic. Modifications of naturally produced penicillins and cephalosporins to yield the nuclei 6-APA and 7-ACA and further modifications of these nuclei to yield the required penicillin or cephalosporin are the preferred commercial routes. These modifications can be chemical or enzymic, and it is the latter that will be emphasized here. The enzymic removal of the aromatic side chain of penicillin G using mycelial extracts of *P. chrysogenum* and *Aspergillus oryzae* was first described by Sakaguchi and Murao (1950). However, the identification of 6-APA and its application to semisynthesis came later (Batchelor et al., 1961). The acylase route to 6-APA is favored over the other possible biological route, namely, direct fermentation of 6-APA by *P. chrysogenum* without added side chain precursor (Batchelor et al., 1959), because the isolation of 6-APA from fermentation broth is complicated.

An extremely wide range of microorganisms has been found to possess acylase activity; penicillin G acylases are mainly produced by bacteria, and penicillin V acylases by molds and actinomycetes (Vandamme and Voets, 1974). A very detailed survey of acylase specificity has been presented by Vandamme and Voets (1974) and Vandamme (1977). Enzymes that efficiently remove the α-aminoadipyl side chain from cephalosporin C to yield 7-ACA have not been found. However, a two-step removal of the α-aminoadipyl side chain by oxidation to glutaric acid (Gilbert et al., 1972) followed by cleavage of the side chain (Shibuya et al., 1981) has been reported. Whether the isopenicillin N acylase (transacylase) of *P. chrysogenum* can remove the cephalosporin C side chain, or can be modified to do so, is yet unclear.

In another degradation, esterase action at position 3 converts cephalosporin C to deacetylcephalosporin C (Jeffery et al., 1961). Abbott and Fukuda (1975a) have described the purification and kinetic properties of a *Bacillus subtilis* cephalosporin acetylesterase. The enzyme was later immobilized by adsorption onto bentonite with little denaturation (Abbott and Fukuda, 1975b). The presence in *C. acremonium* of a cephalosporin C esterase that is repressed by glucose has been reported (Hinnen and Nuesch, 1976). Enzymes that deacylate epithienamycins (Kahan and Kahan, 1980; Cassidy et al., 1981) and PS-5 (Fukagawa et al., 1980) have also been reported.

Following initial indications of β-lactam synthesis by reversal of acylase activity (Rolinson et al., 1960; Claridge et al., 1960), Shimizu et al. (1975) found that most penicillin acylase-producing bacteria could synthesize penicillins from 6-APA. In recent years, very efficient syntheses have been described. Thus Kato et al. (1980) produced amoxicillin with a nonimmobilized whole cell system of *Xanthomonas citri*. More than 90% of the added 6-APA was converted to amoxicillin under optimum conditions. Ampicillin synthesis has been studied with the immobilized enzyme from *E. coli* (Marconi et al., 1975), and cephalexin synthesis has recently been studied by Choi et al. (1981) using immobilized *X. citri* cells. Presumably, this technology could find application in the synthesis of a variety of cephalosporins and penicillins.

b. *β-Lactamases.* β-Lactamases are monomeric proteins that catalyze hydrolysis of the β-lactam ring of penicillins and cephalosporins. They are widespread in nature and probably represent the single most important factor contributing to the resistance of bacteria to β-lactam antibiotics. The genes for β-lactamases may be chromosomal or extrachromosomal and can be mobilized by transposition, conjugation, transduction, or transformation; they therefore spread and survive successfully in the microbial population. Studies of the occurrence, the substrate profiles, and other properties of these enzymes have allowed a classification scheme to be developed (Richmond and Sykes, 1973). Mechanistic studies using synthetic (for example, 6-β-bromopenicillanic acid) and naturally occurring (for example, clavulanic acid) inhibitors of β-lactamase have identified serine at the active site of certain β-lactamases and shown that it participates in forming an acyl-enzyme intermediate (Knott-Hunziker et al., 1979; Fisher et al., 1980). These and other aspects of the subject have recently been reviewed (Sykes and Bush, 1982; Hamilton-Miller and Smith, 1979; Fisher and Knowles, 1980).

2. Penicillin-binding proteins. Bacteria contain proteins that bind penicillins, cephalosporins, and other β-lactams and are referred to as penicillin-binding proteins (PBP's). β-Lactams are believed to act by binding to such proteins, thereby preventing cell wall growth and thus leading to the death of the cell. This model of the mode of action of β-lactams was first suggested by Tipper and Strominger (1965).

The PBP's have been well studied in *E. coli* (Spratt, 1977), and various functions in the development of the cell have been ascribed to them. Thus PBPs 1a and 1b (MW 91,000) are transpeptidases involved in cell elongation; PBP2 (MW 66,000) functions in shape determination; PBP3 (MW 60,000) functions in septation; and PBP4 (MW 49,000), PBP5 (MW 42,000), and PBP6 (MW 40,000) are carboxypeptidases.

The morphological effect of a given β-lactam can therefore be well correlated with its binding to a given PBP. Inhibition of PBP 1a, 1b, 2, and 3 eventually causes cell death. The most important of these is PBP3, through which most β-lactams exert their lethal effect (Spratt, 1980; Curtis, 1981).

III. GENETIC REGULATION OF β-LACTAM PRODUCTION

A. Regulation

1. Growth and Penicillin Production. In a batch culture, the cell is constantly adapting to a new environment, and a steady state may never be achieved. Batch cultures therefore represent a poorly defined system for the study of parameters controlling product formation. With continuous cultures, a steady state can be achieved, and individual variables can be more rationally studied. Observations of continuous cultures can be of great value in interpreting non–steady state phenomena. Application of the chemostat to answer basic questions about growth and β-lactam production is well illustrated by the study of Pirt and Righelato (1967). They found the specific rate of penicillin production (Q_{pen}) to be independent of growth rate between 0.014 h^{-1} and 0.1 h^{-1}. Below 0.014 h^{-1}, Q_{pen} decreases as growth rate decreases. This suggests the need for a growth rate in industrial batch fermentation not lower than 0.014 h^{-1}. The chemostat has also been used to study regulation of penicillin production in lysine auxotrophs of *P. chrysogenum* (C. Y. O'Sullivan and Pirt, 1973).

2. Fermentation of Other β-Lactams. The power of the chemostat to distinguish between growth rate effects and effects due to specific substrates in excess has been demonstrated recently by Lilley et al. (1981) in their studies with *S. cattleya,* a producer of the β-lactams thienamycin and cephamycin C. The specific rate of thienamycin production was diminished by phosphate, whereas production of cephamycin C was reduced at high growth rates and not at low growth rates, irrespective of whether the substrate in excess was carbon, nitrogen, or phosphate.

Low-yielding and high-yielding strains may differ in the way their specific rates of production respond to nutrients or growth rate. With *C. acremonium* and cephalosporin C production, Kuenzi (in a discussion at the Fifth FEMS Symposium in 1977) described optimal cephalosporin C production at high growth rates in a high-yielding strain, yet Matsamura et al. (1978) with a low yielder described optimal production at very low growth rates. However, this difference might be explained if the two groups of workers used different ratios of glucose and phosphate (Kuenzi, 1980).

3. Specific Regulation. While a great deal of data have accumulated from growth studies, further understanding of the processes involved in production can be achieved only when we begin to describe particular features at the enzyme level. One approach has been to determine enzyme activities in washed cell samples from fermentations. Using this technique with protein synthesis inhibitors, Demain and Kennel (1978) found that the glucose effect in the cephalosporin fermentation is due to inhibition and not to repression. In contrast, the uptake pattern of radioactive valine into washed cells of *P. chrysogenum* indicated that the glucose effect in the penicillin fermentation was due to repression and not to inhibition (Martin and Demain, 1980). Other aspects of the regulation of β-lactams in prokaryotes and eukaryotes are discussed in Chapter 3.

4. Industrial Fermentation. There have been some attempts to bring together knowledge of microbial growth and penicillin production and knowledge of the metabolic pathway in an effort to predict whether further improving the yield in penicillin fermentations is feasible. The theoretical estimate of yield is significantly higher than the practically derived value of 0.05 g of penicillin per gram of glucose for a high-yielding strain (Cooney, 1979). The theoretical models of Stouthamer (1979) and Cooney and Acevedo (1977) agreed in principle as to the distribution of glucose to growth, maintenance, and penicillin synthesis. A later model (Cooney, 1979) for fed-batch culture differed in that more glucose went to growth than to maintenance. In all cases, the proportion of glucose diverted to penicillin was much less than that used for other purposes. Thus Cooney (1979) suggested that further yield increases through process and strain improvements are possible.

Culture degeneration is an important issue in commercial production of β-lactams. The study of Righelato (1976) is of particular significance, since it relates growth-limiting substrates not only to the selectability of degenerate (low-yielding) forms but also to the rate at which such types appear. Traditionally, culture stability is assumed in biochemical, genetic, and microbiological studies of organisms producing β-lactams. However, genetic stability should not be assumed, particularly with high-yielding strains where low-yielding types might have a faster growth rate and greater viability on preservation.

B. Genetics

1. Genetic Changes Causing Deregulation. Strain improvement to increase productivity is usually empirical because of the difficulty in determining rate-limiting steps. When attempts are made to pinpoint limiting reactions, by adding intermediates to the fermentation or to washed cell suspensions, the results are often ambiguous. It is not clear whether stimulation is due to increased growth or to a specific effect on β-lactam production. Possibly regulation was more important in the past than now, when higher-yielding, longer-lasting fermentations in industry may well be limited by such factors as carbon supply. However, Martin and Demain (1980) point out that the experimental evidence available, albeit in other systems, does not support this idea. Moreover, studies of regulatory changes in the development of a high-yielding strain might give some clues as to how the strains might be further improved. The problems are not simple; for example, Masurekar and Demain (1974) succeeded in producing a strain resistant to an amino acid analog, only to find that its penicillin yield was lower rather than higher. Ball (1978) did the same with a leucine auxotroph of *P. chrysogenum*.

Genetic deregulation studies have been carried out in many β-lactam producing organisms. For example, Godfrey (1973) obtained evidence for limitation of valine by feedback control of valine synthesis in *S. lipmanii*, when controlled feeding of a leucine auxotroph allowed overproduction of cephamycin C. Niss and Nash (1973) obtained a mutant of *C. acremonium* that no longer needed methionine for cephalosporin C production. The detailed role of methionine as inducer (Drew and Demain, 1973) or sulfur source (Caltrider and Niss, 1966) is still in dispute, despite the elegant work of Treichler et al. (1979) and Demain and co-workers (see Chapter 3).

In *P. chrysogenum*, regulatory mechanisms appear to have been overcome by empirically producing high-yielding mutants. For example, the higher acyl transferase concentration in high-yielding penicillin strains than in low yielders (Pruess and Johnson, 1967) suggests that this activity may have been selected for during strain improvement to increase accretion rate. In addition, acyl transferase has been shown to increase in concentration during the course of a penicillin fermentation. This might have been selected by screening for increasing length of fermentation. In so doing, the decay of this complex may have been reduced.

Sensitivity to feedback inhibition by lysine involves homocitrate synthase (Martin and Demain, 1980). The inhibited enzyme can decay; hence removal of feedback control might be relevant to preserving Q_{pen} even if α-aminoadipate is recycled. Goulden and Chattaway (1969) have shown reduced valine feedback inhibition in a high-yielding strain, and higher-yielding strains have been shown to have a higher rate of sulphate uptake (Segal and Johnson, 1961). Other instances of deregulation mutations in *P. chrysogenum* may include strains in which the regulatory effect of penicillin on its own synthesis has been overcome in high-yielding strains. However, recent data have indicated that penicillin inhibition could be an artifact (Nestaas and Demain, 1981).

2. Blocked Mutants and Recombination Studies. As was indicated in Section II, much of our knowledge of the biosynthesis of β-lactam antibiotics has been gained without the use of blocked (that is, nonproducing, nonauxotrophic) mutants. Nevertheless, in cephalosporin C biosynthesis, such mutants have played a key role in determining the pathway and in generating novel metabolites (Queener et al., 1978). In *P. chrysogenum,* blocked mutants have been isolated and studied by using complementation tests (see Chapter 2). Specific exploitation of such mutants to give improved penicillin productivity by recombination has been suggested by Ball (1978).

Recombination studies with β-lactam producing microorganisms have mainly been carried out in fungi (see Chapter 2), but it can be anticipated that an increasing number of such studies will be carried out with prokaryotes that are known to produce β-lactam antibiotics. The techniques for gene transfer in actinomycetes are now on a firm foundation and include protoplast fusion and transformation with various plasmids modified by using recombinant DNA methodology (see Chapter 1). A more recent development of the long-standing idea of gene duplication is to use recombinant DNA methods to construct plasmid vectors carrying specific regions of DNA. Indeed, a gene bank of approximately 500 plasmids plus inserts could duplicate various regions of the whole genome, and if they were introduced into strains to give appropriate duplications, these strains could have higher yields whether the region duplicated is a structural gene or a regulatory gene. Queener and Baltz (1979) in a recent review discuss many aspects of applied microbial genetics, including the potential application of recombinant DNA to antibiotic production.

ACKNOWLEDGMENTS

We are grateful to Sir Edward Abraham, Professor A. L. Demain, Drs. R. P. Elander and M. Aoki, Drs. S. W. Queener and N. Neuss, and Drs. R. B. Sykes and K. Bush for access to reviews prior to publication.

REFERENCES

Abbott, B. J., and Fukuda, D. S. (1975a) *Appl. Microbiol. 30,* 413–419.

Abbott, B. J., and Fukuda, D. S. (1975b) *Antimicrob. Agents Chemother. 8,* 282–288.

Abraham, E. P., Huddleston, J. A., Jayatilake, G. S., O'Sullivan, J., and White, R. L. (1980) in *Recent Advances in the Chemistry of β-Lactam Antibiotics* (Gregory, G. I., ed.) pp. 125–134, Royal Society of Chemistry, London.

Albers-Schonberg, G., Arison, B. H., Kaczka, E., Kahan, F. M., Kahan, J. S., Lago, B., Maiese, W. M., Rhodes, R. E., and Smith, J. L. (1976) Sixteenth Interscience Conference on Antimicrobial Agents and Chemotherapy.

Aoki, H., Sakai, H., Kohsaka, M., Konomi, T., Hosoda, J., Kubochi, Y., Iguchi, E., and Imanaka, H. (1976) *J. Antibiot. 29,* 492–500.

Asai, M., Haibara, K., Muroi, M., Kintaka, K., and Kishi, T. (1981) *J. Antibiot. 34,* 621–627.

Bahadur, G. A., Baldwin, J. E., Usher, J. J., Abraham, E. P., Jayatilake, G. S., and White, R. L. (1981) *J. Amer. Chem. Soc. 103,* 7650–7651.

Baldwin, J. E., Singh, P. D., Yoshida, M., Sawada, Y., and Demain, A. L. (1980) *Biochem. J. 186,* 889–895.

Ball, C. (1978) in *Antibiotics and Other Secondary Metabolites* (Hütter, R., Leisinger, T., Nuesch, J., and Wehrli, W., eds.) pp. 165–176, Academic Press, New York.

Ball, C. (1982) in *Antibiotics Containing the β-Lactam Structure* (Demain, A. L., and Solomon, N. A., eds.) Springer, Berlin. (in press)

Batchelor, F. R., Doyle, F. P., Nayler, J. H. C., and Rolinson, G. N. (1959), *Nature, London 183,* 257–258.

Batchelor, F. R., Chain, E. B., Hardy, T. L., Mansford, K. R. L., and Robinson, G. M. (1961) *Proc. Roy. Soc. London B 154,* 498–508.

Beecham Ltd. (1979) *British Patent,* Spec. 1,547,222.

Box, S. J., Hood, J. D., and Spear, S. R. (1979) *J. Antibiot. 32,* 1239–1247.

Brewer, S. J., Taylor, P. M., and Turner, M. K. (1980) *Biochem. J. 185,* 555–564.

Brown, A. G., Butterworth, D., Cole, M., Hanscomb, G., Hood, J. D., Reading, C., and Rolinson, G. N. (1976) *J. Antibiot. 29,* 668–669.

Brown, D., Evans, J. R., and Fletton, R. A. (1979) *J. Chem. Soc. Chem. Comm.* 282–283.

Brundidge, S. P., Gaeta, F. C. A., Hook, D. J., Sapino, C., Elander, R. P., and Morin, R. B. (1980) *J. Antibiot. 33,* 1348–1351.

Caltrider, P. G., and Niss, H. F. (1966) *Appl. Microbiol. 14,* 746–753.

Cassidy, P. J., Albers-Schonberg, G., Goegelman, R. T., Arison, B., Stapley, E. O., and Birnbaum, J. (1981) *J. Antibiot. 34,* 637–648.

Choi, W. G., Lee, S. B., and Ryu, D. D. Y. (1981) *Biotech. Bioeng. 23,* 361–371.

Claridge, C. A., Gourevich, A., and Lein, J. (1960) *Nature 187,* 237–238.

Cooney, C. L. (1979) *Process Biochem. 14,* 31–33.

Cooney, C. L., and Acevedo F. (1977) *Biotech. Bioeng. 19,* 1449–1462.

Curtis, N. A. C. (1981) *J. Antimicrob. Chemother. 8,* 85–89.

Demain, A. L., and Kennel, Y. M. (1978) *J. Ferment. Technol. 56,* 323–328.

Drew, S. W., and Demain, A. L. (1973) *Biotechnol. Bioeng. 15,* 743–754.

Elson, S. W., and Oliver, R. S. (1978) *J. Antibiot. 31,* 586–592.

Eriquez, L. A., and Pisano, M. A. (1979) *Antimicrob. Agents Chemother. 16,* 392–397.

Fawcett, P. A., Usher, J. J., and Abraham, E. P. (1975) *Biochem. J. 151,* 741–746.

Fawcett, P. A., Usher, J. J., Huddleston, J. A., Bleaney, R. C., Nisbet, J. J., and Abraham, E. P. (1976) *Biochem. J. 157,* 651–660.

Felix, H. R., Peter, H. H., and Treichler, H. J. (1981) *J. Antibiot. 34,* 567–575.

Fisher, J. F., and Knowles, J. R. (1980) In *Enzyme Inhibitors as Drugs* (Sandler, M., ed.) pp. 209–218, Macmillan, New York.

Fisher, J., Belasco, J. G., Khosla, S., and Knowles, J. R. (1980) *Biochemistry 19,* 2895–2901.

Friedrich, C. G., and Demain, A. L. (1978) *Arch. Microbiol. 119,* 43–47.

Fujisawa, Y., Shirafuji, H., Kida, M., Nara, K., Yoneda, M., and Kanzaki, T. (1973) *Nature New Biol. 246,* 154–155.

Fujisawa, Y., Kitano, K., and Kanzaki, T. (1975) *Agr. Biol. Chem. 39,* 2049–2055.

Fukagawa, Y., Kubo, K., Ishikura, T., and Kuono, K. (1980) *J. Antibiot. 33,* 543–549.

Gilbert, D. A., Arnold, B. H., and Fildes, R. A. (1972) *Brit. Pat. 1,* 272, 769.

Godfrey, O. W. (1973) *Antimicrob. Agents Chemother. 4,* 73–79.

Goulden, S. A., and Chattaway, F. W., (1969) *J. Gen. Microbiol. 59,* 111–118.

Hamilton-Miller, J. M. T., and Smith, J. T., (eds.) (1979) *Beta-Lactamases,* Academic Press, New York.

Harada, S., Shinagawa, S., Nozaki, Y., Asai, M., and Kishi, T. (1980) *J. Antibiot. 33,* 1425–1430.

Higgens, C. E., and Kastner, R. E. (1971) *Int. J. Syst. Bacteriol. 21,* 326–331.

Hinnen, A., and Nuesch, J. (1976) *Antimicrob. Agents Chemother. 9,* 824–830.

Hook, D. J., Chang, L. T., Elander, R. P., and Morin, R. B. (1979) *Biochem. Biophys. Res. Comm. 87,* 258–265.

Hosoda, J., Tani, N., Konomi, T., Ohsawa, S., Aoki, H., and Imanaka, H. (1977) *Agr. Biol. Chem. 41,* 2007–2012.

Imada, A., Kitano, K., Kintaka, K., Muroi, M., and Asai, M. (1981) *Nature 289,* 590–591.

Jayatilake, G. S., Huddleston, J. A., and Abraham, E. P. (1981) *Biochem. J. 194,* 645–647.

Jeffery, J. D'A., Abraham, E. P., and Newton, G. G. F. (1961) *Biochem. J. 81,* 591–596.

Kahan, J. S., and Kahan, F. M. (1980) *British Patent,* Spec. 1,561,108, Feb. 13.

Kahan, J. S., Kahan, F. M., Goegelman, R., Currie, S. A., Jackson, M., Shapley, E. O., Miller, T. W., Miller, A. K., Hendlin, D., Mochales, S., Hernandez, S., Woodruff, H. B., and Birnbaum, J. (1979) *J. Antibiot. 12,* 1–12.

Kato, K., Kawahara, K., Takahashi, T., and Igarasi, S. (1980) *Agr. Biol. Chem. 44,* 821–825.

Kikuchi, T., and Uyeo, S. (1967) *Chem. Pharm. Bull. 15,* 549–570.

Knott-Hunziker, V., Waley, S. G., Orlek, B. S., and Sammes, P. G. (1979) *FEBS Lett. 99,* 59–61.

Kohsaka, M., and Demain, A. L. (1976) *Biochem. Biophys. Res. Comm. 70,* 465–473.

Konomi, T., Herchen, S., Baldwin, J. E., Yoshida, M., Hunt, N. A., and Demain, A. L. (1979) *Biochem. J. 184,* 427–430.

Kuenzi, M. (1980) *Arch. Microbiol. 128,* 78–83.

Kurylowicz, W., Kurzatkowski, W., Woznicka, W., Poloniak-Pracka, H., and Paszkiewicz, A. (1979) *Zbl. Bakt. II. Abt. 134,* 721–732.

Liersch, M., Nuesch, J., and Treichler, H. J. (1976) in *Second International Symposium on the Genetics of Industrial Microorganisms* (MacDonald, K. D., ed.) pp. 179–195, Academic Press, New York.

Lilley, G., Clark, A. E., and Lawrence, G. C. (1981) *J. Chem. Tech. Biotechnol. 31,* 127–134.

Loder, P. B. (1972) *Post. Hig. I Med. Dosw. 26,* 493–500.

Loder, P. B., and Abraham, E. P. (1971a) *Biochem. J. 123,* 471–476.

Loder, P. B., and Abraham, E. P. (1971b) *Biochem. J. 123,* 477–482.

Marconi, W., Bartoli, F., Cecere, F., Galli, G. and Morisi, F. (1975) *Agr. Biol. Chem. 39,* 277–279.

Martin, J. F., and Demain, A. L. (1980) *Microbiol. Rev. 44,* 230–251.

Masurekar, P. S., and Demain, A. L. (1974) *Antimicrob. Agents Chemother. 6,* 366–368.

Matsamura, M., Imanaka, T., Yoshida, T., and Taguchi, H. (1978) *J. Ferment. Technol. 56,* 345–353.

Meesschaert, B., Adriaens, P., and Eyssen, H. (1980) *J. Antibiot. 33,* 722–730.

Nagarajan, R., Boeck, L. D., Gorman, M., Hamill, R. L., Higgens, C. E., Hoehn, M., Stark, W. M., and Whitney, J. G. (1971) *J. Amer. Chem. Soc. 93,* 2308–2310.

Nakayama, M., Kimura, S., Tanabe, S., Mizoguchi, T., Watanabe, I., Mori, T., Miyahara, K., and Kawasaki, T. (1981) *J. Antibiot. 34,* 818–823.

Nestaas, E., and Demain, A. L. (1981) *Eur. J. Appl. Microbiol. Biotechnol. 12,* 170–172.

Neuss, N., Miller, R. D., Affolder, C. A., Nakatsukasa, W., Mabe, J., Huckstep, L. L., De La Higuera, N., Hunt, A. H., Occolowitz, J. L., and Gilliam, J. H. (1980) *Helv. Chim. Acta 63,* 1119–1129.

Neuss, N., and Queener, S. W. (1982) In *β-Lactam Antibiotics: Chemistry and Biology,* Vol. II (Morin, R. B., and Gorman, M., eds.) Academic Press, New York, (in press).

Niss, J. F., and Nash, C. H. (1973) *Antimicrob. Agents Chemother. 4,* 474–478.

Okamura, K., Hiraka, S., Okamura, Y., Fukugawa, Y., Shimauchi, Y., Kouno, K., and Ishikura, I. (1978) *J. Antibiot. 31,* 480–482.

O'Sullivan, C. Y., and Pirt, S. J. (1973) *J. Gen. Microbiol. 76,* 65–75.

O'Sullivan, J., and Abraham, E. P. (1980) *Biochem. J. 186,* 613–616.

O'Sullivan, J., and Abraham, E. P. (1981) in *Antibiotics* Vol. IV, (Corcoran, J. W., ed.) pp. 107–122, Springer, Berlin.

O'Sullivan, J., Aplin, T. R., Stevens, C. M., and Abraham, E. P. (1979a) *Biochem. J. 179,* 47–52.

O'Sullivan, J., Bleaney, R. C., Huddleston, J. A., and Abraham, E. P. (1979b) *Biochem. J. 184,* 421–426.

O'Sullivan, J., Gillum, A. M., Aklonis, C. A., Souser, M. L., and Sykes, R. B. (1982). *Antimicrob. Agents Chemother. 21,* 558–564.

Parker, W. L., and Rathnum, M. (1982) *J. Antibiot. 35,* 300–305.

Parker, W. L., Rathnum, M. L., Scott, J. S., Trejo, W. H., Principe, P. A., and Sykes, R. B. (1982) *J. Antibiot. 35,* 189–195.

Pirt, S. J., and Righelato, R. C. (1967) *Appl. Microbiol. 15,* 1284–1290.

Pruess, D. L., and Johnson, M. J. (1967) *J. Bacteriol. 94,* 1502–1508.

Queener, S. W., and Baltz, R. H. (1979) *Ann. Rep. Ferm. Processes 3,* 5–45.

Queener, S. W., Capone, J. J., Radue, A. B., and Nagarajan, R. (1974) *Antimicrob. Agents Chemother. 6,* 334–337.

Queener, S. W., Sebek, O. K., and Vezina, C. (1978) *Ann. Rev. Microbiol. 32,* 593–636.

Reading, C., and Cole, M. (1977) *Antimicrob. Agents Chemother. 11,* 852–857.

Richmond, M. H., and Sykes, R. B. (1973) *Adv. Microb. Physiol. 9,* 31–88.

Righelato, R. (1976) *J. Appl. Chem. Biotechnol. 26,* 153–159.

Rolinson, G. N., Bachelor, F. R., Butterworth, D., Cameron-Wood, J., Cole, M., Eustace, G. C., Hart, M. V., Richards, M., and Chain E. B. (1960) *Nature, 187,* 236–237.

Sakaguchi, K., and Murao, S. (1950) *J. Agr. Chem. Soc. Japan 23,* 411.

Sankyo Co., Ltd. (1980) Japan Patent No. S550–24129.

Sawada, Y., Hunt, N. A., and Demain, A. L. (1979) *J. Antibiot. 32,* 1303–1310.

Scannell, J. P., Pruess, D. L., Blount, J. F., Ax, H. A., Kellett, M., Weiss, F., Demny, T. C., Williams, J. H., and Stempel, A. (1975) *J. Antibiot. 28,* 1–6.

Segal, I. H., and Johnson, M. J. (1961) *J. Bacteriol. 81,* 91–98.

Shibamoto, N., Koki, A., Nishino, M., Nakamura, K., Kiyoshima, K., Okamura, K., Okabe, M., Okamoto, R., Fukagawa, Y., Shimauchi, Y., and Ishikura, T. (1980) *J. Antibiot. 33,* 1128–1137.

Shibuya, Y., Matsumoto, K., and Fujii, T. (1981) *Agr. Biol. Chem. 45,* 1561–1567.

Shimizu, M., Masuike, T., Fujita, H., Kimura, K., Okachi, R., and Nara, T. (1975) *Agr. Biol. Chem. 39,* 1225–1232.

Spratt, B. G. (1977) *Eur. J. Biochem. 72,* 341–352.

Spratt, B. G. (1980) *Phil. Trans. Roy. Soc. London B 289,* 273–283.

Stapley, E. O., Cassidy, P. J., Tunak, J., Monaghan, R. L., Jackson, M., Hernandez, S., Zimmerman, S. B., Mata, J. M., Currie, S. A., Daoust, D., and Hendlin, D. (1981) *J. Antibiot. 34,* 628–636.

Sterling, I., and Elson, S. W. (1979) *J. Antibiot. 32,* 1125–1129.

Stevens, C. M., Abraham, E. P., Huang, F.-C., and Sih, C. J. (1975) *Fed. Proc. 34,* 625.

Stewart, W. W. (1971) *Nature 229,* 174–178.

Stouthamer, A. J. (1979) in *Genetics of Industrial Microorganisms* (Sebek, O. K., and Laskin, A. I., eds.) pp. 70–76, American Society for Microbiology, Washington, D.C.

Sykes, R. B., and Bush, K. (1982) in *β-Lactam Antibiotics: Chemistry and Biology,* Vol. I (Morin, R. B., and Gorman, M., eds.) pp. 155–207, Academic Press, New York.

Sykes, R. B., Cimarusti, C. M., Bonner, D. P., Bush, K., Floyd, D. M., Georgopapadakou, N. H., Koster, W. H., Liu, W. C., Parker, W. L., Principe, P. A., Rathnum, M. L., Slusarchyk, W. A., Trejo, W. H., and Wells, J. S. (1981) *Nature 291,* 489–491.

Tanaka, K., Shoji, J., Terui, Y., Tsuji, N., Kondo, E., Mayama, M., Kawamura, Y., Hattori, T., Matsumoto, K., and Yoshida, T. (1981) *J. Antibiot. 34,* 909–911.

Tipper, D. J., and Strominger, J. L. (1965) *Proc. Nat. Acad. Sci. U.S.A. 54,* 1133–1141.

Townsend, C. A., and Brown, A. M. (1981) *J. Amer. Chem. Soc. 103,* 2873–2876.

Townsend, C. A., and Nguyen, L. T. (1981) *J. Amer. Chem. Soc. 103,* 4582–4583.

Treichler, H. J., Liersch, M., Nuesch, J., and Dobeli, H. (1979) in *Genetics of Industrial Microorganisms* (Sebek, O. K., and Laskin, A. I., eds.) pp. 97–106, American Society for Microbiology, Washington, D.C.

Troonen, H., Roelants, P., and Boon, B. (1976) *J. Antibiot. 29,* 1258–1267.
Trown, P. W., Smith, B., and Abraham, E. P. (1963) *Biochem. J. 86,* 284–291.
Turner, M. K., Farthing, J. E., and Brewer, S. J. (1978) *Biochem. J. 173,* 839–850.
Vandamme, E. J. (1977) *Adv. Appl. Microbiol. 21,* 89–123.
Vandamme, E. J., and Voets, J. P. (1974) *Adv. Appl. Microbiol. 17,* 311–369.
Wanning, M., Zähner, H., Krone, B., and Zeeck, A. (1981) *Tetrahedron Lett. 22,* 2539–2540.
Wells, J. S., Trejo, W. H., Principe, P. A., Bush, K., Georgopapadakou, N. H., Bonner, D. P., and Sykes, R. B. (1982a) *J. Antibiot. 35,* 184–188.
Wells, J. S., Trejo, W. H., Principe, P. A., Bush, K., Georgopapadakou, N. H., Bonner, D. P., and Sykes, R. B. (1982b) *J. Antibiot. 35,* 295–299.
Whitney, J. G., Brannon, D. R., Mabe, J. A., and Wicker, K. J. (1972) *Antimicrob. Agents Chemother. 1,* 247–251.

CHAPTER
5

Peptides

Horst Kleinkauf
Hans von Döhren

I. INTRODUCTION

Research on peptide antibiotics has contributed valuable insights to biochemistry. Aspects of their chemistry, their biochemical action, and the enzymology of their production have general implications. While we are far from understanding the role these compounds play within their producer's life cycle (Katz and Demain, 1977; Kleinkauf and von Döhren, 1982), their nonribosomal formation has been established, and enzyme studies are in progress. Table 5.1 summarizes currently available information about the biosynthesis of several linear and cyclic compounds, some containing ester bonds. The enzymes involved have been described as multienzyme complexes, but many of them do not contain subunits and consist of multifunctional polypeptide chains. Thus construction of the cyclic decapeptide, tyrocidine, requires only three multienzyme structures catalyzing ten amino acid activations, two epimerizations, nine specific elongations, and one cyclization. These and many other aspects of peptide antibiotics have been reviewed in the last decade (Table 5.2). In this chapter, we discuss recent advances stemming from work on gramicidin S, tyrocidine, linear gramicidin, bacitracin, and polymyxin.

TABLE 5.1 Biosynthetic Data on Peptide Antibiotics[1]

		Multienzyme Systems					
		Minimum Number of Enzymes Required for:					
Compound	*Number of Amino Acids and Structural Type[2]*	*Peptide Sequence*	*Complete Structure*	*Size (Kd) of Multi-enzyme*	*Mimimum Total Number of Enzyme Functions[3]*	*Multi-functional*	*Pantetheine Content*
Alamethicin	19-L, modified	2 (?)	4[4]	150 320	16 (32) 23 (46)	ND[9]	ND
Bacitracin	12 with 7-C	3 (4)	4[5]	150 210 380	11 (20) 5 (9) 13 (24)	+ ND ND	+ + +
Edeine	5-L, modified	3	6[6]	210 180 100	1 (2) 8 (17) ?	ND ND ND	+ + −
Enniatin	6-C, 3 × 2 repeat	1	2[7]	250	5 (9)	+	+
Linear gramicidin	15-L, modified	3 (?)	5	180 350 ?	3 (6) 12 (22)	+ +	+ +
Gramicidin-S	10-C, 2 × 5 repeat	2	2	100 280	2 (3) 10 (18)	+ +	− +
Polymyxin	10 with 7-C	1 (?)	2[8]	300 (?)	22 (42)	ND	ND
Tyrocidine	10-C	3	3	100 230 440	2 (3) 7 (13) 13 (26)	+ + +	− + +

1. Data from Kleinkauf and von Döhren (1981, 1982)

2. L = linear peptide; C = cyclic peptide structure

3. The number of functions includes: amino acid (hydroxy acid) activation, specific elongation (amino acid addition) or peptide bond formation (or ester bond formation), modification (e.g., L → D); note that the currently discussed number of functions usually includes, in addition, (1) formation of a stable adenylate (followed by enzyme aminoacylation) and (2) preceding the specific elongation, transport of the peptide to be elongated to the activated amino acid (Laland et al., 1972; Frøyshov et al., 1978; Kleinkauf and Koischwitz, 1980b). The number with these included is given in brackets.

4. Includes N-terminal acetylation, production of phenylalaninol (Mohr and Kleinkauf, 1978)

5. Includes the fourth multienzyme proposed by Vitković and Pfaender (1982)

6. Includes production of β-tyrosine (Parry and Kurylo-Borowska, 1980) and diaminohydroxyazelaic acid as well as C-terminal modification (Kurylo-Borowska and Heaney-Kieras, 1982; Kurylo-Borowska, 1975)

7. Includes S-adenosylmethionine production (Zocher and Kleinkauf, 1978; Zocher et al., 1982a, b)

8. Includes N-terminal modification (Komura and Kurahashi, 1980b)

9. ND = not determined

TABLE 5.2 Review Articles and Books

Author(s)	*Title*
Lipmann et al. (1971)	Polypeptide Synthesis on Protein Templates: the Enzymatic Synthesis
Laland and Zimmer (1973)	The Protein Template Mechanism of Synthesis for the Peptide Antibiotics Produced by *Bacillus brevis*
Kurahashi (1974)	Biosynthesis of Small Peptides of Gramicidin S and Tyrocidine
Ramachandran (1975)	The Gramicidins
Suzuki and Kimura (1976)	Chemistry of Polymyxin
Izumiya (1976)	Chemistry of Gramicidin S
Katz and Demain (1977)	The Peptide Antibiotics of *Bacillus:* Chemistry, Biogenesis and Possible Functions
Vining and Wright (1977)	Biosynthesis of Oligopeptides
Storm et al. (1977)	Polymyxin and Related Peptide Antibiotics
Kleinkauf and Koischwitz (1978)	Peptide Bond Formation in Non-ribosomal Systems
Perlman (1978)	Antibiotics (Peptides)
Frøyshov et al. (1978)	Biosynthesis of Microbial Peptides by the Thiotemplate Mechanism
Izumiya et al. (1979)	Synthetic Aspects of Biologically Active Cyclic Peptides, Gramicidin S and Tyrocidines
Kleinkauf (1979)	Antibiotic Polypeptides-Biosynthesis on Multifunctional Protein Templates
Zimmer et al. (1979)	Peptide Antibiotics
Brewer (1979)	Gramicidin
Lipmann (1980)	Bacterial Production of Antibiotic Polypeptides by Thiol-Linked Synthesis on Protein Templates
Kleinkauf and von Döhren (1981)	Nucleic-Acid-Independent Synthesis of Peptides
Kurahashi (1981)	Biosynthesis of Peptide Antibiotics
Vandamme (1981)	Properties, Biogenesis and Fermentation of the Cyclic Decapeptide Antibiotic Gramicidin S
Kleinkauf and von Döhren (1982)	Peptide Antibiotics, Biosynthesis and Functions

II. REGULATION OF PRODUCTION

A. Secondary Metabolism

Little is known about the regulation of peptide antibiotic formation; information on the enzyme "families" involved is just being evaluated. Since enzyme production and enzyme properties may vary with growth conditions and are not always directly related to formation and release of the peptide, the system is undoubtedly complex. Growth curves can usually be divided into the growth phase (trophophase) and the production phase (idiophase) (Bu'Lock and Powell, 1965). Enzymes for producing most peptide antibiotics are repressed during growth and derepressed during transition to idiophase. They may also be present during the trophophase but inactive because of inhibition, compartmentation, or lack of precursors. For a general discussion of the regulation of antibiotic biosynthesis, the reader is referred to Chapter 3. Since growth conditions for peptide production may differ from those selected for the isolation and study of enzyme systems (Kleinkauf, 1979; Matteo et al., 1976; Frøyshov, 1977), we will discuss these two topics separately.

B. Peptide Production

1. Bacitracins a. *Structures*. Bacitracins (Fig. 5.1) are formed from a linear dodecapeptide by two cyclization reactions leading to a thiazoline moiety and a seven-membered ring (Ressler and Kashelikar, 1966). Of eight compounds isolated from fermentation broths (Tsuji and Robertson, 1975), the structures of bacitracin A and two preparative artifacts, F formed by oxidation (Konigsberg and Craig, 1962; Hirotsu et al., 1979) and "low-potency A" formed by spontaneous epimerization of the N-terminal Ile to D-allo-Ile (Craig and Konigsberg, 1957), have been described; the position of the valine replacement in bacitracin B has not yet been established.

b. *Producer strains*. The original bacitracin-producing organism was called *Bacillus subtilis* Tracy I (Hickey, 1964; Zimmer et al., 1979). Later, it was classified as a variant of *Bacillus licheniformis* (Meleney and Johnson, 1949) and designated *B. licheniformis* ATCC 10716 (Snoke, 1960). Other producing strains are *B. licheniformis* ATCC 9945, 11945, 11946, 14580, and *B. subtilis* 14593.

c. *Fermentation improvement*. Since bacitracins or zinc bacitracins are widely used as growth promotors in animal production, many fermentation procedures have been devised. Generally, with batch procedures, yields of 500 units/ml are obtained on soybean meal media (Hickey, 1964). To avoid spore contamination of equipment, sporulation-deficient strains have been used (Tyc et al., 1976; Kuzovnikova et al., 1978). That there is no correlation between sporulation and peptide production is shown by normal sporulation of the *B. licheniformis* nonproducer mutant, SB 319 (Haavik and Thomassen,

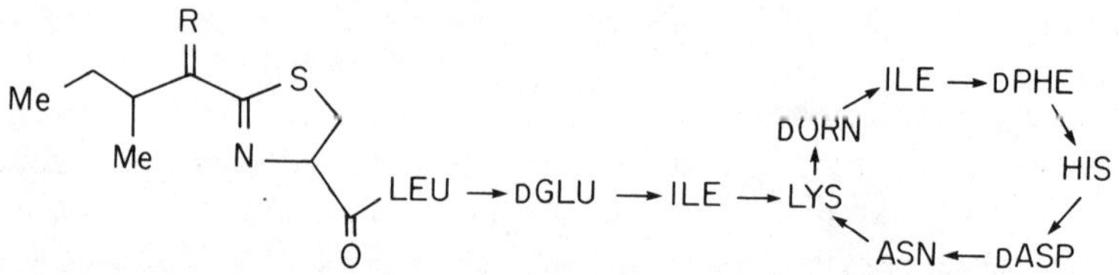

FIGURE 5.1 Structures of bacitracins A (R = H, NH_2), F (R = O) and B (one Ile in bacitracin A replaced by Val).

1973). Some stimulation of peptide production has been achieved by mutational procedures (Tyc and Kadzikiewicz, 1976b), for example, those used to obtain phage-resistant nonsporulator strains (Tyc et al., 1979); spontaneous autolysin-resistant mutants produce up to twice the amount of peptide but are unstable (Lukin et al., 1981). Addition of glutamate (Szczesniak et al., 1976), pantothenate, β-alanine, and/or cystine (Smekal et al., 1979) to fermentations has given significant increases. Oxygen-enriched aeration, which decreases carbohydrate uptake and growth rate, increases specific productivity. Flickinger and Perlman (1979) have suggested that O_2 enrichment be used to provide the growth rate control needed to optimize antibiotic production.

In search of continuous procedures, Morikawa et al. (1979, 1980) and Suzuki and Karube (1979) used polyacrylamide gel-immobilized cells. Suspensions in 0.5–1% peptone showed similar productivity to free cell suspensions, and the activity was stabilized (half-life: 10 days). Continuous production in an aerated reactor proved superior to continuous fermentation.

d. *Control mechanisms.* A study of bacitracin production in a glutamate–citrate medium where the antibiotic is produced during growth showed no correlation between enzyme activity and formation of the antibiotic (Fig. 5.2). Thus additional mechanisms appear to regulate production (Haavik, 1974b; Frøyshov, 1977). An apparent catabolite repression by glucose (Laishley and Bernlohr, 1968) actually arose from a pH drop due to production of organic acids (Haavik, 1974c). An apparent phosphate control was also found to depend on pH changes (Haavik, 1974a). With pH control, high peptide yields were obtained even with 2% phosphate in the medium.

Significant effects of amino acids on bacitracin formation have been described recently for *B. licheniformis* ATCC 10716 and for a high-yielding mutant AL (Haavik and Vessia, 1978; Haavik, 1979, 1981). Studies on the comparable (twofold to threefold) stimulation of gramicidin S yield in *Bacillus brevis* ATCC 9999 cultures given by phenylalanine suggest that the amino acids may be acting as precursors; no change was observed in the level of synthetase in the *B. brevis* cells (Demain and Matteo, 1976). A precursor function is reasonable in the case of bacitracin, since some D-amino acids, notably DOrn, Dglu and DPhe, inhibit peptide biosynthesis and all D-isomers are incorporated at reduced rates. However, interpretation of in vivo ex-

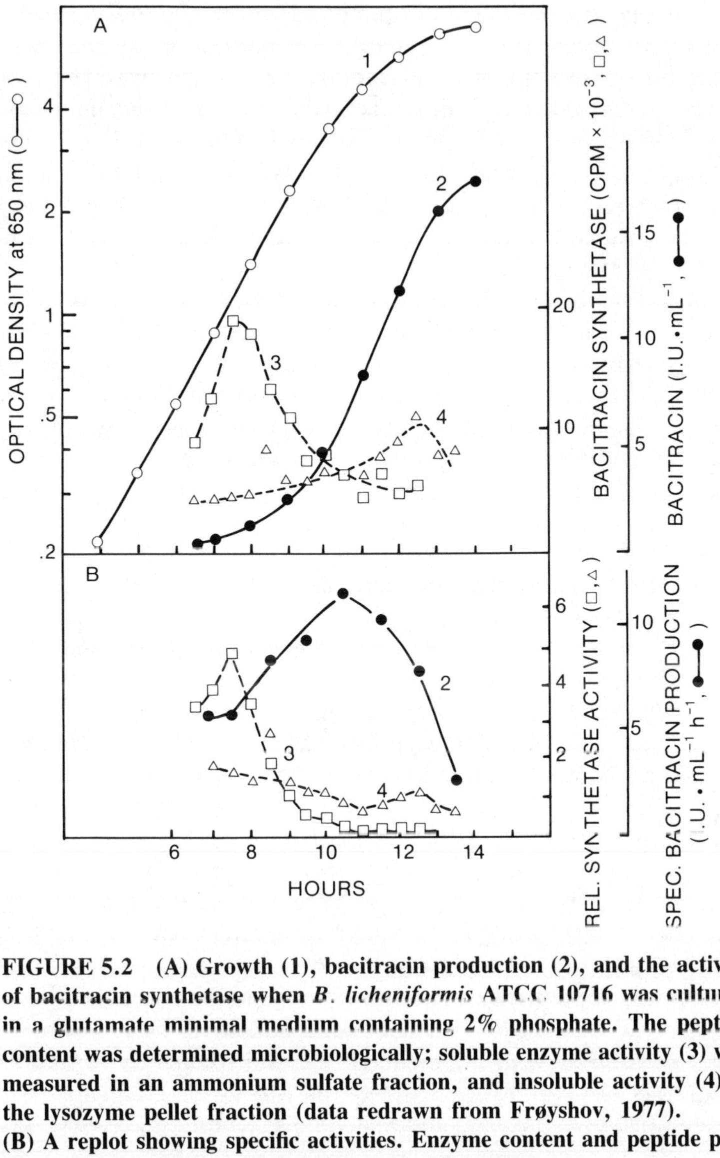

FIGURE 5.2 (A) Growth (1), bacitracin production (2), and the activity of bacitracin synthetase when *B. licheniformis* ATCC 10716 was cultured in a glutamate minimal medium containing 2% phosphate. The peptide content was determined microbiologically; soluble enzyme activity (3) was measured in an ammonium sulfate fraction, and insoluble activity (4) in the lysozyme pellet fraction (data redrawn from Frøyshov, 1977). (B) A replot showing specific activities. Enzyme content and peptide productivity are related to the absorbance of the culture.

periments is hazardous without pool estimations or at least uptake studies, since the results deviate in some respects from those obtained in vitro using protoplasts (Cornell and Snoke, 1964) or cell-free systems (Ishihara et al., 1968; Pfaender, 1972; Frøyshov and Laland, 1974).

A possible regulatory role for leucine has been proposed from its effect on bacitracin and its synthetase in mutant AL of *B. licheniformis*, but although leucine stimulated growth threefold and production 15- to 30-fold, it had no effect on an in vitro system or on production by the wild type. Almost twofold stimulation was achieved in the wild type when both leucine and ornithine were added to a glutamate medium, or when leucine auxotrophs were used. However, a similar leucine-dependent stimulation of gramicidin S formation by *Bacillus brevis* ATCC 9999 originated in growth stimulation, since the specific activity of the cells remained unchanged and the effect was not observed with glycerol instead of fructose as the carbon source (Vandamme, 1981).

A mechanism by which bacitracin synthetase could be regulated by feedback from a bactracin–metal complex has been proposed by Frøyshov et al. (1980). A crude enzyme system was inhibited at peptide concentrations of several mg/ml. There is a speculative connection between bacitracin-mediated metal ion uptake (Haavik, 1976a, Haavik and Frøyshov, 1975) and metal ion complexes between substrate and product.

2. Gramicidin S. The gramicidin S field has recently been reviewed by Vandamme (1981), so in the following short account, we concentrate on new information.

a. *Structure*. Gramicidin S is a symmetrical cyclic decapeptide (Fig. 5.3). Its structure has been known since 1951 (Izumiya et al., 1979; Battersby and Craig, 1951). There are no homologies to the linear gramicidins, but there is a close homology to tyrocidines. A unique property of gramicidin S is the lack of isomers with amino acid replacements in the product formed in vivo, despite the broad specificity of gramicidin S synthetase indicated by the numerous replacements carried out in vitro (Lipmann et al., 1971; Aarstad et al., 1979; Kleinkauf and Koischwitz, 1980a; H. von Döhren and H. Kleinkauf, unpublished). However, recent mass spectrometric studies of the in vivo product do show a minor component with an additional 14 mass units, indicating a Val→Leu or Orn→Lys replacement (Höhne and H. von Döhren, unpublished). Also, ornithine can be replaced by lysine in vivo by directed biosynthesis in minimal medium (H. von Döhren et al., unpublished). Studies with gramicidin-J are still frequently reported. This compound was thought to be a structural analog of gramicidin-S but is actually identical with it (Otani and Saito, 1964; Kurahashi, 1964). A variety of other peptides have been described from strains of *B. brevis*. The data

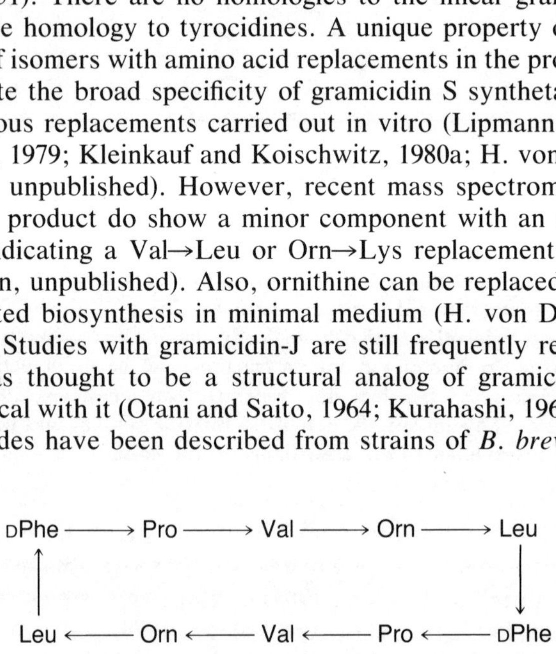

FIGURE 5.3 Structure of gramicidin S.

on some (for example, on brevolin, brevin) were obtained many years ago, and their status is uncertain.

b. *Producer strains*. In various aspects of current work on gramicidin S, three strains of *B. brevis* are being used: the original Gause-Brazhnikova strain GB; derived from this, ATCC 9999; and the Nagano strain. Much attention has been paid to strain GB. Four variants of types R, P^+, S, and P^- are formed reversibly (Zharikova et al., 1964, 1965, 1966; Mitronova and Zharikova, 1965; Stoletov et al., 1965; Zharikova and Lukin, 1966; Rautenshtein et al., 1967; Peshkov et al., 1968; Surguchev et al., 1969). The latter two do not produce gramicidine S but make the related peptides, gracellin (in P^-), esein, and bresein (in S) (Zharikova et al., 1975). The possible involvement of extrachromosomal DNA in dissociation of the GB strain, or in peptide production in general, has been suggested (A. P. Dobritsa et al., 1978; S. V. Dobritsa and Dobritsa, 1978; Marahiel et al., 1981b). Marahiel et al. (1979) based their suggestion on the high-frequency formation of gramicidin S nonproducer mutants with ethidium bromide or acridine orange. Covalently closed circular ccc DNA's have been isolated from all four variants of the GB strain (A. P. Dobritsa et al., 1978) and from gramicidin S producer and nonproducer mutants of the ATCC strain (Marahiel et al., 1981a, b). The ccc DNA's from the ATCC strains are chromosomal- or membrane-associated, have sizes of 47.1 and 44.6 Md, respectively, and have only slightly different restriction cleavage maps (Marahiel et al., 1981). So far no evidence for their involvement in peptide formation has been obtained. The sensitivity of gramicidin-S formation to intercalating dyes is apparently not related to elimination of ccc DNA but rather to a property of certain genes involved in nongrowth events like sporulation (Rogolsky and Nakamura, 1974; Burke and Spizizen, 1977).

c. *Production*. Gramicidin S normally occurs only intracellularly, although C. W. Chiu et al. (unpublished results) have found it to accumulate in the fermentation broth on a fumarate minimal medium (Chiu et al., 1982). While gramicidin S from *B. brevis* ATCC 9999 as well as tyrothricin from *B. brevis* ATCC 8185 were originally produced in surface culture, fermentation procedures are now used (Perlman, 1974). Vandamme (1981) has reviewed these in detail. Efforts to improve the productivity of the ATCC strain to equal that reported in the 1960's for the GB strain have led to the use of pH-controlled fermentations (Vandamme, 1981; Vandamme et al., 1981, 1982). This prevents the lowering of external and internal pH which inhibits peptide formation, and thus prolongs the phase of gramicidin S production (compare bacitracin production, Haavik 1974b, c). Values for specific gramicidin S production reach 22% of the dry cell weight. Under these conditions (Fig. 5.4), gramicidin S was formed only when the growth rate decreased. Constant pH apparently prolonged this phase. In an amino acid–supplemented fructose medium, constant pH conditions increased gramicidin S production four-fold and doubled biomass. Growth was biphasic. In the initial growth phase, arginine and glutamine were consumed, but gramicidin S was not

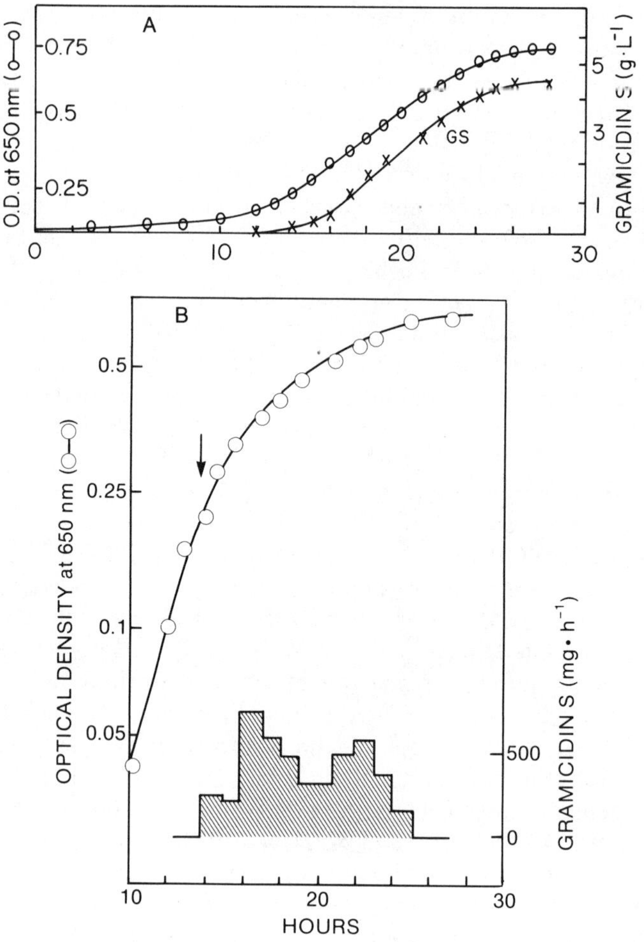

FIGURE 5.4 (A) Apparent formation of gramicidin S (GS) during "active growth" in a batch fermentation of *B. brevis* cultured in a complex yeast extract peptone medium controlled at pH 7.3 (reproduced from Vandamme, 1981; Vandamme et al., 1982).
(B) A replot of the data shows that production starts at the end of the log phase and continues during subsequent slower growth. True log phase production of bacitracin has been reported recently (Hanlon and Hodges, 1981) using carbon or nitrogen substrates imposing reduced growth.

formed. Production occurred, following a diauxic lag, during subsequent growth while fructose and histidine were being used. Analysis of various fermentations, comparing growth rate and specific production of gramicidin S, reveals that maximum product formation occurs at a growth rate 75–80% below the maximum (H. von Döhren, unpublished).

3. Polymyxins. a. *Structures*. The structures of polymyxins and related peptides have been discussed recently by Shoji (1978). Polymyxins are acylated undecapeptides containing a seven-membered ring structure (Fig. 5.5); the related octapeptin group of acylated nonapeptides have partially identical or identical seven-membered rings (Table 5.3). The sequence similarities indicate that closely related enzyme systems are involved in their biosynthesis (Kleinkauf and von Döhren, 1981). The structural differences reflect strain-dependent variations in amino acid substrate specificity. We find exchanges of Dab→DSer, DLeu→DPhe, Leu→Phe, and Ile→Leu→Thr.

b. *Producer strains and fermentations*. Production of specific polymyxins, circulins, colistins, or octapeptins is associated with *Bacillus* species. Each strain produces a particular peptide sequence, and variations occur only in the acyl moiety (Table 5.4). The earlier fermentation procedures have been summarized by Nakamura (1972). A most extensive search for producer strains in Japanese soil samples has been carried out by Shoji and coworkers (Shoji, 1978). Mutational improvement of peptide yield has not been significant so far (Matselyukh et al., 1978). As with bacitracin, colistin has been produced by polyacrylamide-immobilized cells. Here, *B. polymyxa* ATCC 10401 was used (Suzuki et al., 1979).

c. *Control mechanisms*. Pavlyuk et al. (1979b) showed that a rich source of organic nitrogen stimulates polymyxin B production more than do amino acids or combinations of amino acids. Whereas for gramicidin S, a high phosphate concentration of 50 mM gives optimal production, 0.5 mM of phosphate has been found to be optimal for polymyxin B (Ermakova et al., 1976; Pavlyuk et al., 1979c). This corresponds well with the optimum for formation of gramicidin S synthetase multienzymes. Polymyxin B formation by *B. polymyxa*, but not cell growth, is inhibited by increasing K^+ concentrations (Nefelova et al., 1980a).

In glucose-containing media where *B. polymyxa* fails to sporulate, peptides are produced during logarithmic growth (Nefelova et al., 1979). Also, two mucoidless log-phase-producer mutants show delayed and inhibited sporulation (Emilianowicz-Czerska et al., 1973). These results suggest that peptide

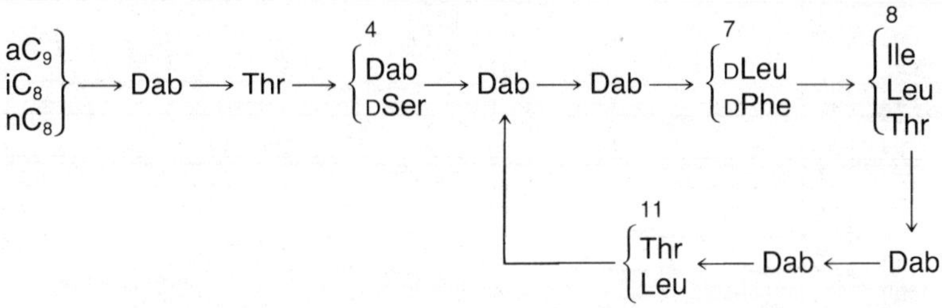

FIGURE 5.5 Structures of polymyxins. The acyl moieties are either 6-methyl-heptanoic acid (iC_8), 6-methyloctanoic acid (aC_9), or octanoic acid (nC_8).

TABLE 5.3 Structures of polymyxins and octapeptins

Name(s) Polymyxin	*Composition* FA—Dab —Thr—	3 —Dab —Dab —	6—	7—Dab —Dab —	10	*Reference*
–(circulin A)	aC_9[1]	Dab	DLeu	Ile	Thr	Fujikawa et al. (1965)
–(circulin B)	iC_8[2]	Dab	DLeu	Ile	Thr	Hayashi et al. (1968)
E_1 (colistin A)	aC_9	Dab	DLeu	Leu	Thr	Suzuki et al. (1963, 1964a)
E_2 (colistin B)	iC_8	Dab	DLeu	Leu	Thr	
A_1 (M?)	aC_9	Dab	DLeu	Thr	Thr	Silaev et al., (1975);
A_2	iC_8	Dab	DLeu	Thr	Thr	Wilkinson and Lowe, (1966)
B_1	aC_9	Dab	DPhe	Leu	Thr	Suzuki et al. (1964b);
B_2	iC_8	Dab	DPhe	Leu	Thr	Wilkinson and Lowe (1964)
B_3	nC_8[3]	Dab	DPhe	Leu	Thr	Elverdam et al. (1981)
C_1 (P_1)	aC_9	Dab	Phe	Thr	Thr	Kimura et al. (1969)
C_2 (P_2)	iC_8	Dab	Phe	Thr	Thr	Jones (1949)
D_1	aC_9	DSer	DLeu	Thr	Thr	Hayashi et al. (1966)
D_2	iC_8	DSer	DLeu	Thr	Thr	
S_1	aC_9	DSer	DPhe	Thr	Thr	Shoji et al. (1978)
T_1	aC_9	Dab	DPhe	Leu	Leu	
K_1	?	(Dab)	(DLeu)	Thr	Thr	Kimura (1971)
K_2	?	(Dab)	(DLeu)	Thr	Thr	
F_1	aC_9					
F_2	iC_8	(Dab_5 Thr Leu_2 Ser Ile)				Parker et al. (1977)
F_3	nC_8					
I_1	aC_9	(Dab_6 Thr_2 Phe Ile)				Elverdam et al. (1981)
O_1	aC_9	(Dab_6 Thr_2 Leu Nva)				
L_2	iC_8	(Dab_6 Thr_2 Leu Val)				

TABLE 5.3 (Continued)

Name(s) *Octapeptin*	*FA*—DDab—Dab—Dab—	*Composition* 4—	5—Dab—Dab—Leu	*Reference*
A_1	$aC_{11}h$[4]	DLeu	Leu	
A_2	$iC_{10}h$[5]	DLeu	Leu	
A_3	$nC_{10}h$[6]	DLeu	Leu	Meyers et al. (1973)
B_1	$aC_{11}h$	DLeu	Phe	
B_2	$iC_{10}h$	DLeu	Phe	
B_3	$nC_{10}h$	DLeu	Phe	
C_1	aC_9h[7]	DPhe	Leu	Shoji et al. (1976)
Bu 1880 (B_1)	$aC_{11}h$	Dab_5 Leu_2 Phe		Japanese Patent (1975)
TM-743 (C_1)	C_8h[3]/C_9h			Japanese Patent (1976); Meyers et al. (1976)

1. 6-methyloctanoic acid
2. 6-methylheptanoic acid
3. octanoic acid
4. 8(*S*)-methyl-3(*R*)-hydroxydecanoic acid
5. 8-methyl-3(*R*)-hydroxynonanoic acid
6. 3(*R*)-hydroxydecanoic acid
7. 3-hydroxyanteisononanoic acid

TABLE 5.4 Properties of GS2 from gramicidin S non-producing mutants of *B. brevis*[1]

Group	*Mutant*	*Activation[2] of*				*Pantothenic Acid[3]*	*Initiation Reaction[4]*	*Apparent Molecular Weight*	
		Pro	*Val*	*Orn*	*Leu*			*Native (S^5;Kd^6)*	*Denaturated[7] (Kd)*
IV[8]									
	BI-4	+	+	+	+	–	–		
	C-3	+	+	+	+	–	–		
	E-1	+	+	+	+	–	–		
	E-2	+	+	+	+	–	–		
VI	BII-3	–	+	+	+	+	–	12.2	280
	BI-3	+	–	+	+	+	+	12.2	280
	BI-6	+	–	+	+	+	+		
	BII-1	+	–	+	+	+	+		
	E-4	+	–	+	+	+	+		
	E-5	+	–	+	+	+	+		
	BI-9	+	+	+	–	+	+	12.2	280
	BI-2	+	+	+	+	+	+		
	h7	+	–	–	–		+	6.1; 110	
	hh	+	+	+	–		+	11.3	260

1. Data from Kambe et al., 1974; Shimura et al., 1974; Hori et al., 1978, 1981, 1982; Saito, 1982; Kleinkauf and von Döhren, 1981.
2. Detected by ATP–PPi exchange reaction.
3. Content determined microbiologically.
4. Formation of cyclo-DPhePro with complementary GS1.
5. Determined by active enzyme centrifugation in sucrose gradient (GS2, 12.2 s).
6. Determined by gel filtration (GS2, 280 Kd).
7. SDS polyacrylamide gel electrophoresis.
8. GS1 of these mutant strains is deficient in aminoacylation and racemization reactions (Hori et al., 1981).

production and sporulation are independent processes. However, a different conclusion has been reached from other observations. A linkage between polymyxin B production and sporulation was proposed to account for DLeu inhibition of peptide production and sporulation in vivo (Jayaraman and Paulus, 1971; Jayaraman and Kannan, 1972); both polymyxin formation and sporulation are also inhibited by valine and threonine (Nefelova et al., 1980b). Also, revertants from rif^r antibiotic$^-$ spo$^-$-mutants acquired all three properties together (Kaur et al., 1978). Sporulating cells contained polymyxin synthetase enzymes associated with membrane DNA (Kaur et al., 1978, 1982; Balakrishnan et al., 1980; Vasantha et al., 1980); these enzymes were solubilized during active polymyxin formation. It was suggested that polymyxin plays a regulatory role in transcription.

4. Tyrocidines and Linear Gramicidins. a. *Structures.* In a study of metabolic products from soil organisms, Dubos, in 1939, isolated tyrothricin, a mixture of tyrocidines and linear gramicidins (Figs. 5.6 and 5.7). The structures of tyrocidines A–C were established by Craig and co-workers, while tyrocidines D and E were obtained by directed biosynthesis in vivo (Ruttenberg and Mach, 1966) and in vitro (Fujikawa et al., 1968a). Other replacements in vitro are Orn→Lys and Leu→Ile. The sequence similarities between gramicidin-S, tyrocidines, and gratisin (Fig. 5.8) are surprising, and their biosynthetic implications will be discussed later.

The *linear* gramicidins, as they are often called to stress the difference from gramicidin S, are modified pentadecapeptides. Their structures were elucidated mainly by Witkop and co-workers in the early 1960's (see Izumiya et al., 1979; Ramachandran, 1975).

b. *Producer strains.* Both tyrocidines and gramicidins are produced by the original Dubos strain (*B. brevis* ATCC 8185) or by *B. brevis* ATCC 10068. Although mutants producing only tyrocidine have been isolated (Mukherjee and Paulus, 1977; Sarkar et al., 1978), no stable gramicidin producers

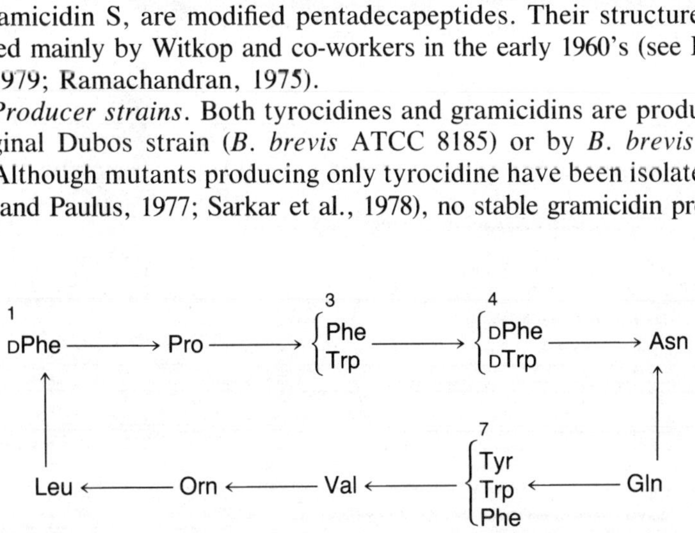

FIGURE 5.6 Structures of tyrocidines. Terminology of tyrocidines is related to the residues 3-4-7. Thus Phe-Phe-Tyr is A, Trp-Phe-Tyr is B, Trp-Trp-Tyr is C, all Trp is D, and all Phe is E. E has been obtained only in vitro, and no structural proof has been carried out.

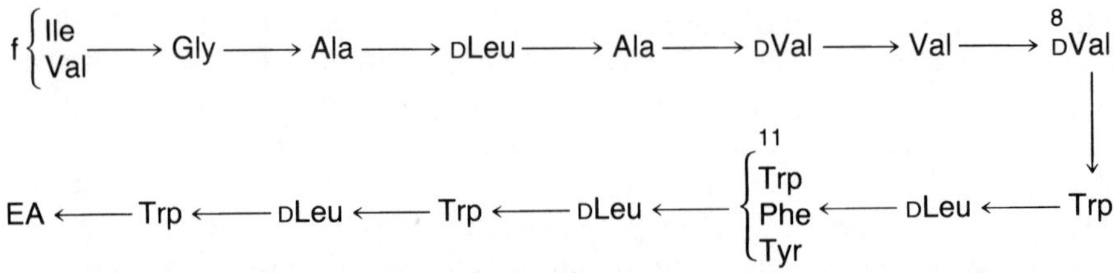

FIGURE 5.7 Structures of linear gramicidins (f = formyl, EA = ethanolamine). Terminology A, B, or C identifies position 11 as Trp, Phe, or Tyr. If Val is replaced by Ile, the gramicidins are A_1, B_1, or C_1 (Ramachandran, 1975).

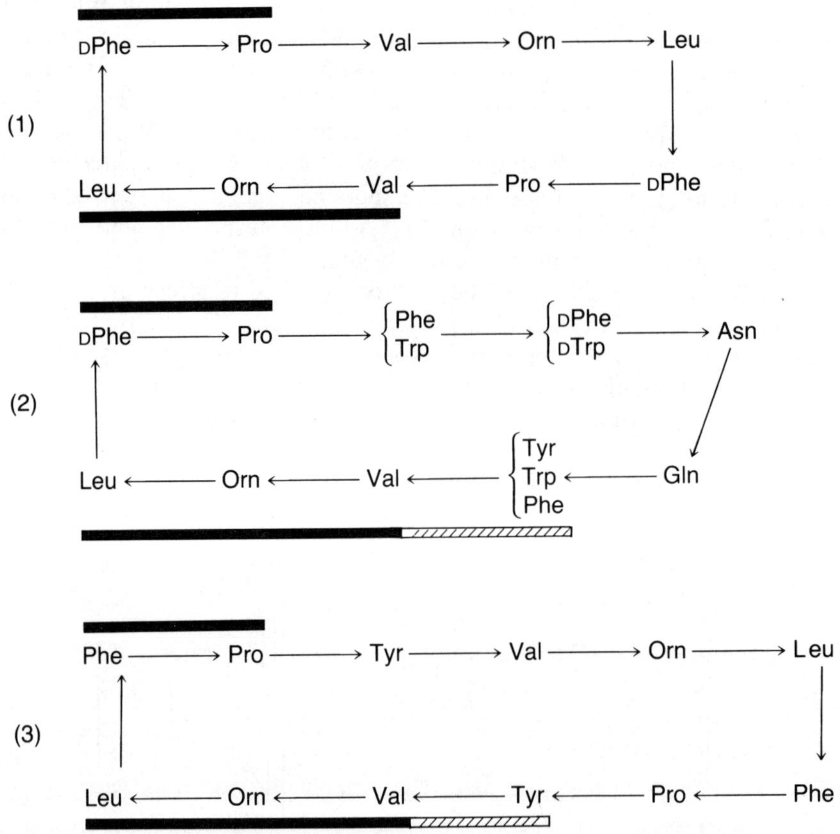

FIGURE 5.8 Sequence homologies in gramicidin S (1), tyrocidines (2), and gratisin (3). The gratisin producer is a spontaneous mutant derived from the gramicidin S producer *B. brevis* var. GB. The optical configuration of gratisin amino acids remains to be established.

have been found. Some evidence for dependence of gramicidin formation on tyrocidine will be discussed later.

C. Production of Enzyme Systems

1. Introduction. As was pointed out in a recent review (Kleinkauf, 1979), the yield of a secondary metabolite is an unreliable indicator of the production of enzyme systems. Cells may contain gramicidin S synthetase even though no peptide production is observed (Matteo et al., 1976). Nevertheless, several studies comparing peptide production in vivo and enzyme activity measured in vitro have shown a direct correlation (Fig. 5.9). To obtain high

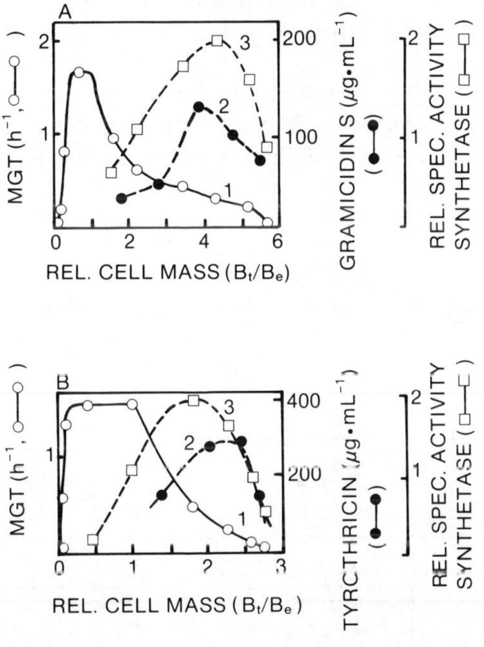

FIGURE 5.9 Correlation of enzyme activity and peptide production in batch fermentations (A) of *B. brevis* ATCC 9999 and (B) of *B. brevis* ATCC 8185 (data from Kurahashi et al., 1968; Fujikawa et al., 1968a). The reciprocal mean generation time (MGT, 1) has been correlated with the relative cell mass (end of log phase = 1.0), specific gramicidin S or tyrothricin production (2), and enzyme activity (3) measured in cell extracts. Note that the growth rate declined from its log phase value and was about 20% of the peak rate during highest productivity. Cell numbers increased threefold to sixfold after log phase ended. Enzyme activity and productivity correlated well.

enzyme yields, high specific activities are required. These can be obtained by specific induction, by synchronization, or in chemostat cultures. Of these methods, only continuous fermentation has been used so far (Matteo et al., 1976; Chiu et al., 1982). Several attempts to establish cell-free systems have failed. Soluble enzyme forms may become membrane-associated (bacitracin, gramicidin S, tyrocidine), while other enzymes are solubilized (polymyxin) during peptide production (Fig. 5.10). Generally, peptide synthetases are active for a limited time and are inactivated by yet unknown mechanisms. Enzyme systems can be assayed by partial reactions (for example, amino acid–dependent ATP-PPi exchange, amino acid epimerization, amino acid–dependent phosphate or pyrophosphate release from ATP, protein aminoacylation) or by peptide formation. In multicomponent systems, characterization of the enzymes may be difficult owing to interference by other enzyme activities (aminoacyl-tRNA-ligases, ATPase, pyrophosphatase). In general, crude extracts have been found to catalyze product formation and can be used for preparative purposes (Wang et al., 1977; von Döhren, 1982).

A general difficulty is the quantitative assay of enzyme systems in crude preparations. Thus gramicidin S formation by gramicidin S synthetase shows a nonlinear dependence on protein concentration (Tzeng et al., 1975). This

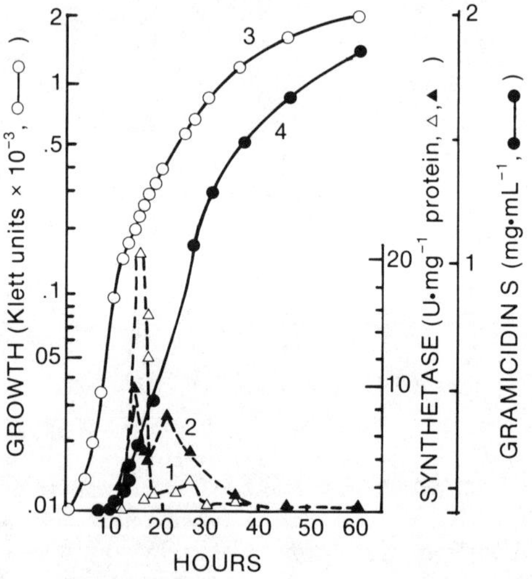

FIGURE 5.10 Formation of soluble (1) and insoluble (2) types of gramicidin S synthetase during growth (3) of *B. brevis* ATCC 9999 in a fructose-based medium (Vandamme, 1976; Vandamme et al., 1982). Gramicidin S production (4) correlated with increasing amounts of the insoluble form that can be solubilized by Triton X-100.

apparently reflects the interaction of multienzymes not associated as stable complexes (Koischwitz and Kleinkauf, 1976a), and a kinetic treatment has been developed (Wang and Hamilton, 1977). To determine the specific activity of gramicidin S synthetase, measurements must be made at different protein concentrations until a constant value is obtained. Since gramicidin-S synthetase has been selected as a model system for preparative enzymatic reactions, its production is being studied extensively. Other synthetases have been purified in small amounts only.

2. Fermentations. All enzyme systems are produced within a defined period of the growth cycle, usually characterized by a change in the kinetics of growth and in the pH of the culture. Cells producing little (bacitracin, Frøyshov, 1977) or no (gramicidin A, Matteo et al., 1976; Aust and von Döhren, 1982) product sometimes contain highly active enzymes. This has been correlated with either the lack of a precursor (notably phenylalanine or ornithine) or immobilization of the enzymes by attachment to a membrane (Frøyshov, 1977; Vandamme, 1981; Vandamme et al., 1982). It is well known that phosphate significantly influences gramicidin-S formation, the optimum level varying from 10 to 60 mM, depending on the medium used (Vandamme, 1981). In a glutamate minimal medium, the enzyme yield has been increased at least tenfold by lowering the phosphate concentration from 1.36% to 0.1%, corresponding to an optimal level of only 7 mM of phosphate (Aust and von Döhren, 1982). In chemostat experiments using a fumarate minimal medium, Chiu et al. estimated an optimal phosphate concentration of about 1 mM (Chiu et al., 1982). Pulsing a culture growing at a dilution rate of 0.27 h^{-1} with phosphate caused an immediate decrease in activity, with recovery at a phosphate concentration below 5 mM (Fig. 5.11). This suggests that phosphate represses enzyme formation. However, in a glycerol-based medium, gramicidin S synthetase was produced under carbon, nitrogen, phosphate, or sulfate limitation, implicating growth rate, rather than the nature of limitation, as the important factor.

Enzyme activities decline at low dilution rates, and limited sporulation is observed under carbon limitation in glycerol- as well as fumarate-grown cultures (Matteo et al., 1976; Chiu et al., 1982). This decline of activity may be due to in vivo inactivation of the synthetase, for which both oxidative and hydrolytic mechanisms (cofactor cleavage, limited proteolysis) have been proposed (Friebel and Demain, 1977a, b; Kleinkauf and Koischwitz, 1974, 1980a, Vandamme, 1981).

Chemostat experiments demonstrating that sporulation conditions deviate from those favoring multienzyme formation indicate that the two processes are unlinked. This was also the conclusion from studies with peptide non-producer mutants of *B. brevis* ATCC 9999 (Kambe et al., 1974), *B. licheniformis* (Haavik and Thomassen, 1973), and *B. brevis* ATCC 8185 (H.

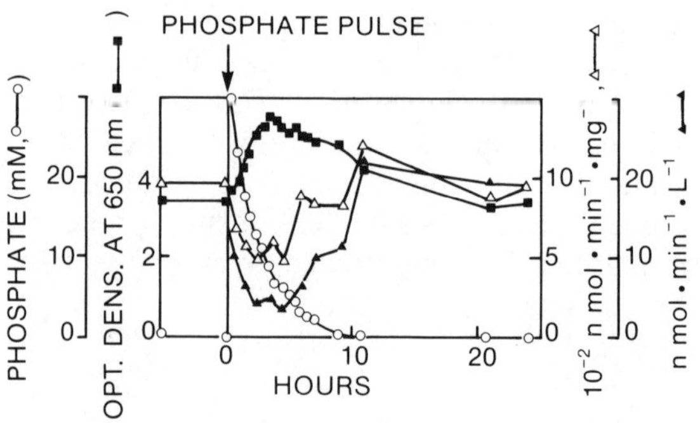

FIGURE 5.11 Response of a phosphate-limited chemostat culture of *B. brevis* ATCC 9999 during growth at a dilution rate of 0.27 h^{-1} to a phosphate pulse. The pulse leads to an almost exponential decay of enzyme activity with recovery as soon as the phosphate concentration is below 5 mM (Chiu et al., 1982). Enzyme activity is plotted as specific (nmol·min^{-1}·mg^{-1}) and volumetric (nmol·min^{-1}·L^{-1}) activity.

Ristow, personal communication). Although similar mechanisms may control both processes, they operate independently. This is also evident from studies with a yeast extract/peptone (5%/5%) medium where high enzyme levels were obtained but cells did not sporulate (Tzeng et al., 1975; D. I. C. Wang, personal communication). The well-known *B. subtilis* sporulation inducer, decoynine (Mitani et al., 1977), causes enzyme formation in midlog fumarate-grown cells, but activity declines rapidly, indicating inactivation (H. I. Aust, unpublished results).

In summary, gramicidin-S synthetase is formed in batch as well as continuous cultures. Enzyme levels as well as peptide production can be improved by optimization of fermentation parameters. In enzyme production, severe problems are posed by in vivo inactivation mechanisms. Enzymes remain active for only a limited period of time. To isolate the enzymes, cultures are cooled and flushed with nitrogen while cells are rapidly separated from the medium (Friebel and Demain 1977a, b). Recognizing optimal enzyme activity in batch fermentation is difficult. Since enzyme assays take approximately one hour, growth rate and pH are usually used to estimate the best time of harvest. At maximum enzyme concentration, a drop followed by a rise in the respiratory rate has been observed (Aust and von Dohren, 1982; Vandamme et al., 1982). Its cause remains to be established, but the associated peak in dissolved oxygen tension can also be used to predict maximum gramicidin synthetase production (Fig. 5.12).

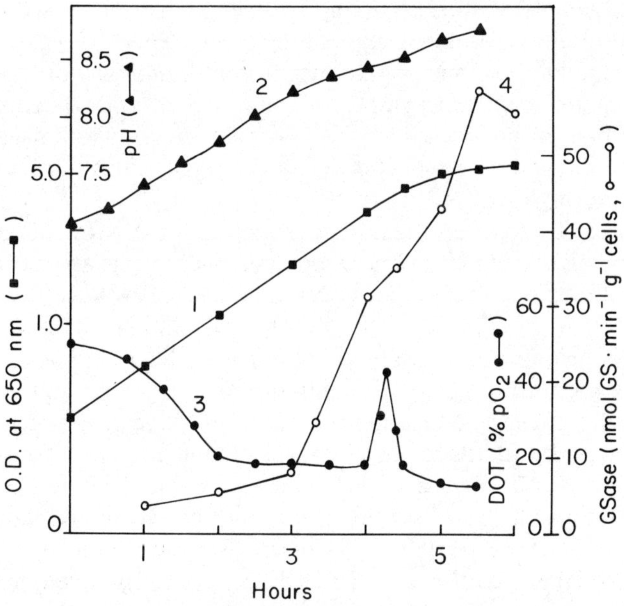

FIGURE 5.12 Growth rate (1), pH (2), and dissolved oxygen tension (DOT, 3) during gramicidin S synthetase formation (GSase, 4) in a batch culture in a 1% glutamate medium. DOT was kept at 20% saturation during growth. Note the reproducible peak associated with enzyme synthesis (data from Aust and von Döhren, 1982).

III. PEPTIDE SYNTHETASES

A. Introduction

1. Multifunctional Proteins. The term *multifunctional* is often used to describe not only catalytic functions, but also noncatalytic association phenomena (binding of small ligands, protein–protein interactions). The current definition of *multifunctional proteins* as proteins combining several *autonomous functions* on one peptide chain and to which distinct *regions* or *domains* can be assigned (Bisswanger and Schmincke-Ott, 1980) reflects the need for precision in describing complicated new enzyme structures. However, it appears to be too simplified an approach. Function, defined as a chemical property of a protein, cannot always be correlated with a region or domain (Wetlaufer 1973). A region or domain defined on a functional basis (functional "unit") is not always *one* distinct region of a peptide chain but may be composed of several structural domains. We would suggest using the term multifunctional not only to refer to the protein, but to further specify its connection. A stable complex of single-function subunits as well as a stable complex of multifunctional polypeptide chains such as yeast fatty acid synthetase could then both be termed "multifunctional proteins." Well-characterized peptide synthetases are complicated enzyme systems containing multifunctional polypeptide chains. The number of functions assigned to one peptide chain will remain unclear (Table 5.1) until uncertainty about the definition of "function" is resolved.

The standard procedure for detecting subunits is dodecylsulfate treatment followed by polyacrylamide gel electrophoresis. Problems arising from limited proteolysis in enzyme preparations containing proteases have been discussed by Weber et al. (1972). If sufficient dodecylsulfate is present (>1.4 g/g protein), the protein can be fully denaturated by heat (95°C); disulfides can be reduced by adding up to 20% dithiothreitol (Phillips and Agin, 1977). The procedure appears to be quite reliable. Other denaturing reagents (urea, guanidine) may be included, or chemical modifications (sulfonation, citraconylation) may be carried out (Kleinkauf, 1979). The size of the GS2 component of gramicidin S synthetase in both native and denaturated state has been estimated to be 280 Kd (Koischwitz and Kleinkauf, 1976b). The sizes of tyrocidine synthetase (TY2, TY3), linear gramicidin synthetase (LG1, LG2), enniatin synthetase, and bacitracin synthetase (BA1) components have been estimated by active enzyme centrifugation and dodecylsulfate gel electrophoresis (Lee and Lipmann, 1975; Zocher et al., 1982b; Ogawa et al., 1981; H. von Döhren, unpublished). Some of the data are compiled in Table 5.1. Apparently, the multifunctional structure is quite common, since extensive purification with convincing structure/activity correlations has pointed to covalent structures rather than subunit-containing systems.

2. Functions of Peptide Synthetases. The biosynthetic process for forming bacitracins, gramicidin S, linear gramicidin, polymyxins, and tyrocidines has been called a *protein template mechanism* (Lipmann, 1971), a *thiotemplate mechanism* (Laland and Zimmer, 1973), and recently a *multienzyme thiotemplate mechanism* (Kurahashi, 1981). The reactions involved are amino acid activation, enzyme aminoacylation, amino acid racemization, and peptide bond formation with transport of peptides by 4′-phosphopantetheine. These reactions have been equated with functions, and multienzyme systems with 50 and more functions have been described. To compare this type of synthetase to other multienzymes, the number of functions should be reduced as shown in Table 5.1. Each function is described below. (For reviews, see Lipmann, 1980, 1982; Kleinkauf and Koischwitz, 1978; Kleinkauf, 1979; Kleinkauf and von Döhren, 1981, 1982; Kurahashi, 1981.)

a. *Amino acid activation and enzyme aminoacylation.* Activation reactions proceed generally by random addition of the substrate amino acids (a) and ATP (as $MATP^{2-}$), followed by aminoacyl adenylate formation, its enzymatic stabilization, and subsequent aminoacylation of specific enzyme thiol groups (1).

(1) 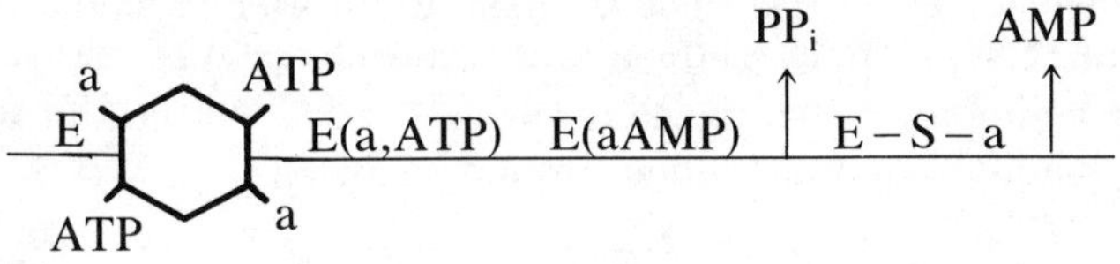

Kinetic analysis has been carried out by using the amino acid–dependent ATP–PP exchange reaction (2), formation of [^{32}P]-phosphate from [γ-^{32}P]ATP (3), the (^{14}C)-AMP-ATP exchange reaction (4), and isolation of enzyme-bound [^{14}C]- and [^{3}H]- amino acids (5).

(2) $$E(aAMP) + {}^{*}PP_i \rightleftharpoons E + a + ATP^{*}$$

(3) $$a + ATP^{*} + E \rightleftharpoons E(aAMP) + PP_i{}^{*} \xrightleftharpoons{PP_i ase} P_i + P_i{}^{*}$$

(4) $$E-S-a + {}^{*}AMP \rightleftharpoons Ea^{*}AMP + PP_i \rightleftharpoons E + a + {}^{*}ATP$$

(5) $$E + 2a^{*} + 2\ ATP \rightleftharpoons E^{a^{*}AMP}_{-S-a^{*}} + 2PP_i + AMP$$

It has been proposed that two moles of each amino acid (*a) bind per mole of enzyme, as is shown in Eq. (5). Although this has not yet been investigated in detail, complexes isolated by gel filtration contain labile aminoacyl adenylates, together with stable amino acids (thioesters) in equal amounts. No more than 1 mol/mol has been detected for these intermediates. An interpretation is given in (6).

(6) 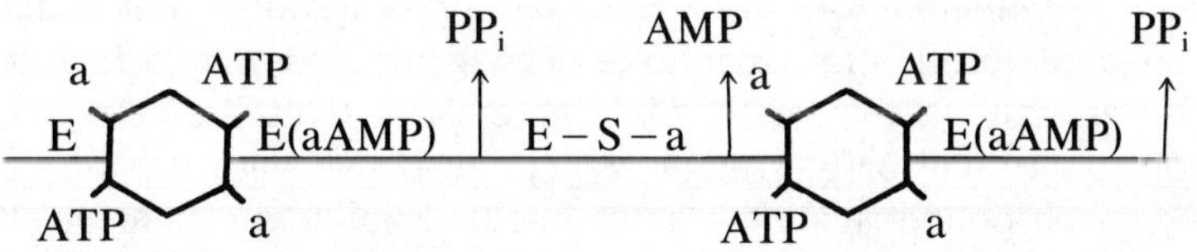

The uncoupling of enzyme aminoacylation with retention of aminoacyl adenylate formation has been shown for GS1 (DTNB or mutant enzyme, Kanda et al., 1981; Saito, 1982) and GS2 (Pro-chloromethylketone, Leung and Baxter, 1972).

b. *Amino acid racemization* Few details are available on the racemization of amino acids in peptide biosynthesis. Only the system in gramicidin S synthetase GS1 has been partially characterized (Kanda et al., 1981). Although the multienzyme has been classified as an ATP-dependent Phe-racemase, the enzyme-bound thiolester is not a racemate. It has about 70% D-configuration, regardless of which isomer has formed the aminoacyl adenylate (Takahashi et al., 1971). The configuration after reaction with a suitable acceptor (NH_3 or GS2 charged with Pro-thioester) is exclusively D, so we have an epimerase rather than a racemase (Eq. 7). The lack of stereospecificity in Phe-binding is not a unique property of GS1.

(7)

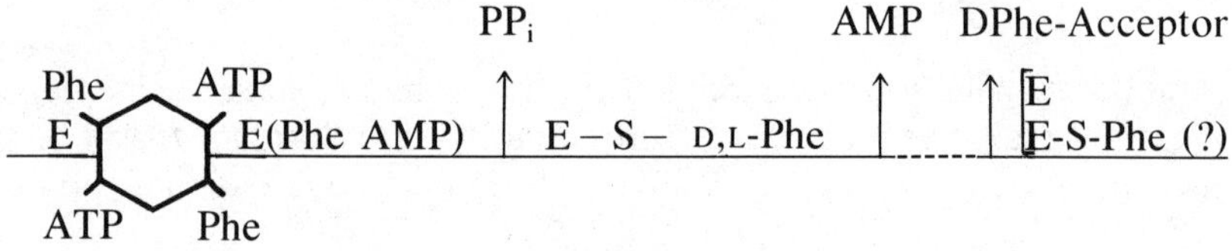

In a preliminary study, Lee (1976) showed that ^3H in [α-^3H]Phe is exchanged to the enzyme in an acid-stable, alkali-labile form, presumably attached to a base. D-Isomers of the aromatic amino acids (Phe, Tyr, Trp) can be incorporated into tyrocidine and gramicidin S. In linear gramicidin synthetase, the multienzyme LG2 activates DVal and DLeu (Akers et al., 1977). In bacitracin formation, DGlu, DPhe, and DAsp are incorporated, but only the L form of Orn is used (Frøyshov and Laland, 1974; Frøyshov, 1979). In polymyxin synthetase, activation of DLeu is detectable but by no means comparable to Leu activation (S. Komura, personal communication). Even when the L configuration is present in the final product, the D form may be activated. This has been demonstrated for the Val and Leu activation site of GS2. Also, these two binding sites activate not only Leu, Val, Ile, norleucine, and norvaline, but also, at detectable rates, Phe and cyclohexylalanine. The D forms of the corresponding amino acids are activated but not incorporated into gramidicin S (Aarstad et al., 1979; Laland et al., 1982; Lipmann et al., 1971; Saxholm et al., 1972).

c. *Peptide bond formation.* Peptide bonds are formed in initiation, elongation, and termination reactions. The general principle in elongation is "head growth" (Lipmann, 1968) where the activated carboxyl group of a donor amino acid or intermediate peptide reacts with the amino group of the enzyme-bound acceptor amino acid. Similar mechanisms have been proposed in ribosomal, enzymatic peptide, and fatty acid elongation (Kleinkauf and von Döhren, 1982). The central role in elongation reactions is attributed to 4′-phosphopantetheine (Fig. 5.13). This cofactor fulfills a transport function in many polymerizing multienzyme systems involving activated carboxyl groups. Although its origin in peptide synthetases has not been established, it is ususally derived from CoA by specific transferase action. 4′-Phosphopantetheine is covalently attached to the multienzyme complex. However, the enzyme-bound cofactor may not be the only type of transporting agent in peptide formation, since bacitracin synthetase contains a soluble form of pantetheine (Roland et al., 1975). As in polyketide biosynthesis, inter-

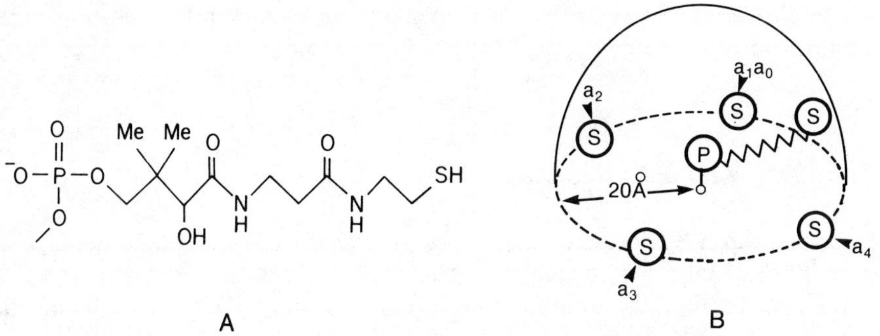

FIGURE 5.13 (A) Structure, and (B) model showing operation of the enzyme-bound cofactor 4′-phosphopantetheine.

mediates are attached to the cofactor as activated thiolesters. Transport is thought to proceed by transthiolation (8):

$$(8)\quad E\begin{cases} S^{P}H \\ S\text{-donor} \end{cases} \rightleftharpoons E\begin{cases} S^{P}\text{-donor} \\ SH \end{cases}$$

The donor compound attached to pantetheine may be modified (for example, by dehydration or reduction of polyketides) or react dircetly with an acceptor compound (9).

$$(9)\quad E\begin{cases} S^{P}\text{-donor} \\ S\text{-acceptor} \end{cases} \longrightarrow E\begin{cases} S^{P}H \\ S\text{-}\underbrace{\text{acceptor-donor}}_{\text{donor*}} \end{cases}$$

The effective concentration of enyzyme-attached substrates is high, and a diffusion-controlled reaction would account for the observed reaction rates.

The observation that all amino acids may be attached to a multienzyme before initiation suggests an *exclusion principle*: peptides or N-modified amino acids are transported because of their closer structural homology to carboxylic acids. Amino acids themselves are not transported because the extra charge of an amino group would prevent transthiolation by protonation of a postulated thiolate pantetheine acceptor (Koischwitz, 1978).

The gramicidin S initiation reaction is absent in multienzymes from pantetheine-free mutants (Hori et al., 1981; Saito, 1982), indicating that activated amino acids are transported during initiation (10).

$$(10)\quad E^{1}\text{-S-D,LPhe} + E^{2}\text{-S-Pro} \longrightarrow E^{1}\text{-SH} + E^{2}\text{-S-ProD-Phe}$$

It remains to be established how the peptide sequence and thus the sequential addition of amino acids is controlled.

B. Enzymology of Synthetases

1. Gramicidin S Synthetase. a. *Enzyme organization.* The two multienzymes GS1 (100 Kd) and GS2 (280 Kd) have been purified and partially characterized, but problems in handling and stability have been reported (Kleinkauf and Koischwitz, 1974, 1980a, b; Zimmer and Laland, 1975; Koischwitz and Kleinkauf, 1976a, b; Kleinkauf and von Döhren, 1980, 1981; von Döhren, 1982). Several methods have been devised for preparing crude enzymes (Augenstein et al., 1974; Wang et al., 1977), GS1 (Yamada and Kurahashi, 1969; Vater and Kleinkauf, 1976; Kanda et al., 1978), and GS2 (Koischwitz and Kleinkauf, 1976a; Christiansen et al., 1977; Hori et al., 1978, 1981). Some of them use PMFS as a proteinase inhibitor, although GS2 is inactivated by this serine-directed reagent (Altmann et al., 1982). The structure of GS2 is highly susceptible to proteolysis (Kleinkauf and Koischwitz, 1980b, c; Kleinkauf and von Döhren, 1981; Altmann et al., 1982) and oxidation (Friebel and Demain, 1977a, b). The integrity of its structure is required for catalytic functioning of enzymes of the reaction sequence. Various efforts to disintegrate the multienzyme structures by physical methods (Astrup, 1972; Kleinkauf and Koischwitz, 1974; Koischwitz and Kleinkauf, 1976b; Christiansen et al., 1977; Kleinkauf, 1979) have indicated that multifunctional polypeptide chains are present. This is in accord with genetic data (Table 5.4) but proof of a single polypeptide chain awaits more complete genetic analysis. Gramicidin S nonproducer mutants of *B. brevis* ATCC 9999 or the Nagano strain have been classified into (1) multienzyme nonproducers, (2) GS1 nonproducers, (3) GS2 nonproducers, and (4) those with pantetheine-free GS2, (5) altered GS1 structures, and (6) altered GS2 structures (Iwaki et al., 1972; Shimura et al., 1974; Kambe et al., 1974; Kanda et al., 1978, 1981; Hori et al., 1978, 1981; Marahiel et al., 1979; Saito, 1982).

The chromosomal location of genes participating in peptide synthesis has not been established, but those for gramicidin S formation can reasonably be divided into four sets controlling: expression of GS genes, GS1, GS2, and 4′-phosphopantetheine-transferase. Distinct genes for amino acid–activating subunits later assembled to multienzyme structures have not been found, although separate enzymes activating Pro and Orn, with a size of 80–100 Kd, respectively, have been described (Kurahashi, 1981). The first of these was characterized as a GS2 fragment by its cross-reaction with GS2 antibodies and similar substrate specificity profile (Bothe et al., 1982). The altered imino acid affinity, together with altered Mg^{2+} dependence of the Pro activation reaction of this fragment, has also been observed with a GS2 fragment derived by limited trypsinolysis (H. von Döhren et al., unpublished). The sequence of amino acid activation centers has been concluded from fragment analysis to correspond to the sequence in gramicidin-S (Kleinkauf and Koischwitz, 1980a, b, c; Koischwitz, 1979).

b. *Reaction sequence*. The sequence proposed in 1971 (Kleinkauf et al., 1971) is summarized in Scheme 5.1. Its elucidation has been reviewed comprehensively (Lipmann et al., 1971; Laland and Zimmer, 1973; Kurahashi, 1974, 1981; Kleinkauf and Koischwitz, 1978; Lipmann, 1980, 1982). The activation reactions in which an amino (imino) acid and the mono-Mg (or Mn, Ca) complex of ATP form an aminoacyl adenylate (Reactions 1–5) proceed by random binding of substrates (Kittelberger et al., 1982; M. Altmann, unpublished results). Binding patterns of ATP analogs are similar at all sites but different from those of known families of aminoacyl-tRNA synthetases (Freist et al., 1981). The low specificity of amino-acid binding is remarkable. Instead of Phe, other aromatic amino acids, or even cyclohexylalanine, are accepted (Lipmann et al., 1971; Aarstad et al., 1979, 1980; Laland et al., 1982; Kleinkauf and von Döhren, 1980). Adenylates are then accepted by specific thiol groups (enzyme aminoacylation, Reactions 6–10). A mutationally altered GS1 lacking an essential DTNB-rapid-reacting thiol group is defective in ester binding and epimerization (Kanda et al., 1981; Saito, 1982).

Activated Phe is bound in L,D configuration but only the D form is transferred to a suitable acceptor. Initiation requires interaction of both multienzymes. Loading with Pro is a prerequisite for transfer of Phe to GS2, indicating the possible acceptor (Reaction 11). The dipeptide DPhePro may be lost by cyclization (Fig. 5.14) in an early termination reaction (19). This side reaction has been used to prepare various cyclodipeptides from imino and aromatic amino acids (von Döhren, 1982). Another side reaction is the Orn side chain cyclization (Reactions 20 and 21, Fig. 5.14; Gadow, 1981; J. Vater et al., unpublished results). Pantetheine-mediated transport of peptides to acceptor amino acids (Reactions 12–17) leads to enzyme-bound pentapeptides, which react by head-to-tail cyclization (Reaction 18 a, b). The mechanism of this reaction (intermolecular or intramolecular) has not yet been established (Stoll et al., 1970; Kurahashi, 1974; Kleinkauf and Koischwitz, 1980b).

c. *Applications*. Attempts to stabilize crude preparations of the synthetase for peptide production in enzyme reactors by adding soluble polymers gave only slight improvements that depended on the degree of purification. At temperatures above 42°C, purified GS2 has a half-life in the range of minutes. High activation energies indicate predominantly denaturation processes. At the usual operating temperature of 30–37°C, where these processes should be negligible, crude and pruified preparations were inactivated by approximately first-order reactions with low activation energies corresponding to hydrolytic processes (von Döhren, 1982). In the presence of substrates, gramicidin S production persists for several hours at 37°C. In a model reaction, 80% of labeled Leu added was incorporated into gramicidin S

Scheme 5.1

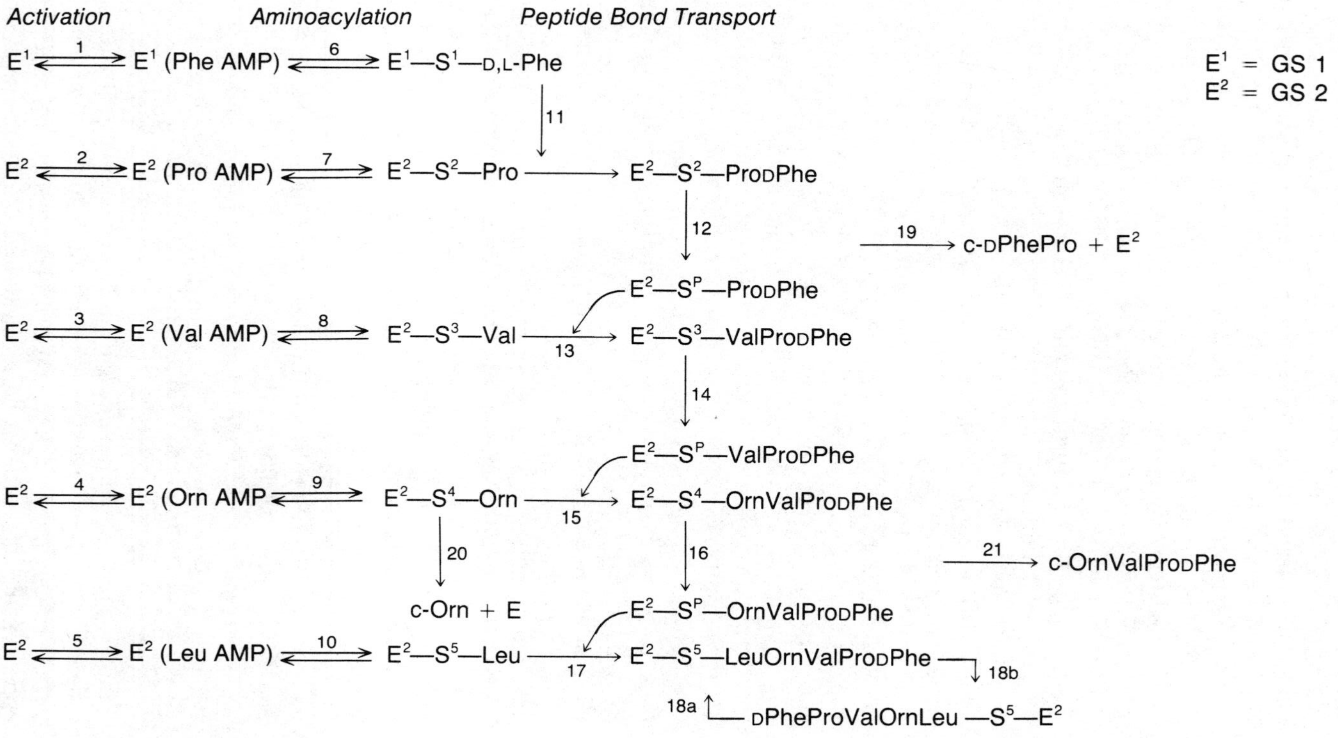
Activation
Aminoacylation
Peptide Bond Transport
E¹ = GS 1
E² = GS 2
E¹ 1 E¹ (Phe AMP) 6 E¹—S¹—D,L-Phe
E² 2 E² (Pro AMP) 7 E²—S²—Pro 11 E²—S²—ProDPhe
12 E²—Sᴾ—ProDPhe 19 c-DPhePro + E²
E² 3 E² (Val AMP) 8 E²—S³—Val 13 E²—S³—ValProDPhe
14 E²—Sᴾ—ValProDPhe
E² 4 E² (Orn AMP 9 E²—S⁴—Orn 15 E²—S⁴—OrnValProDPhe
20 c-Orn + E
16 E²—Sᴾ—OrnValProDPhe 21 c-OrnValProDPhe
E² 5 E² (Leu AMP) 10 E²—S⁵—Leu 17 E²—S⁵—LeuOrnValProDPhe 18b
18a DPheProValOrnLeu —S⁵—E²

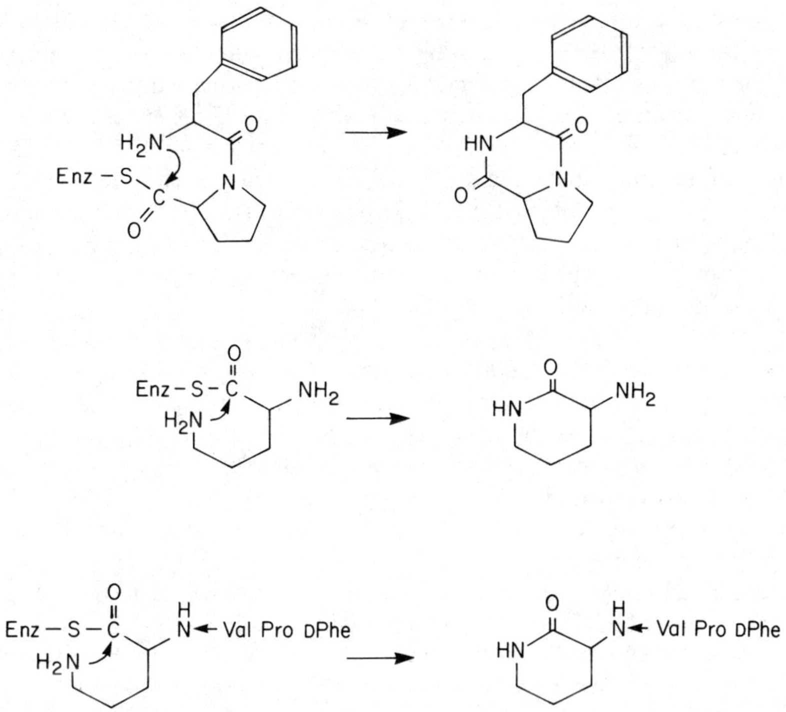

FIGURE 5.14 Early termination reactions in gramicidin S formation. (1) cyclization of enzyme-bound DPhePro (2,5-piperazinedione), (2) cyclization of thiolester-bound Orn (β-amino-α-piperidone), and (3) tetrapeptide loss by cyclization of Orn.

(Hamilton et al., 1973; Demain and Wang, 1976). Since stabilization of the enzyme system has not been accomplished, cell-free production of gramicidin-S cannot compete with fermentation procedures. However, the broad substrate specificity of the synthetase can be used to advantage to make various peptide analogs. Their rates of production can be roughly estimated from single analog substitutions (Kleinkauf and Koischwitz, 1980c, von Döhren, 1982).

2. Tyrocidine Synthetase. a. *Enzyme organization.* The tyrocidine system has been investigated mainly in the laboratories of Kurahashi and Lipmann and has been reviewed recently (Lipmann, 1980, 1982; Kurahashi, 1981). It consists of three multienzymes (100, 230, and 440 Kd). The multienzymes (or *polyenzymes*) have been purified (Lee et al., 1973; Lee and Lipmann, 1975). Treatment with dodecylsulfate only partially dissociated multienzymes TY2 and TY3. Further evidence of their single polypeptide structure was later provided by H. von Döhren et al. (unpublished results). Incubating crude extracts at 37°C or treating purified enzymes with a Triton X-100-soluble factor from the producer caused dissociation into a pantetheine-containing protein (17–20 Kd) and subunits (about 70 Kd), which activated amino acids (Lee et al., 1973; Lee and Lipmann, 1974, 1975, 1977). The nature of the enzymatic degradation has yet not been established. The functional exchange of GS1 and TY1 has been demonstrated (Fujikawa et al., 1971). From the amino acid sequence homology of gramicidin S and tyrocidines, structural homologies between multienzymes GS2, TY2, and TY3 can be anticipated.

Several unexpected results have come from studies of fragmented subunits. The pantetheine-containing carrier protein has a strong tendency to self-aggregate. The activating proteins are similar in size and are not separated by ion exchange or adsorption chromatography; the carrier protein is associated with individual subunits as 90-Kd complexes during gel filtration; affinity of Pro for the 90-Kd complex is higher than for the 70-Kd subunit (Lee and Lipmann, 1974, 1975, 1977). This finding has been accommodated in an alternative interpretation of the elongation process (Lee and Lipmann, 1977; Lipmann, 1980, 1982). However, this interpretation does not account for the accumulation of intermediates as has been discussed by Kurahashi (1981) and Laland et al. (1978).

b. *Reaction sequence*. In Scheme 5.2 the reaction sequence leading to tyrocidine E (all-Phe) is given. At all aromatic amino acid positions (Reactions 1, 3, 4, 7), Phe can be replaced by Tyr, Trp, and other analogs (Mach and Tatum, 1964; Fujikawa et al., 1968b; Rao et al., 1968; Roskoski et al., 1970a, b; Lee and Lipmann, 1975). Amino acid activation (reactions 1–10) is followed by amino acylation of specific thiol groups (SH^1–SH^{10}, Reactions 11–20). The enzyme–substrate complexes that have been isolated contain both thiol esters and aminoacyladenylates, indicating a second activation reaction.

Initiation, as in gramicidin S formation, is a transfer of DPhe to the Pro acceptor on TY2 (21). The dipeptide formed is transferred to pantetheine (22) and accepted by the third amino acid at thiol site 3 (23). It may be lost as the piperazinedione (39) either at the Pro site or as the pantetheine-bound intermediate. With the addition of Phe and DPhe (23, 25), the tetrapeptide is formed on TY2 and finally transferred to Asn on TY3 (27). The occurrence of DPhe in initiation and peptide transfer suggests similarities between at least the activation sites of TY1 and TY2. Indeed, both isomers of Phe are activated; DPhe, however, at lower rates in TY2 (Lee et al., 1973; Lee and Lipmann, 1975). Elongations proceed to the decapeptide (29–37), detection of which indicates that the cyclization (38) is slow in comparison to gramicidin S cyclization. Two side reactions detected in the gramicidin S sequence (40, 41) can be predicted to occur also in the tyrocidine system.

3. Linear Gramicidins. a. *Enzyme organization.* Although complete formation of linear gramicidin in vitro has not been achieved, the pentadecapeptide has been formed by four enzyme fractions, two of which have been purified and partially characterized (Akers et al., 1977; Akashi, 1977; Akashi et al., 1977; Akashi and Kurahashi, 1977, 1978). Such structural features of gramicidin as N-terminal formylation, the repeated TrpDLeu sequence and the C-terminal ethanolamine imply some difference from the tyrocidine type of biosynthesis. The initiating enzyme LG1 (160–180 Kd) activates Val (Ile) and Gly and also formylates the starter. The formyl dipeptide is then elongated by LG2 (300–350 Kd) to the heptapeptide. Additions of Val and the dipeptide elements TrpDLeu are apparently performed by smaller enzymes (Akashi, 1977). LG1 and LG2 contain pantetheine (Akers et al., 1977) and are stable in SDS, indicating multifunctional polypeptide chains (H. von Döhren, unpublished).

b. *Reaction sequence.* The partially elucidated sequence of reactions is summarized in Scheme 5.3. Activation of amino acids as aminoacyladenylates (1–15) is followed by enzyme aminoacylation (16–30). Reactions have been listed only to the decapeptide. Formylation of enzyme-bound Val has been demonstrated by Akashi and Kurahashi (1977, 1978). In the absence of formyltetrahydrofolic acid, peptide is synthesized, but dipeptide hydrolysis (63) and cyclization (64) are favored (Akers et al., 1977). The transport of formylVal by 4′-phosphopantetheine is in accord with the exclusion principle of peptide transport. Elongation is thought to proceed as described in Section III.A.2. Involvement of pantetheine beyond the heptapeptide stage has not been shown. D-Amino acids in the alternating D,L sequence are thought to be derived by epimerization of activated Val and Leu, although DLeu and, to a small extent, DVal can be activated directly (Akers et al., 1977). Ethanolamine may be introduced either directly (as in the chemical termination of enzyme-bound pentadecapeptide, Bauer et al., 1972) or, as has been suggested by Kurahashi (1981), as serine followed by decarboxylation.

4. Bacitracin. a. *Enzyme organization.* Although in vitro synthesis of bacitracin has been accomplished (e.g. Ishihara and Shimura, 1974; Frøyshov and Laland, 1974; Wang et al., 1977), purified enzyme systems have been found to be inactive (Roland et al., 1977) or inefficient. Vitković and Pfaender (1982) recently proposed that a fourth multienzyme is needed in addition to the three (BA1–3) already known (Frøyshov, 1974, 1975, 1979), but insufficient data are available as yet for evaluation. Multienzymes BA1, BA2, and BA3 activate residues 1–5, 6 and 7, and 8–12 respectively. A structural study (Ogawa et al., 1981; Ishihara et al., 1982) established that BA1 is a 140-Kd multienzyme. The other multienzymes have not been characterized well. Pantetheine is present in all three components, but there is some disagreement about the amount and state of the cofactor in BA2

Scheme 5.2

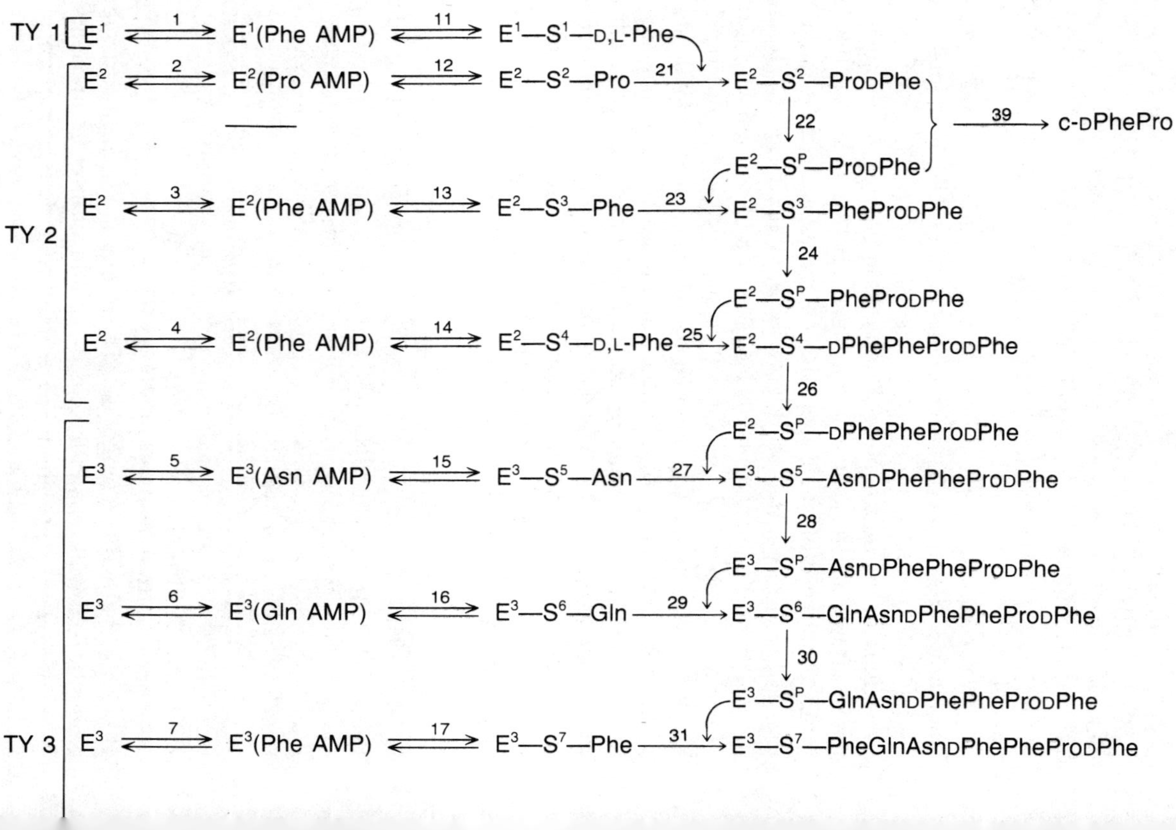
TY 1
TY 2
TY 3
E¹ ⇌(1) E¹(Phe AMP) ⇌(11) E¹—S¹—D,L-Phe
E² ⇌(2) E²(Pro AMP) ⇌(12) E²—S²—Pro —(21)→ E²—S²—ProDPhe
E²—S^P—ProDPhe
—(39)→ c-DPhePro
E² ⇌(3) E²(Phe AMP) ⇌(13) E²—S³—Phe —(23)→ E²—S³—PheProDPhe
E²—S^P—PheProDPhe
E² ⇌(4) E²(Phe AMP) ⇌(14) E²—S⁴—D,L-Phe —(25)→ E²—S⁴—DPhePheProDPhe
E²—S^P—DPhePheProDPhe
E³ ⇌(5) E³(Asn AMP) ⇌(15) E³—S⁵—Asn —(27)→ E³—S⁵—AsnDPhePheProDPhe
E³—S^P—AsnDPhePheProDPhe
E³ ⇌(6) E³(Gln AMP) ⇌(16) E³—S⁶—Gln —(29)→ E³—S⁶—GlnAsnDPhePheProDPhe
E³—S^P—GlnAsnDPhePheProDPhe
E³ ⇌(7) E³(Phe AMP) ⇌(17) E³—S⁷—Phe —(31)→ E³—S⁷—PheGlnAsnDPhePheProDPhe
22
24
26
28
30

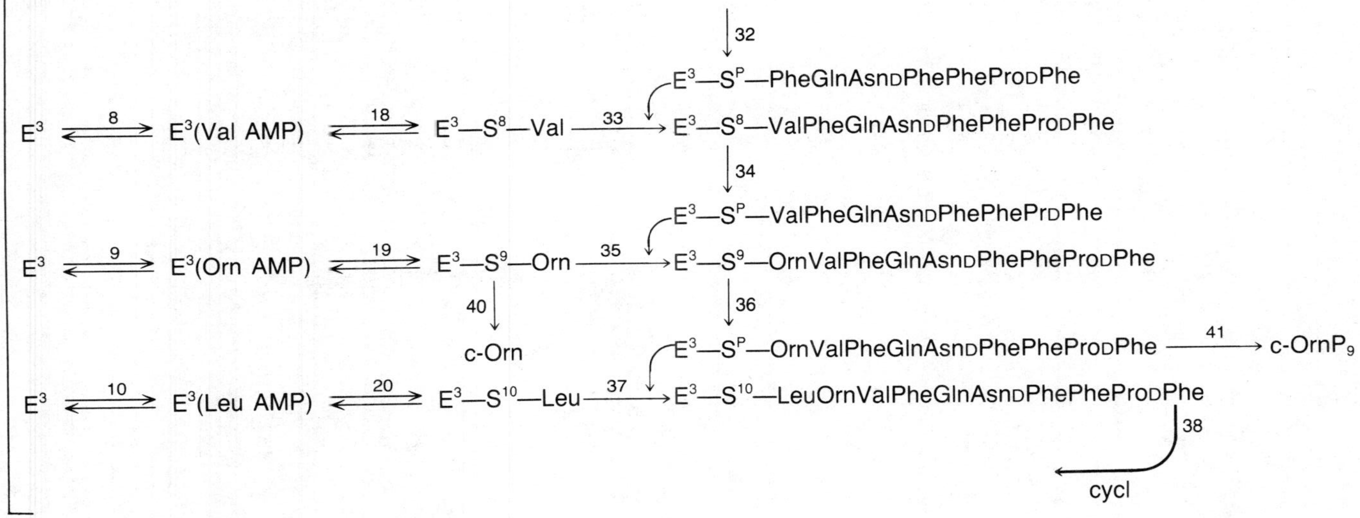
E³ ⇌ E³(Val AMP) ⇌ E³—S⁸—Val → E³—S⁸—ValPheGlnAsnDPhePheProDPhe
E³—Sᴾ—PheGlnAsnDPhePheProDPhe
32
8
18
33
34
E³ ⇌ E³(Orn AMP) ⇌ E³—S⁹—Orn → E³—S⁹—OrnValPheGlnAsnDPhePheProDPhe
E³—Sᴾ—ValPheGlnAsnDPhePhePrDPhe
9
19
35
40
c-Orn
36
E³—Sᴾ—OrnValPheGlnAsnDPhePheProDPhe
41
c-OrnP₉
E³ ⇌ E³(Leu AMP) ⇌ E³—S¹⁰—Leu → E³—S¹⁰—LeuOrnValPheGlnAsnDPhePheProDPhe
10
20
37
38
cycl

Scheme 5.3

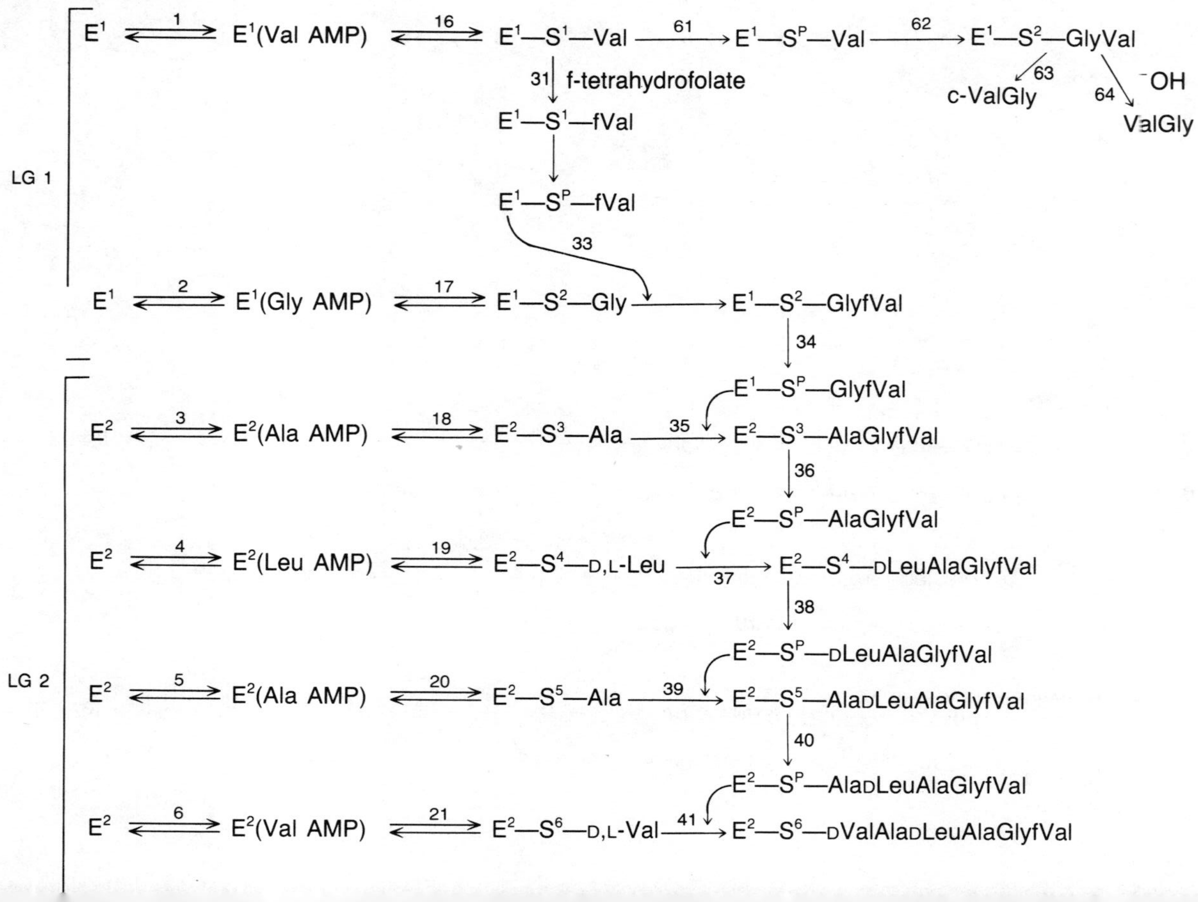
LG 1
LG 2
E¹ ⇌ E¹(Val AMP) ⇌ E¹—S¹—Val → E¹—Sᴾ—Val → E¹—S²—GlyVal
f-tetrahydrofolate
E¹—S¹—fVal
E¹—Sᴾ—fVal
c-ValGly
⁻OH
ValGly
E¹ ⇌ E¹(Gly AMP) ⇌ E¹—S²—Gly → E¹—S²—GlyfVal
E¹—Sᴾ—GlyfVal
E² ⇌ E²(Ala AMP) ⇌ E²—S³—Ala → E²—S³—AlaGlyfVal
E²—Sᴾ—AlaGlyfVal
E² ⇌ E²(Leu AMP) ⇌ E²—S⁴—D,L-Leu → E²—S⁴—DLeuAlaGlyfVal
E²—Sᴾ—DLeuAlaGlyfVal
E² ⇌ E²(Ala AMP) ⇌ E²—S⁵—Ala → E²—S⁵—AlaDLeuAlaGlyfVal
E²—Sᴾ—AlaDLeuAlaGlyfVal
E² ⇌ E²(Val AMP) ⇌ E²—S⁶—D,L-Val → E²—S⁶—DValAlaDLeuAlaGlyfVal

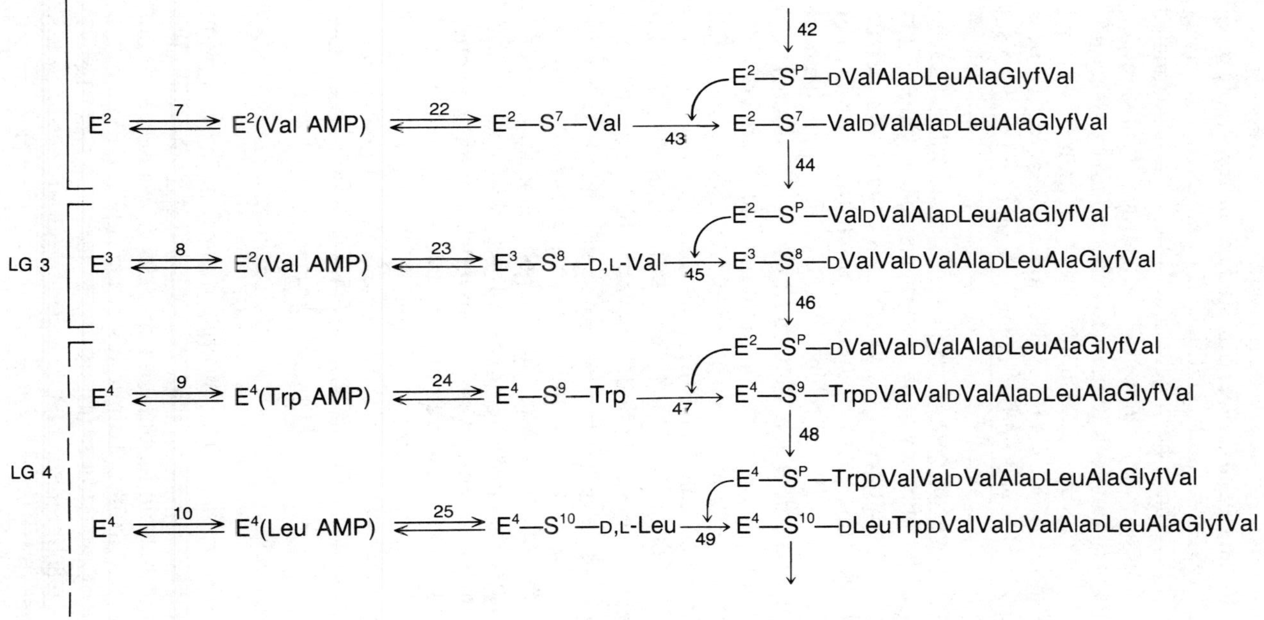

LG 3
LG 4
E² ⇌(7) E²(Val AMP) ⇌(22) E²—S⁷—Val →(43) E²—S⁷—ValᴅValAlaᴅLeuAlaGlyfVal
E²—Sᴾ—ᴅValAlaᴅLeuAlaGlyfVal
E³ ⇌(8) E²(Val AMP) ⇌(23) E³—S⁸—ᴅ,ʟ-Val →(45) E³—S⁸—ᴅValValᴅValAlaᴅLeuAlaGlyfVal
E²—Sᴾ—ValᴅValAlaᴅLeuAlaGlyfVal
E⁴ ⇌(9) E⁴(Trp AMP) ⇌(24) E⁴—S⁹—Trp →(47) E⁴—S⁹—TrpᴅValValᴅValAlaᴅLeuAlaGlyfVal
E²—Sᴾ—ᴅValValᴅValAlaᴅLeuAlaGlyfVal
E⁴ ⇌(10) E⁴(Leu AMP) ⇌(25) E⁴—S¹⁰—ᴅ,ʟ-Leu →(49) E⁴—S¹⁰—ᴅLeuTrpᴅValValᴅValAlaᴅLeuAlaGlyfVal
E⁴—Sᴾ—TrpᴅValValᴅValAlaᴅLeuAlaGlyfVal
42
44
46
48

(Ishihara et al., 1975; Roland et al., 1975). In studies on trypsinolysis of bacitracin enzymes (Frøyshov and Mathiesen, 1982), amino acid–activating fragments of 70 Kd have been detected. BA1, with an estimated size of 335 Kd, is considered to be a dimer of 2 × 140 Kd (Ogawa et al., 1981). Tryptic fragments may represent dimers or other complex structures and need to be characterized further. In a nonproducing mutant (Frøyshov, 1975; Haavik and Frøyshov, 1975), BA1 and BA3 are not separated during purification. A pentapeptide formed on this complex cannot be transferred to BA2.

b. *Reaction sequence*. (Scheme 5.4). Activation of the constituent amino acids (1–12) is followed by aminoacylation of enzyme thiols (S^1–S^{12}, 13–24). Initiation generates enzyme-bound Ile-Cys (25), which may be modified to a Δ-2 thiazoline ring structure (46). In attempts to characterize nascent dipeptide structures, Frøyshov (1975) obtained three compounds; Ishihara and Shimura (1979) failed to detect a thiol group in their product and concluded that the thiazoline ring was present. Vitković and Pfaender (1982) described a possible incorporation of the peptides IleCys, CysLeu, and IleCysLeu into bacitracin and concluded that ring formation occurs later, possibly by the action of a protease. This is based on the observation that a specific seryl protease produces a peptide, believed to be bacitracin, from antibiotic-free extracts of *B. licheniformis* (Vitković and Sadoff, 1977a, b). It has been proposed that these extracts contain a protein-associated bacitracin precursor devoid of the thiazoline moiety. The dipeptide elongation process has been concluded to be a pantetheine-mediated peptide transport with accumulation of covalent intermediates (Frøyshov, 1975). A pentapeptide (33) and a heptapetide (37) are transferred in intermolecular reactions. Cyclization (47) proceeds by intramolecular reaction of the δ-NH_2 group of DOrn with the activated carboxyl (Asn) of the dodecapeptide. A heptapeptide ring similar in size to those in polymyxin or octapeptin is formed (Kleinkauf and von Döhren, 1981).

5. Polymyxin. a. *Enzyme organization*. Several attempts have been made to isolate a polymyxin-synthesizing system. As with linear gramicidin, the N-terminal modification controlling initiation introduces problems. From the first report of in vitro colistin formation in a crude cell-free preparation of *Bacillus colistinus* Koyama (Ito et al., 1970), a decade passed before Komura and Kurahashi (1980b) reported the complete synthesis of polymyxin E with enzymes isolated from *Aerobacillus polyaerogenes*. This system consisted of a multienzyme of 300 Kd activating the constituent amino acids Dab, Leu, and Thr (Komura and Kurahashi, 1980a) together with a crude acylating enzyme transferring an octanoyl residue from CoA to the initiating Dab enzyme complex (S. Komura and K. Kurahashi, submitted for publication). A similar multienzyme, characterized by Dab activation, has been described by others (Jayaraman et al., 1969; Paulus, 1975; Kaur et al., 1978, 1982; Vasantha et al., 1980; Balakrishnan et al., 1980). In *B. polymyxa* 2459, a polymyxin B producer, a membrane-associated multienzyme complex activating the constituents Dab, Leu, Thr, and Phe is solubilized by polymyxin in parallel with enhancement of peptide biosynthesis. Both the membrane-attached and soluble forms from stationary phase cells contain two main active fractions, of 300 Kd and 170 Kd, activating all constituents. Analysis by electrophoresis with sodium dodecyl sulfate present reveals at least six protein components. Their role within a possible complex synthetase remains to be established. Analysis of mutants provides evidence for a subunit structure, unlike enzymes from gramicidin S nonproducing mutants. Membrane fractions missing either one, two, or three activating functions have been obtained, whereas in gramicidin S synthetase preparations, only one such function is missing at a time.

b. *Reaction sequence*. No scheme has been included for polymyxin biosynthesis. As in the other examples discussed, amino acid activation and binding of the amino acids to the enzyme are observed. However, accumulation of intermediates and the participation of pantetheine remain to be established.

IV. POSSIBLE FUNCTIONS OF PEPTIDE ANTIBIOTICS WITHIN THEIR PRODUCER ORGANISMS

Although the peptides discussed here share antibiotic activity, their structures and properties are diverse (Table 5.5). We shall discuss the possible role of only our selected peptides in the life cycle of their producers (Table 5.6). There has been much speculation about the role of peptide antibiotics, but as was pointed out by Demain (1974), it would be wiser to concentrate on evaluating the function of one compound than to seek a generalized explanation. In a review of peptide antibiotics, Katz and Demain (1977) critically evaluated experimental data and speculations. Several contributions made since then are summarized below.

A. Gramicidin S

Nonproducer mutants of *Bacillus brevis* ATCC 9999 form heat-sensitive spores with reduced levels of dipicolinic acid (Marahiel et al., 1979, 1982). Gramicidin S apparently regulates the outgrowth of spores without effecting germination. A lag phase in outgrowth of several hours was lost in nonproducer and washed spores but could be restored by exogenously added peptide (Nandi and Seddon, 1978; Seddon and Nandi, 1978; Marahiel et al., 1979; Lazaridis et al., 1980). Spore defects have been partially restored by adding

Scheme 5.4

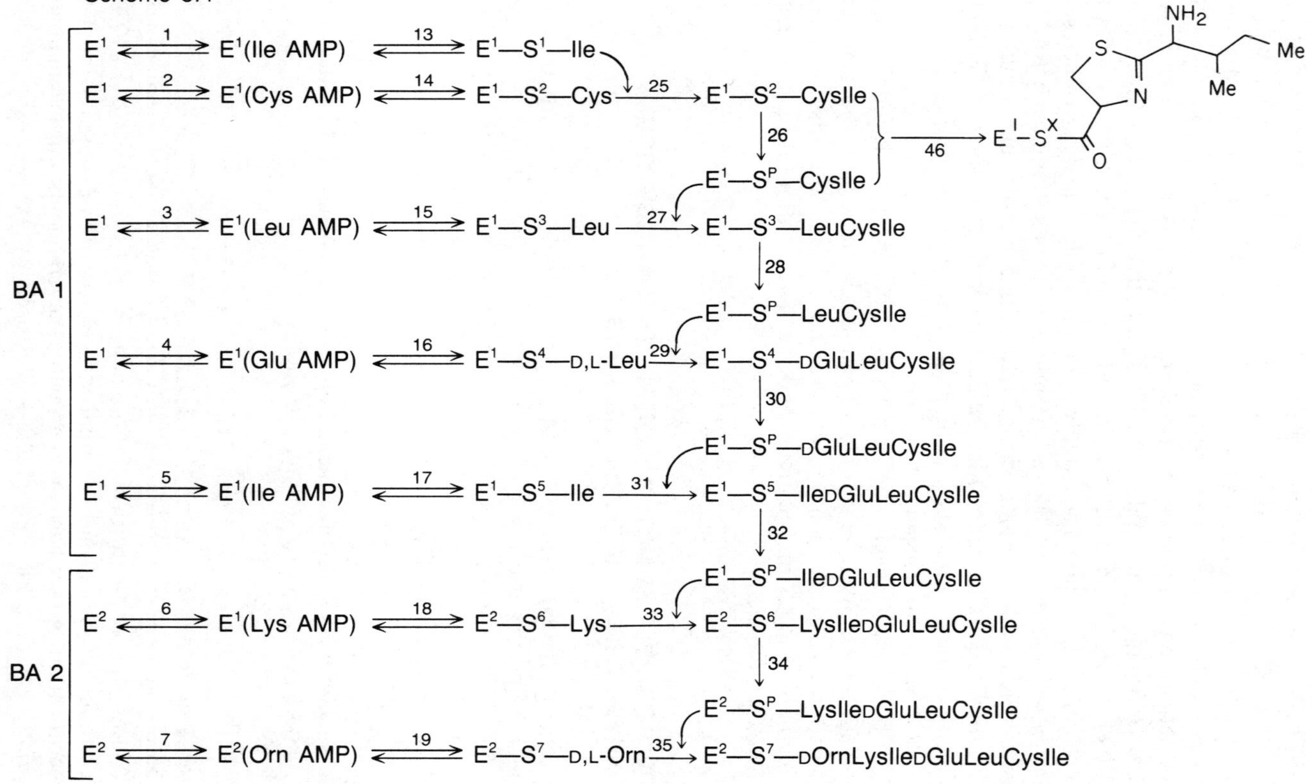
BA 1
BA 2
E¹
E¹(Ile AMP)
E¹—S¹—Ile
E¹(Cys AMP)
E¹—S²—Cys
E¹—S²—CysIle
E¹—Sᴾ—CysIle
E¹(Leu AMP)
E¹—S³—Leu
E¹—S³—LeuCysIle
E¹—Sᴾ—LeuCysIle
E¹(Glu AMP)
E¹—S⁴—D,L-Leu
E¹—S⁴—DGluLeuCysIle
E¹—Sᴾ—DGluLeuCysIle
E¹—S⁵—Ile
E¹—S⁵—IleDGluLeuCysIle
E¹—Sᴾ—IleDGluLeuCysIle
E²
E¹(Lys AMP)
E²—S⁶—Lys
E²—S⁶—LysIleDGluLeuCysIle
E²—Sᴾ—LysIleDGluLeuCysIle
E²(Orn AMP)
E²—S⁷—D,L-Orn
E²—S⁷—DOrnLysIleDGluLeuCysIle
E¹—Sˣ
NH2
S
N
O
Me
1
2
3
4
5
6
7
13
14
15
16
17
18
19
25
26
27
28
29
30
31
32
33
34
35
46

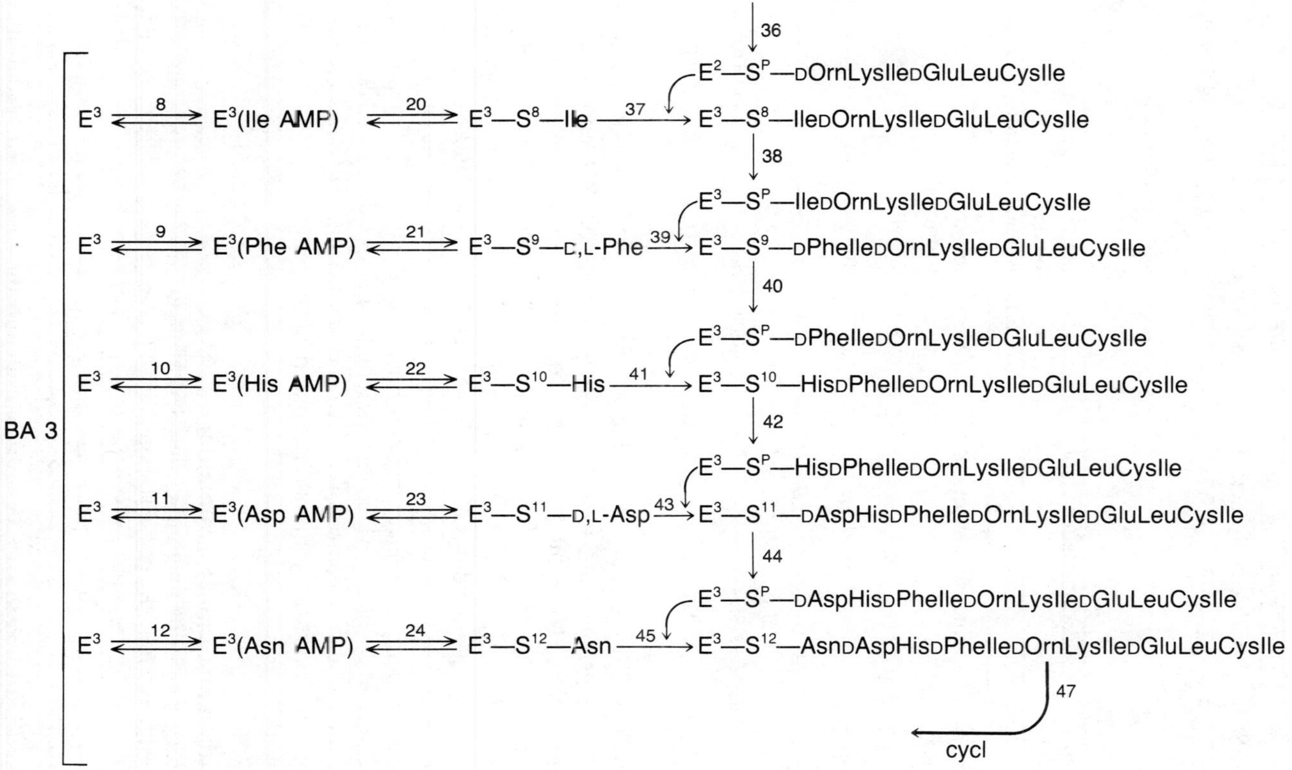
36
E²—Sᴾ—DOrnLysIleDGluLeuCysIle
E³ 8 E³(Ile AMP) 20 E³—S⁸—Ile 37 E³—S⁸—IleDOrnLysIleDGluLeuCysIle
38
E³—Sᴾ—IleDOrnLysIleDGluLeuCysIle
E³ 9 E³(Phe AMP) 21 E³—S⁹—D,L-Phe 39 E³—S⁹—DPheIleDOrnLysIleDGluLeuCysIle
40
E³—Sᴾ—DPheIleDOrnLysIleDGluLeuCysIle
E³ 10 E³(His AMP) 22 E³—S¹⁰—His 41 E³—S¹⁰—HisDPheIleDOrnLysIleDGluLeuCysIle
42
E³—Sᴾ—HisDPheIleDOrnLysIleDGluLeuCysIle
E³ 11 E³(Asp AMP) 23 E³—S¹¹—D,L-Asp 43 E³—S¹¹—DAspHisDPheIleDOrnLysIleDGluLeuCysIle
44
E³—Sᴾ—DAspHisDPheIleDOrnLysIleDGluLeuCysIle
E³ 12 E³(Asn AMP) 24 E³—S¹²—Asn 45 E³—S¹²—AsnDAspHisDPheIleDOrnLysIleDGluLeuCysIle
47
cycl
BA 3

TABLE 5.5 Properties of Some Peptide Antibiotics

Compound	*General Properties*	*Biochemical Properties*
Gramicidin S	Cationic surfactant	Membrane active Inhibition of transport functions Spore component Germination effector Enzyme inhibitor
Tyrocidines	Cationic surfactant	Membrane active DNA-binding Transcriptional effector
Linear gramicidin S	Neutral surfactant	Membrane active Channel forming Cation transport (K^+, Na^+, Li^+)
Bacitracins	Cation binding	Cation transport(Mn^{2+},Co^{2+},Zn^{2+}) Enzyme inhibitor
Polymxins	Cationic surfactant	Membrane active

References: Perlman (1978), Storm et al., (1977), Ramachandran (1975), Katz and Demain (1977).

gramicidin-S to growing cultures. During germination, the first RNA synthesis was sensitive to gramicidin S in vivo but not in vitro. Spore RNA polymerase was not inhibited at 80 μg/ml, while RNA formation in vivo was suppressed at less than 5 μg/ml (Danders and Marahiel, 1981; Marahiel et al., 1982). This inhibition has been related to reversibly altered transport properties of membranes by gramicidin S binding. The membrane-associated respiratory electron transport system was susceptible to exogenous peptide only in the germination and outgrowth phases (Nandi et al., 1981). Activities of membrane-associated enzymes like NADH and succinate oxidase decreased with increasing gramicidin S concentration in vivo and in vitro and may be reactivated during germination after gramicidin S has been degraded.

B. Tyrocidine and Linear Gramicidin

A regulatory function for gramicidin (Sarkar et al., 1979) or both peptides (Ristow et al., 1979) at the transcriptional level has been suggested. Tyrocidine forms a complex with native DNA, preventing initiation of RNA synthesis (Chakraborty et al., 1978). RNA synthesis can be partially restored by adding gramicidin and *B. brevis* RNA polymerase (Ristow, 1977; Schazschneider, 1978); this occurs because gramicidin interacts with tyrocidine, weakening the tyrocidine–DNA complex. Inhibition of RNA synthesis by gramicidin is competitive with respect to the duplex DNA template. Apparently, by interacting with the DNA-binding site, gramicidin can selectively

TABLE 5.6 Possible Functions of Some Peptides in Their Producer Organisms

Peptide	*Organism*	*Proposed Functions*	*Evidence*	*References*
Bacitracins	*B. licheniformis* ATCC 10716 ATCC 14580	M^{2+} carrier	Reduced uptake of M^{2+} (e.g., Mn^{2+}) ions in nonproducer mutants, enhanced uptake with exogenously added peptide	Haavik and Frøyshov (1975) Haavik (1976a, b)
Gramicidin S	*B. brevis* ATCC 9999 GB, Nagano	Germination effector, inhibition of membrane-associated functions of the spore membrane	Extractable spore component effect on outgrowth of non-producer mutants and washed spores, inhibition of NADH and succinate oxidases, inhibition of RNA synthesis during outgrowth	Nandi and Seddon (1978) Seddon and Nandi (1978) Marahiel et al. (1979, 1982), Lazaridis et al. (1980), Nandi et al. (1981), Danders and Marahiel (1981)
Tyrocidine and linear gramicidin	*B. brevis* ATCC 8185	Transcriptional regulators	Tyrocidine–DNA complex with blocked initiation, partial reactivation by gramicidin; transcriptional effects induction of sporulation in nitrogen-starved cells	Sarkar and Paulus (1972), Sarkar et al. (1977, 1978, 1979), Ristow et al. (1975a, b, 1979, 1982), Mukherjee and Paulus (1977), Chakraborty et al. (1978), Paulus et al. (1979), Ristow (1977), Hansen (1980), Schazschneider (1978)
Polymyxin	*B. polymyxa* 2459	Solubilization of membrane-associated polymyxin synthetase	Solubilization of multienzymes only in cells producing maximal amounts of peptides. In vitro solubilization from membrane complex	Vasantha et al. (1980), Balakrishnan et al. (1980), Kaur et al. (1982)

modulate the affinity of RNA polymerase for different promoters (Sarkar et al., 1979). If growing cells of *B. brevis* ATCC 8185 are starved of nitrogen, sporulation may be induced by tyrothricin (a mixture of gramicidins and tyrocidines) or tyrocidine but not by gramicidin (Ristow et al., 1979, 1982). Studies with mutants producing only one of the peptides (Mukherjee and Paulus, 1977) have been hampered by their instability (H. Ristow, personal communication). If, in a sporulation medium, tyrocidine formation is suppressed by phenylserine, sporulation is reduced, and no gramicidin is formed. Adding tyrocidine to this culture then induces sporulation, gramicidin biosynthesis, and protease formation (H. Ristow et al., unpublished results).

C. Bacitracins

Growth of a sporulating nonproducer of *B. licheniformis* (Haavik and Thomassen, 1973) is inhibited by EDTA, while bacitracin or excess Mn^{2+} promotes growth (Haavik and Frøyshov, 1975). The peptide inhibits its producer strain in the presence of excess divalent metal ions (Mn^{2+}, Co^{2+}, Zn^{2+}) (Haavik, 1975, 1976a, b), suggesting a transport function.

D. Polymyxins

Vegetative cells of the polymyxin B producer *B. polymyxa* 2459 contain the peptide-forming multienzyme complex in a membrane-bound form (Balakrishnan et al., 1980; Vasantha et al., 1980). During sporulation, delocalization parallels peptide production (Kaur et al., 1982). This complex, together with other polypeptides apparently associated with membrane DNA, may thus regulate gene expression. Polymyxin may be involved in specific dissociation of membrane DNA-binding proteins.

REFERENCES

Aarstad, K., Zimmer, T.-L., and Laland, S. G. (1979) *FEBS Lett. 103,* 118–121.

Aarstad, K., Zimmer, T.-L., and Laland, S. G. (1980) *Eur. J. Biochem. 112,* 335–338.

Akashi, K., Kubota, K., and Kurahashi, K. (1977) *J. Biochem. 81,* 269–272.

Akashi, K., and Kurahashi, K. (1977) *Biochem. Biophys. Res. Comm. 77,* 259–267.

Akashi, K., and Kurahashi, K. (1978) *J. Biochem. 83,* 1219–1229.

Akashi, K. (1977) Ph.D. thesis, Osaka University, Osaka, Japan.

Akers, H. A., Lee, S. G., and Lipmann, F. (1977) *Biochemistry 16,* 5722–5729.

Altmann, M., Döhren, H. von, El-Samaraie, A., Kittelberger, R., Pore, M. S., and Kleinkauf, H. (1982) in *Peptide Antibiotics—Biosynthesis and Functions* (Kleinkauf, H., and Döhren, H. von, eds.) pp. 243–252, de Gruyter, Berlin.

Astrup, E. G. (1972) Ph.D. thesis, University of Oslo, Oslo, Norway.

Augenstein, D. C., Thrasher, K. D., Sinskey, A. J., and Wang, D. I. C. (1974) *Biotechnol. Bioeng. 16,* 1433–1441.

Aust, H. J., and Döhren, H. von (1982) in *Peptide Antibiotics—Biosynthesis and Functions* (Kleinkauf, H., and Döhren, H. von, eds.) pp. 137–148, de Gruyter, Berlin.

Balakrishnan, R., Kaur, S., Goel, A. K., Padmavathy, S., and Jayaraman, K. (1980) *Arch. Biochem. Biophys. 200,* 45–54.

Battersby, A. R., and Craig, L. C. (1951) *J. Amer. Chem. Soc. 73,* 1887–1888.

Bauer, K., Roskoski, R., Jr., Kleinkauf, H., and Lipmann, F. (1972) *Biochemistry 11,* 3266–3271.

Bisswanger, H., and Schmincke-Ott, E. (eds.) (1980) *Multifunctional Proteins,* John Wiley, New York.

Bothe, D., Döhren, H. von, Zschiedrich, H., El-Samaraie, A., Krause, M., and Kleinkauf, H. (1982) in *Peptide Antibiotics—Biosynthesis and Functions* (Kleinkauf, H., and Döhren, H. von, eds.) pp. 233–241, de Gruyter, Berlin.

Brewer, G. A. (1979) *Anal. Profiles Drug Substr. 8,* 179–218.

Bu'Lock, J. D., and Powell, A. J. (1965) *Experientia 21,* 55–56.

Burke, W. F., and Spizizen, J. (1977) *J. Bacteriol. 129,* 1215–1221.

Charkraborty, T., Hansen, J., Schazschneider, B., and Ristow, H. (1978) *Eur. J. Biochem. 90,* 261–270.

Chiu, C. W., Bernhard, T., and Dellweg, G. (1982) in *Peptide Antibiotics—Biosynthesis and Functions* (Kleinkauf, H., and Döhren, H. von, eds.) pp. 149–154 de Gruyter, Berlin.

Christiansen, C., Aarstad, K., Zimmer, T.-L., and Laland, S. G. (1977) *FEBS Lett. 81,* 121–124.

Cornell, N., and Snoke, J. E. (1964) *Biochim. Biophys. Acta 91,* 533–536.

Craig, L. C., and Konigsberg, W. (1957) *J. Org. Chem. 22,* 1345–1353.

Danders, W., and Marahiel, M. A. (1981) *FEMS Microbiol. Lett. 10,* 277–283.

Demain, A. L. (1974) *Ann. N.Y. Acad. Sci. 235,* 601–612.

Demain, A. L., and Matteo, C. C. (1976) *Antimicrob. Agents Chemother. 9,* 1000–1003.

Demain, A. L., and Wang, D. I. C. (1976) in *Second International Symposium on the Genetics of Industrial Microorganisms.* (Macdonald, K. D., ed.) pp. 115–128, Academic Press, London.

Dobritsa, A. P., Dobritsa, S. V, and Tanyahin, V. I. (1978) *Molec. Gen. Genet. 164,* 195–204.

Dobritsa, S. V., and Dobritsa, A. P. (1978) *Eksperim. Izuch. Razvitiya Mikroorganizmov, Pushchino,* pp. 7–10.

Döhren, H. von (1982) in *Peptide Antibiotics—Biosynthesis and Functions* (Kleinkauf, H., and Döhren, H. von, eds.) pp. 169–182, de Gruyter, Berlin.

Elverdam, I., Larsen, P., and Lund, E. (1981) *J. Chromatogr. 218,* 653–669.

Emilianowicz-Czerska, W., Szczepankowska, M., Niemcyzk, H., and Wolkowicz, M. (1973) *Acta Microbiol. Pol. Ser. B. 5,* 29–34.

Ermakova, G. N., Nefelova, M. V., and Egorov, N. S. (1976) *Antibiotiki 21,* 596–600.

Flickinger, M. C., and Perlman, D. (1979) *Antimicrob. Agents Chemother. 15,* 653–661.

Friest, W., Sternbach, H., and Cramer, F. (1981) *Hoppe-Seyler's Z. Physiol. Chem. 362,* 1247–1254.

Friebel, T. E., and Demain, A. L. (1977a) *FEMS Microbiol. Lett. 1,* 215–218.

Friebel, T. E., and Demain, A. L. (1977b) *J. Bacteriol. 130,* 1010–1016.

Frøyshov, Ø. (1974) *FEBS Lett. 44,* 75–78.

Frøyshov, Ø. (1975) *Eur. J. Biochem. 59,* 201–206.
Frøyshov, Ø. (1977) *FEBS Lett. 81,* 315–318.
Frøyshov, Ø. (1979) in *Regulation of Secondary Product and Plant Hormone Metabolism* (Luckner, M., and Schreiber, K., eds.) pp. 189–198, Pergamon, Elmsford, N.Y.
Frøyshov, Ø., and Laland, S. G. (1974) *Eur. J. Biochem. 46,* 235–242.
Frøyshov, Ø., and Mathieson, A. (1982) in *Peptide Antibiotics—Biosynthesis and Functions* (Kleinkauf, H., and Döhren, H. von, eds.) pp. 307–314, de Gruyter, Berlin.
Frøyshov, Ø., Zimmer, T. L., and Laland, S. G. (1978) in *International Review of Biochemistry, Amino Acid and Protein Biosynthesis II,* Vol. 18 (Arnstein, H. R. V., ed.) pp. 49–78, University Park Press, Baltimore, Md.
Frøyshov, Ø., Mathiesen, A., and Haavik, H. I. (1980) *J. Gen. Microbiol. 117,* 163–167.
Fujikawa, K., Suketa, Y., Hayashi, K., and Suzuki, T. (1965) *Experientia 21,* 307–308.
Fujikawa, K., Suzuki, T., and Kurahashi, K. (1968a) *Biochim. Biophys. Acta 161,* 232–246.
Fujikawa, K. Sakamoto, Y., Suzuki, T., and Kurahashi, K. (1968b) *Biochim. Biophys. Acta 169,* 520–533.
Fujikawa, K., Sakamoto, Y., and Kurahashi, K. (1971) *J. Biochem. 69,* 869–879.
Gadow, A. (1981) Diplomarbeit, Technical University, Berlin.
Haavik, H. I. (1974a) *J. Gen. Microbiol. 84,* 226–230.
Haavik, H. I. (1974b) *J. Gen. Microbiol. 84,* 321–326.
Haavik, H. I. (1974c) *J. Gen. Microbiol. 81,* 383–390.
Haavik, H. I. (1975) *Acta Pathol. Microbiol. Scand. Sect. B: Microbiol. 83B,* 513–518.
Haavik, H. I. (1976a) *Acta Pathol. Microbiol. Scand. Sect. B: Microbiol. 84B,* 117–124.
Haavik, H. I. (1976b) *J. Gen. Microbiol. 96,* 393–399.
Haavik, H. I. (1979) *Folia Microbiol. 24*(3), 234–239.
Haavik, H. I. (1981) *FEMS Microbiol. Lett. 10*(2), 111–114.
Haavik, H. I., and Frøyshov, Ø. (1975) *Nature 254,* 79–82.
Haavik, H. I., and Thomassen, S. (1973) *J. Gen. Microbiol. 76,* 451–454.
Haavik, H. I., and Vessia, B. (1978) *Acta Pathol. Microbiol. Scand. Sect. B, 86,* 67–70.
Hamilton, B. K., Montgomery, J. P., and Wang, D. I. C. (1973) in *Enzyme Engineering,* Vol. II (Pye, E. K., and Wingard, L. B., Jr., eds.) pp. 153–159, Plenum Press, New York.
Hanlon, G. W., and Hodges, N. A. (1981) *J. Bacteriol. 147,* 427–431.
Hansen, J. (1980) Ph.D. thesis, Free University, Berlin.
Hayashi, K., Suketa, Y., Tsukamoto, K., and Suzuki, T. (1966) *Experientia 22,* 354–355.
Hayashi, K., Suketa, Y., and Suzuki, T. (1968) *Experientia 24,* 656–657.
Hickey, R. J. (1964) *Progr. Ind. Microbiol. 5,* 93–150.
Hirotsu, Y., Nishiuchi, Y., and Shiba, T. (1979) *Pept. Chem. 16,* 171–176.
Hori, K., Kanda, M., Kurotsu, T., Miura, S., Yamada, Y., and Saito, Y. (1981) *J. Biochem. 90,* 439–447.
Hori, K., Kurotsu, T., Kanda, M., Miura, S., Nozoe, A., and Saito, Y. (1978) *J. Biochem. 84,* 425–434.

Hori, K., Kurotsu, T., Kanda, M., Miura, S., Yamada, Y., and Saito, Y. (1982) *J. Biochem. 91,* 369–379.
Ishihara, H., and Shimura, K. (1974) *Biochim. Biophys. Acta 338,* 588–600.
Ishihara, H., and Shimura, K. (1975) *Seikagaku Jikken Koza 7,* 689–670.
Ishihara, H., and Shimura, K. (1979) *FEBS Lett. 99,* 109–112.
Ishihara, H., Sasaki, T., and Shimura, K. (1968) *Biochim. Biophys. Acta 166,* 496–504.
Ishihara, H., Endo, Y., Abe, S., and Shimura, K. (1975) *FEBS Lett. 50,* 43–46.
Ishihara, H., Ogawa, I., and Shimura, K. (1982) in *Peptide Antibiotics—Biosynthesis and Functions* (Kleinkauf, H., and Döhren, H. von, eds.) pp. 289–297, de Gruyter, Berlin.
Ito, M., Koyama, Y., Aida, K., and Uemura, T. (1970) *Biochim. Biophys. Acta 215,* 418–420.
Iwaki, M., Shimura, K., Kanda, M., Kaji, E., and Saito, Y. (1972) *Biochem. Biophys. Res. Comm. 48,* 113–118.
Izumiya, N. (1976) *Tampakushitsu Kakusan Koso Bessatsu 76,* 166–190.
Izumiya, N., Kato, T., Aoyagi, H., Waki, M., and Kondo, M. (1979) *Synthetic Aspects of Biologically Active Cyclic Peptides—Gramicidin S and Tyrocidines* Kodansha, Tokyo.
Japanese Patent (1975) 50-25795.
Japanese Patent (1976) 51-9789.
Jayaraman, K., and Kannan, R. (1972) *Biochem. Biophys. Res. Commun. 48,* 1235–1239.
Jayaraman, K., and Paulus, H. (1971) *Cell Processes Growth, Develop. Differentiation, Proc. Symp.,* 85–91.
Jayaraman, K., Monreal, J., and Paulus, H. (1969) *Biochim. Biophys, Acta 185,* 447–457.
Jones, T. S. G. (1949) *Ann. N.Y. Acad. Sci. 51,* 909–916.
Kambe, M., Imae, Y., and Kurahashi, K. (1974) *J. Biochem. 75,* 481–493.
Kanda, M., Hori, K., Kurotsu, T., Miura, S., Nozoe, A., and Saito, Y. (1978) *J. Biochem. 84,* 435–441.
Kanda, M., Hori, K., Kurotsu, T., Miura, S., Yamada, Y., and Saito, Y. (1981) *J. Biochem. 90,* 765–771.
Katz, E., and Demain, A. L. (1977) *Bacteriol. Rev. 41,* 449–474.
Kaur, S., Balakrishnan, R., and Jayaraman, K. (1978) *Biochem. Biophys, Res. Comm. 81,* 50–57.
Kaur, S., Chandrasekaran, G. N., Vasantha, N., and Jayaraman, K. (1982) in *Peptide Antibiotics—Biosynthesis and Functions* (Kleinkauf, H., and Döhren, H. von, eds.) pp. 453–466, de Gruyter, Berlin.
Kimura, Y. (1971) Japanese Patent 71-16,152.
Kimura, Y., Murai, E., Fujisawa, M., Tatsuki, T., and Nobue, F. (1969) *J. Antibiot. 22,* 449–450.
Kittelberger, R., Altmann, M., and Döhren, H. von (1982) in *Peptide Antibiotics—Biosynthesis and Functions* (Kleinkauf, H., and Döhren, H. von, eds.) pp. 209–218, de Gruyter, Berlin.
Kleinkauf, H. (1979) *Planta Medica 35,* 1–18.
Kleinkauf, H., and Döhren, H. von (1980) *6th Int. Fermentation Symp. London* (Canada).
Kleinkauf, H., and Döhren, H. von (1981) *Curr. Topics Microbiol. Immunol. 91,* 129–177.

Kleinkauf, H., and Döhren, H. von (eds.) (1982) *Peptide Antibiotics—Biosynthesis and Functions* de Gruyter, Berlin.
Kleinkauf, H., and Koischwitz, H. (1974) in *Lipmann Symposium: Energy Biosynthesis and Regulation in Molecular Biology* (Richter, D., ed.) pp. 336–344, de Gruyter, Berlin.
Kleinkauf, H., and Koischwitz, H. (1978) *Progr. Mol. Subcell. Biol. 6,* 59–112.
Kleinkauf, H., and Koischwitz, H. (1980a) in *Cell Compartmentation and Metabolic Channeling* (Nover, L., Lynen, F., and Mothes, K., eds.) pp. 147–158, Fischer/Elsevier, Stuttgart, West Germany.
Kleinkauf, H., and Koischwitz, H. (1980b) in *Multifunctional Proteins* (Bisswanger, H., and Schmincke-Ott, E., eds.) pp. 217–233, Wiley, Elmsford, N.Y.
Kleinkauf, H., and Koischwitz, H. (1980c) in *Chemical Recognition in Biology* (Chapeville, F., and Haenni, A.-L., eds.) pp. 205–216, Springer, Berlin.
Kleinkauf, H., Roskoski, R., Jr., and Lipmann, F. (1971) *Proc. Nat. Acad. Sci. 68,* 2069–2072.
Koischwitz, H. (1978) Ph.D. thesis, Technical University, Berlin.
Koischwitz, H. (1979) *Hoppe Seyler's Z. Physiol. Chem. 360,* 307.
Koischwitz, H., and Kleinkauf, H. (1976a) *Biochim. Biophys. Acta 429,* 1041–1051.
Koischwitz, H., and Kleinkauf, H. (1976b) *Biochim. Biophys. Acta 429,* 1052–1061.
Komura, S., and Kurahashi, K. (1980a) *J. Biochem. 88,* 285–288.
Komura, S., and Kurahashi, K. (1980b) *Biochem. Biophys. Res. Comm. 95,* 1145–1151.
Konigsberg, W., and Craig, L. C. (1962) *J. Org. Chem. 27,* 934–938
Kurahashi, K. (1964) *J. Biochem 56,* 101–102.
Kurahashi, K. (1974) *Ann. Rev. Biochem. 43,* 445–459.
Kurahashi, K. (1981) in *Antibiotics,* Vol. *IV, Biosynthesis* (Corcoran, J. W., ed.) pp 325–352, Springer, Berlin.
Kurahashi, K., Yamada, M., Mori, K., Fujikawa, K., Kambe, M., Imae, Y., Sato, E., Takahashi, H., and Sakamoto, Y. (1968) *Cold Spring Harb. Symp. Quant. Biol. 34,* 815–826.
Kurahashi, K., Komura, S., Akashi, K., and Nishio, C. (1982) in *Peptide Antibiotics—Biosynthesis and Functions* (Kleinkauf, H., and Döhren, H. von, eds.) pp. 275–288, de Gruyter, Berlin.
Kurylo-Borowska, Z. (1975) *Meth. Enzymol. 43,* 559–567.
Kurylo-Borowska, Z., and Heaney-Kieras, J. (1982) in *Peptide Antibiotics—Biosynthesis and Functions* (Kleinkauf, H., and Döhren, H. von, eds.) pp. 315–324, de Gruyter, Berlin.
Kuzovnikova, T. A., Chulkova, A. P., and Zhdanov, V. G. (1978) *Khim. Mutagen. Gibrid.* 165–170.
Laishley, E. J., and Bernlohr, R. W. (1968) *J. Bacteriol. 96,* 322–329.
Laland, S. G., and Zimmer, T-L. (1973) *Essays Biochem. 9,* 31–57.
Laland, S. G., Frøyshov, Ø., Gilhuus-Moe, C. C., and Zimmer, T.-L. (1972) *Nature 239,* 43–44.
Laland, S. G., Zimmer, T.-L., and Frøyshov, Ø. (1978) in *Bioactive Peptides Produced by Microorganisms* (Umezawa, H., Takita, T., and Shiba, T., eds.) pp. 7–34, Kodansha, Tokyo, and Wiley, New York.
Laland, S. G., Aarstad, K., and Zimmer, T.-L. (1982) in *Peptide Antibiotics—Biosynthesis and Functions* (Kleinkauf, H., and Döhren, H. von, eds.) pp. 185–194, de Gruyter, Berlin.

Lazaridis, I., Frangov-Lazaridis, M., McCuish, F. C., Nandi, S., and Seddon, G. (1980) *FEMS Microbiol. Lett. 7,* 229–232.

Lee, S. G. (1976) *10th Int. Congr. Biochem. Hamburg 1976, Abstr. 4-3-352* and personal communication.

Lee, S. G., and Lipmann, F. (1974) *Proc. Nat. Acad. Sci. 71,* 607–611.

Lee, S. G., and Lipmann, F. (1975) *Meth. Enzymol. 43,* 585–607.

Lee, S. G., and Lipmann, F. (1977) *Proc. Nat. Acad. Sci. 74,* 2343–2347.

Lee, S. G., Roskoski, R., Jr., Bauer, K., and Lipmann, F. (1973) *Biochemistry 12,* 398–405.

Leung, D. C., and Baxter, R. M. (1972) *Biochim. Biophys. Acta 279,* 34–47.

Lipmann, F. (1968) *Essays Biochem. 4,* 1–23.

Lipmann, F. (1971) *Science 173,* 875–884.

Lipmann, F. (1980) *Adv. Microb. Physiol. 21,* 227–266.

Lipmann, F. (1982) in *Peptide Antibiotics—Biosynthesis and Functions* (Kleinkauf, H., and Döhren, H. von, eds.) pp. 23–45, de Gruyter, Berlin.

Lipmann, F., Gevers, W., Kleinkauf, H., and Roskoski, R., Jr. (1971) *Adv. Enzymol. 35,* 1–34.

Lukin, G. A., Zvenigorodskii, V. I., and Zhdanov, V. G. (1981) *Antibiotiki 26,* 87–89.

Mach, B., and Tatum, E. L. (1964) *Proc. Nat. Acad. Sci. 52,* 876–884.

Marahiel, M. A., and Döhren, H. von (1982) in *Peptide Antibiotics—Biosynthesis and Functions* (Kleinkauf, H., and Döhren, H. von, eds.) pp. 375–380, de Gruyter, Berlin.

Marahiel, M. A., Danders, W., Krause, M., and Kleinkauf, H. (1979) *Eur. J. Biochem. 99,* 49–55.

Marahiel, M. A., Lurz, R., and Kleinkauf, H. (1981a) in *Advances in Biotechnology,* Vol. III, *Fermentation Products* (Moo-Young, M., Vezina, C., and Singh, K., eds.) pp. 43–49, Pergamon Press, Toronto.

Marahiel, M. A., Lurz, R., and Kleinkauf, H. (1981) *J. Antibiot. 34,* 323–330.

Marahiel, M. A., Danders, W., Kraepelin, G., and Kleinkauf, H. (1982) in *Peptide Antibiotics—Biosynthesis and Functions* (Kleinkauf, H., and Döhren, H. von, eds.) pp. 389–397, de Gruyter, Berlin.

Matselyukh, B. P., Strizhkova, A. M., Matselyukh, D. Y., Sitalo, N. A., Preobrazhenskii, L. N., and Evseeva, P. N. (1978) *Tsitol. Genet. 12,* 441–444.

Matteo, C. C., Cooney, C. L., and Demain, A. L. (1976) *J. Gen. Microbiol. 96,* 415–422.

Meleney, F. L., and Johnson, B. A. (1949) *Am. J. Med. 7.* 794–806.

Meyers, E., Brown, W. E., Principe, P. A., Rathnum, M. C., and Parker, W. L. (1973) *J. Antibiot. 26,* 437–443.

Meyers, E., Parker, W. L., Brown, W. E., Shoji, J., and Wakisaka, Y. (1976) *J. Antibiot. 29,* 1241–1242.

Mitani, T., Heinze, J. E., and Freese, E. (1977) *Biochem. Biophys, Res. Comm. 77,* 1118–1125.

Mitronova, T. N., and Zharikova, G. G. (1965) *Mikrobiologiya 34,* 835–839.

Mohr, H., and Kleinkauf, H. (1978) *Biochim. Biophys, Acta 526,* 375–386.

Morikawa, Y., Ochiai, K., Karube, I., and Suzuki, S. (1979) *Antimicrob. Agents Chemother. 15,* 126–130.

Morikawa, Y., Karube, I., and Suzuki, S. (1980) *Biotechnol. Bioeng. 22,* 1015–1023.

Mukherjee, P. K., and Paulus, H. (1977) *Proc. Nat. Acad. Sci. U.S.A. 74,* 780–784.

Nakamura, S. (1972) *Iyakuhin Kaihatsu Kiso Koza 14,* 678–681.

Nandi, S., and Seddon, B. (1978) *Biochem. Soc. Trans. 6,* 409–411.

Nandi, S., Lazaridis, I., and Seddon, B. (1981) *FEMS Microbiol. Lett. 10,* 71–75.

Nefelova, M. V., Filippova, M. S., Ermakova, G. N., and Egorov, N. S. (1979) *Biol. Nauki (1),* 74–78.

Nefelova, M. V., Ermakova, G. N., and Egorov, N. S. (1980a) *Antibiotiki 25,* 326–330.

Nefelova, M. V., Filippova, M. S., and Egorov, N. S. (1980b) *Mikrobiologiya 49,* 294–297.

Ogawa, I., Ishihara, H., and Shimura, K. (1981) *FEBS Lett. 124,* 197–201.

Otani, S., and Saito, Y. (1964) *J. Biochem. 56,* 103–105.

Parry, R. J., and Kurylo-Borowska, Z. (1980) *J. Amer. Chem. Soc. 102,* 836–837.

Paulus, H. (1975) *Meth. Enzymol. 43,* 579–584.

Paulus, H., Sarkar, N., Mukherjee, P. K., Langley, D., Ivanov, V. T., Sheppel, E. N., and Veatch, W. (1979) *Biochemistry 18,* 4532–4536.

Pavlyuk, Y. V., Bogatskii, M. A., Orlova, N. V., and Anakhova, V. A. (1979a) *Antibiotiki 24,* 566–570.

Pavlyuk, Y. V., Bogatskii, M. A., Orlova, N. V., and Anakhova, V. A. (1979b) *Antibiotiki 24,* 815–820.

Pavlyuk, Y. V., Bogatskii, M. A., Orlova, N. V., and Anakhova, V. A. (1979c) *Antibiotiki 24,* 723–727.

Perlman, D. (1974) *Chem. Techn.* 210–216.

Perlman, D. (1978) in *Encyclopedia of Chemical Technology* (Kirk-Othmer), pp. 991–1036, Wiley, New York.

Peshkov, M. A., Kapitonova, A. V., Lukin, A. A., Polyakova, I. N., Skortsova, E. B., and Blagoslavova, E. I. (1968) *Mikrobiologiya 37,* 92–94.

Pfaender, P. (1972) *Zentralbl. Bakterol., Parasitenk., Infektionskr., Hyg. Abt. 1: Orig., Reihe A 220,* 319–321.

Philips, D. R., and Agin, P. P. (1977) *J. Biol. Chem. 252,* 2121–2126.

Ramachandran, L. K. (1975) *J. Scient. Ind. Res. 34,* 249–265.

Rao, R. K., Bhagavan, N. V., Rao, K. R., and Hall, J. B. (1968) *Biochemistry 7,* 3072–3077.

Rautenshtein, Y. I., Pochtareva, V. I., and Zharikova, G. G. (1967) *Mikrobiologiya 36,* 1060–1065.

Ressler, C., and Kashelikar, D. V. (1966), *J. Amer. Chem. Soc. 88,* 2025–2035.

Ristow, H. (1977) *Biochim. Biophys. Acta 477,* 177–184.

Ristow, H., Schazschneider, B., Bauer, K., and Kleinkauf, H. (1975a) *Biochim. Biophys Acta 390,* 246–252.

Ristow, H., Schazschneider, B., Vater, J., and Kleinkauf, H. (1975b) *Biochim. Biophys. Acta 414,* 1–8.

Ristow, H. Pschorn, W., Hansen, J., and Winkel, U. (1979) *Nature 280,* 165–166.

Ristow, H., Russo, J., Stochaj, E., and Paulus, H. (1982) in *Peptide Antibiotics—Biosynthesis and Functions* (Kleinkauf, H., and Döhren, H. von, eds.) pp. 381–388, de Gruyter, Berlin.

Rogolsky, M., and Nakamura, H. T. (1974) *J. Bacteriol. 119,* 57–61.

Roland, I., Frøyshov, Ø., and Laland, S. G. (1975) *FEBS Lett. 60,* 305–308.

Roland, I., Frøyshov, Ø., and Laland, S. G. (1977) *FEBS Lett. 84,* 22–24.

Roskoski, R., Jr., Gevers, W., Kleinkauf, H., and Lipmann, F. (1970a) *Biochemistry 9,* 4839–4845.

Roskoski, R., Jr., Kleinkauf, H., Gevers, W., and Lipmann, F. (1970b) *Biochemistry 9,* 4846–4851.

Ruttenberg, M. A., and Mach, B. (1966) *Biochemistry 5,* 2864–2869.

Saito, Y. (1982) in *Peptide Antibiotics—Biosynthesis and Functions* (Kleinkauf, H., and Döhren, H. von, eds.) pp. 195–207, de Gruyter, Berlin.

Sarkar, N., and Paulus, H. (1972) *Nature 239,* 228–230.

Sarkar, N., Langley, D., and Paulus, H. (1977) *Proc. Nat. Acad. Sci. 74,* 1478–1482.

Sarkar, N., Mukherjee, P. K., Langley, D., and Paulus, H. (1978) *Spores 7,* 226–231.

Sarkar, N., Langley, D., and Paulus, H. (1979) *Biochemistry 18,* 4536–4541.

Saxholm, H., Zimmer, T.-L. and Laland, S. G. (1972) *Eur. J. Biochem. 30,* 138–144.

Schazschneider, B. (1978) Ph.D. thesis, Technical University, Berlin.

Seddon, B., and Nandi, S. (1978) *Biochem. Soc. Trans. 6,* 412–413.

Shimura, K., Iwaki, M., Kanda, M., Hori, K., Kaji, E., Hasegawa, S., and Saito, Y. (1974) *Biochim. Biophys. Acta 338,* 577–587.

Shoji, J. (1978) *Adv. Appl. Microbiol. 24,* 187–214.

Shoji, J., Hinoo, H., Wakisaka, Y., Koizumi, K., Mayama, M., Matsuura, S., and Matsumoto, K. (1976) *J. Antibiot. 29,* 516–520.

Silaev, A. B., Maevskaya, S. N., Trifonova, Zh. P., Yolikava, E. P., Vasileva, N. M., and Katrukha, S. S. (1975) *J. Gen. Chem. USSR* (Engl. transl.) *45,* 2287–2291; *Zh. Obshch. Khim. 45,* 2331–2337.

Smekal, F., Paleckova, F., and Vinter, V. (1979) Czech. Patent 175,992.

Snoke, J. E. (1960) *J. Bacteriol. 80,* 522–527.

Stoletov, V. N., Glazer, V. M., and Shestakov, S. V. (1965) *Mikrobiologiya 34,* 584–589

Stoll, E., Frøyshov, Ø., Holm, H., Zimmer, T. L., and Laland, S. G. (1970) *FEBS Lett. 11,* 348–352.

Storm, D. R., Rosenthal, K. S., and Swanson, P. E. (1977) *Ann. Rev. Biochem. 46,* 723–763.

Surguchev, A. P., Zaitseva, G. N., and Zharikova, G. G. (1969) *Mikrobiologiya 38,* 984–998.

Suzuki, S., and Karube, I. (1979) *ACS Symp. Ser.,* 106.

Suzuki, T., and Kimura, Y. (1976) *Tampakushitsu Kakusan Koso Bessatsu 76,* 114–130.

Suzuki, T., Hayashi, K., and Fujikawa, K. (1963) *J. Biochem. 54,* 412–418.

Suzuki, T., Hayashi, K., Fujikawa, K., and Tsukamoto, K. (1964a) *Biochemistry 56,* 182–189.

Suzuki, T., Hayashi, K., Fujikawa, K., and Tsukamoto, K. (1964b) *Biochemistry 56,* 335–343.

Suzuki, S., Karube, I., and Morikawa, Y. (1979) *Jap. Kokai Tokkyo Koho* 79, 138, 194.

Szczesniak, T., Karabin, L., Krzemien-Brodowicz, M., Dziegielewski, K., Korona, A., Siwek, U., Wawrzyniewicz, M., Dutkiewicz, M., Fabijanski, M., et al. (1976) Polish Patent 79,673.

Takahashi, H., Sato, E., and Kurahashi, K. (1971) *J. Biochem. 69,* 973–976.

Tsuji, K., and Robertson, J. H. (1975) *J. Chromatogr. 112,* 663–672.
Tyc, M., and Kadzikiewicz, T. (1976) Polish Patent 90,102.
Tyc, M., Kadzikiewicz, T., Karabin, L., Wyrzuc, J., Taflinska, K., Michalak, D., and Komorowska, C. (1976) Polish Patent 79,336.
Tyc, M., Kadzikiewicz, T., Tomczyk, A., Zakrzewska, J., Galewski, P., Taflinska, K., Zaorska, H., and Gajowiak, E. (1979) Polish Patent 102,636.
Tzeng, C. H., Thrasher, K. D., Montgomery, J. P., Hamilton, B. K., and Wang, D. I. C. (1975) *Biotechnol. Bioeng. 17,* 143–151.
Vandamme, E. J. (1976) D.Sc. thesis, University of Ghent, Ghent, Belgium.
Vandamme, E. J. (1981) in *Topics in Enzyme and Fermentation Biotechnology* (Wiseman, A., ed.) pp. 187–261, Eilis Horwood, Chichester, England.
Vandamme, E. J., Leyman, D., De Visscher, P., De Buyser, D., and Vansteenkiste, G. (1981) *J. Chem. Tech. Biotechnol. 31,* 247–257.
Vandamme, E. J., Leyman, D., De Buyser, D., De Visscher, P., Spriet, J., Vansteenkiste, G. C., Nimi, O., Poirier, A., and Demain, A. L. (1982) in *Peptide Antibiotics—Biosynthesis and Functions* (Kleinkauf, H., and Döhren, H. von, eds.) pp. 117–136. de Gruyter, Berlin.
Vasantha, N., Balakrishnan, R., Kaur, S., and Jayaraman, K. (1980) *Arch. Biochem. Biophys. 200,* 40–44.
Vater, J., and Kleinkauf, H. (1976) *Biochim. Biophys. Acta 429,* 1062–1072.
Vining, L. C., and Wright, J. L. C. (1977) in *Biosynthesis* Vol. V, pp. 240–305, Chemical Society, London.
Vitković, L., and Pfaender, P. (1982) in *Peptide Antibiotics—Biosynthesis and Functions* (Kleinkauf, H., and Döhren, H. von, eds.) pp. 297–306, de Gruyter, Berlin.
Vitković, L., and Sadoff, H. L. (1977a) *J. Bacteriol. 131,* 891–896.
Vitković, L., and Sadoff, H. L. (1977b) *J. Bacteriol. 131,* 897–905.
Wang, D. I. C., and Hamilton, B. K. (1977) *Biotechnol. Bioeng. 19,* 1225–1232.
Wang, D. I. C., Stramondo, J., and Fleischaker, R. (1977) in *Biotechnological Applications of Proteins and Enzymes* (Bohak, Z., and Sharon, N., eds.) pp. 183–201, Academic Press, New York.
Weber, K., Pringle, J. R., and Osborn, M. (1972) *Meth. Enzymol. 26,* 3–27.
Wetlaufer, D. B. (1973) *Proc. Nat. Acad. Sci. 70,* 697–701.
Wilkinson, S., and Lowe, L. A. (1964) *Nature 204,* 185.
Wilkinson, S., and Lowe, L. A. (1966) *Nature 211,* 311.
Zharikova, G. G., and Lukin, A. A. (1966) *Mikrobiologiya 35,* 100–106.
Zharikova, G. G., Sachenko, G. V., and Mitronova, T. N. (1964) *Mikrobiologiya 33,* 605–609.
Zharikova, G. G., Pogliazova, M. N., and Makarova, G. Y. (1965) *Mikrobiologiya 34,* 995–998.
Zharikova, G. G., Makarova, G. Y., Poglazova, M. N., and Zarubina, A. P. (1966) *Mikrobiologiya 35,* 647–650.
Zharikova, G. G., Zarubina, A. P., Kherat, D. M., Myaskovskaya, S. P., and Maksimov, V. N. (1975) in *Antibiotiki ikh Produtsenty* (Silaev, A. B., ed.), pp. 163–186, Nauka, Moskow.
Zimmer, T.-H., and Laland, S. G. (1975) *Meth. Enzymol. 43,* 567–579.
Zimmer, T.-L., Frøyshov, Ø., and Laland, S. G. (1979) in *Economic Microbiology* Vol. III, *Secondary Products of Metabolism,* (Rose, A. H., ed.) pp. 123–150, Academic Press, New York.

Zocher, R., and Kleinkauf, H. (1978) *Biochim. Biophys. Res. Comm. 81,* 1162–1167.

Zocher, R., Keller, U., and Kleinkauf, H. (1982a) in *Peptide Antibiotics—Biosynthesis and Functions* (Kleinkauf, H., and Döhren, H. von, eds.) pp. 269–274, de Gruyter, Berlin.

Zocher, R., Keller, U., and Kleinkauf, H. (1982b) *Biochemistry 21,* 43–48.

CHAPTER
6

Peptidolactones

Yasushi Okumura

I. INTRODUCTION

Many kinds of depsipeptide antibiotics have been discovered among microbially synthesized secondary metabolites. The producers are multifarious, and production of each antibiotic is a strain rather than a species character. Within the antibiotic group are various kinds of chemical structures. In this chapter, only the peptidolactones are considered.

Fundamentally, peptidolactone antibiotics consist of two parts. One is a peptide chain, and the other is a chromophore or acyl group that binds to the N-terminal amino group of the peptide. The C-terminal carboxyl group of the peptide or the carboxyl group of a dicarboxylic amino acid within the peptide is linked to a hydroxyl group elsewhere in the molecule by an ester bond. The hydroxyl groups participating in ester formation are derived from serine, threonine, or a hydroxy fatty acid. Most peptidolactone antibiotics, like the peptide antibiotics, contain one or more unusual amino acids, such as D-isomers and N-methyl derivatives not present in proteins.

Generally, the biosynthesis of peptide antibiotics is not directed by ribosomes but by a protein thiotemplate mechanism (see Chapter 5). Although the detailed mechanisms of peptidolactone synthesis are not yet known, available evidence indicates that it, too, is directed by a protein thiotemplate. It is noteworthy that inhibitors of protein synthesis, such as chloramphenicol and puromycin, do not inhibit peptidolactone antibiotic synthesis (Katz and Weissbach, 1962a; Kamal and Katz, 1976). Because protein thiotemplate mechanisms exhibit broader substrate specificities than those permitted by the ribosomal mechanism, it is possible to replace some constituent amino acids with analogs. This has been exploited to produce novel and useful antibiotics.

Peptidolactone antibiotics exhibit a considerable range of biological activities. Their practical use is limited by their toxicity in animals or by low water solubility. If these limitations can be overcome, the value of peptidolactone antibiotics will increase. This chapter reviews present knowledge of the structure and biochemistry of this group as a basis for the development of their commercial potential.

II. ACTINOMYCINS

Actinomycins are composed of a phenoxazinone chromophore and two pentapeptide chains. Their general structure is shown in Fig. 6.1. Individual actinomycins differ at amino acid sites in their peptides. The first actinomycin was reported by Waksman and Woodruff (1940). Since then, many others have been isolated, usually as complex mixtures. The nomenclature of actinomycins is confusing because several different systems have been adopted. Meienhofer and Atherton (1973) proposed an excellent general nomenclature, but the specific designations are longer than those in common use and have not gained wide acceptance.

Actinomycins are active against gram-positive bacteria and are highly toxic in animals. Because of the toxicity, practical use is limited to certain cancer chemotherapies. Many analogs have been produced by controlled (directed) biosynthesis and by chemical modification for testing as improved anticancer agents. This "trial and error" method has not yielded any compounds with more desirable properties, and precise structure–activity relationships need to be better understood for the design of superior antibiotics.

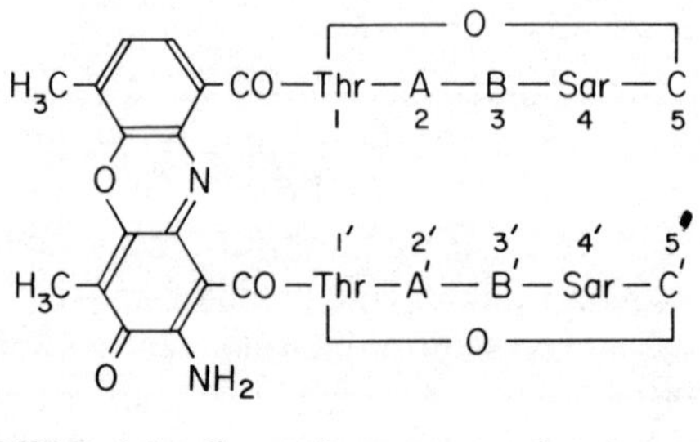

FIGURE 6.1 General structure of actinomycin.

A. Producing Organisms and Structures

Actinomycin-producing organisms are listed in Table 6.1. The structures of antibiotics from natural sources, excluding those prepared by controlled biosynthesis, are shown in Table 6.2. It is noticeable that the actinomycins produced by *Streptomyces fradiae* and *Micromonospora floridensis* have several amino acids that are distinctly different from those synthesized by most *Streptomyces* species. They contain 5-methylproline and its oxidized forms and, additionally, a threonine analog such as hydroxythreonine or N-methylthreonine. A substantial number of different actinomycins have been obtained by chemical modification. These are described by Meienhofer and Atherton (1973) and will not be covered here. Isolation, purification, and structure elucidation of the actinomycins have been reviewed by Mauger and Katz (1978).

B. Biochemistry

The biosynthesis of actinomycin has been studied by many groups, and the following facts have been established:

1. Actinomycin synthesis is not directed by the ribosomal system of protein synthesis (Katz and Weissbach, 1962a).
2. Each amino acid in actinomycin peptides is derived from a corresponding L-amino acid (Katz and Weissbach, 1963; Katz et al., 1962; Ciferri et al., 1962).
3. Tryptophan is a precursor of the actinomycin chromophore (Sivak et al., 1962).
4. Most methyl groups in actinomycin are derived from methionine (Katz and Weissbach, 1963; Ciferri et al., 1962).

There remain many areas of uncertainty, such as the mechanism of peptide chain synthesis and lactone formation. Some of these are being clarified by current studies of actinomycin biochemistry.

TABLE 6.1 Actinomycin-Producing Organisms

Microorganism		*Actinomycin Complex*	*Reference*
Streptomyces	*antibioticus*	A	Waksman and Woodruff (1940)
	aureofaciens	P_2	Rao and Renn (1960)
	chrysomallus	C	Brockmann and Grubhofer (1949)
	flavus	X	Waksman and Gregory (1954)
	fradiae	Z	Bossi et al. (1958)
	michiganensis	X	Corbaz et al. (1958b)
	parvulus	A	Manaker et al. (1955)
Micromonospora	*floridensis*	Schering compounds	Wagman et al. (1976)

TABLE 6.2 Amino Acid Sequences of Actinomycins

Actinomycin	*Site of Actinomycin Peptides* 1	2	3	4	5
$I(=A_I=B_I=X_0)$	Thr	Val	Pro	Sar	MeVal
	Thr	Val	HyPro	Sar	MeVal
X_0	Thr	Val	Pro	Sar	MeVal
	Thr	Val	*a*HyPro	Sar	MeVal
$II(=A_{II}=B_{II}=F_8)$	Thr	Val	Sar	Sar	MeVal
	Thr	Val	Sar	Sar	MeVal
$III(=A_{III}=B_{III}=F_9=X_0)$	Thr	Val	Pro	Sar	MeVal
	Thr	Val	Sar	Sar	MeVal
$IV(=A_{IV}=B_{IV}=D$ $=C_1=I_1=X_1)$	Thr	Val	Pro	Sar	MeVal
	Thr	Val	Pro	Sar	MeVal
X_{1a}	Thr	Val	Sar	Sar	MeVal
	Thr	Val	KetoPro	Sar	MeVal
$V(=A_V=B_V=X_2)$	Thr	Val	Pro	Sar	MeVal
	Thr	Val	KetoPro	Sar	MeVal
$VI(=C_2=I_2)$	Thr	Val	Pro	Sar	MeVal
	Thr	*a*Ileu	Pro	Sar	MeVal
C_{2a}	Thr	*a*Ileu	Pro	Sar	MeVal
	Thr	Val	Pro	Sar	MeVal
$VII(=C_3)$	Thr	*a*Ileu	Pro	Sar	MeVal
	Thr	*a*Ileu	Pro	Sar	MeVal
Z_1	Thr	Val	4-Keto-5-MePro	Sar	MeVal
	HyThr	Val	3-Hy-5-MePro	Sar	MeAla
Z_5	Thr	Val	4-Keto-5-MePro	Sar	MeVal
	HyThr	Val	5-MePro	Sar	MeAla
Schering Compound 1_a	Thr	Val	4-Keto-5-MePro	Sar	MeVal
	Thr	Val	5-MePro	Sar	MeVal
1_b	Thr	Val	4-Keto-5-MePro	Sar	MeVal
	Thr	Val	5-MePro	Sar	MeAla
1_c	Thr	Val	4-Keto-5-MePro	Sar	MeVal
	Thr	Val	3-Hy-5-MePro	Sar	MeVal
6_a	Thr	Val	4-Keto-5-MePro	Sar	MeVal
	Thr	Val	3-Hy-5-MePro	Sar	MeAla
10	Thr	Val	4-Keto-5-MePro	Sar	MeVal
	HyThr	Val	5-MePro	Sar	MeAla

Abbreviations: Thr = threonine, Val = valine, Sar = sarcosine, MeVal = N-methylvaline, HyPro = hydroxyproline, *a*HyPro = *allo*-hydroxyproline, KetoPro = ketoproline, *a*Ileu = *allo*-isoleucine, 4-Keto-5-MePro = 4-keto-5-methylproline, HyThr = hydroxythreonine, 3-Hy-5-MePro = 3-hydroxy-5-methylproline, MeAla = N-methylalanine, 5-MePro = 5-methylproline.

From Mauger and Katz (1978). Reproduced with permission of Elsevier Scientific Publishing Company, Amsterdam.

1. Biosynthesis of the Constituents. a. *Threonine moiety*. Uniformly labeled L-threonine added to cultures of *Streptomyces antibioticus* was incorporated into both actinomycin and cell proteins during short-term incubations. Addition of chloramphenicol decreased incorporation of the amino acid into proteins and increased its incorporation into the antibiotic (Katz et al., 1965). These results indicate that there is a common threonine pool for the actinomycin and protein-synthesizing systems. Furthermore, the two systems are different from each other. The biosynthesis of hydroxythreonine or N-methylthreonine in the actinomycins produced by *M. floridensis* and *S. fradiae* has not been examined.

b. *D-Valine or D-allo-isoleucine moiety*. Under normal fermentation conditions (no supplementation with amino acids), *S. antibioticus* produces mainly actinomycin IV in which the second amino acid position is occupied by D-valine. Addition of D-valine to the culture inhibited antibiotic synthesis; the inhibition was reversed by adding L-valine (Katz, 1960). which also selectively stimulated actinomycin IV synthesis (Katz et al., 1961). The implication that L-valine was the precursor of D-valine in actinomycin was confirmed by the incorporation of L-[1-^{14}C]valine into both actinomycin and cell proteins. The labeled amino acid in actinomycin was equally divided between D-valine and N-methyl-L-valine. D-[1-^{14}C]Valine added to the culture was immediately taken up by the organism but was poorly utilized for antibiotic synthesis. The small amount of radioisotope incorporated was again distributed equally between the D-valine and the N-methyl-L-valine moieties (Katz and Weissbach, 1962a, 1963).

Katz et al. (1961) reported that addition of a high concentration of L-isoleucine, L-threonine, or D-threonine to *Streptomyces chrysomallus* greatly enhanced the synthesis of actinomycin VII (containing D-*allo*-isoleucine) and inhibited the synthesis of actinomycins IV and VI (containing D-valine). Supplementation with D-*allo*-isoleucine or D-isoleucine did not stimulate synthesis of actinomycin VII. These results indicated that D-*allo*-isoleucine in actinomycin came from L-isoleucine, and this was confirmed by experiments with radioactive L-isoleucine and L-threonine. The latter is a normal precursor of isoleucine and was incorporated into both the L-threonine and D-*allo*-isoleucine moieties of actinomycin (Albertini et al., 1964). Studies on the mechanism by which L-valine and L-isoleucine are converted to the corresponding D-enantiomers during actinomycin synthesis (Salzman et al., 1964) showed that both the amino group and the carbon skeleton of L-valine were retained. When L-[2,3-^{3}H]valine was added to the culture medium of *Streptomyces parvulus,* radioactivity was incorporated into actinomycin and cell proteins (Mason et al., 1977). In the D-valine of actinomycin, the tritium label was detected only at C-3, whereas in the valine from cell protein, it was present at both C-2 and C-3. Specific loss of the C-2 hydrogen atom of L-valine during antibiotic synthesis indicated that conversion of L-valine to the D-enantiomer is a racemase-type reaction and α,β-dehydrovaline is not an intermediate.

c. *Proline moiety*. Proline or its analogs occupy the third amino acid position in actinomycin peptides. Under normal fermentation conditions, *S. antibioticus* produces actinomycins I (containing L-proline and 4-hydroxy-L-proline) and V (containing L-proline and 4-keto-L-proline) together with actinomycin IV (containing only L-proline). [^{14}C-(U)]Proline added to the culture was immediately incorporated into both the proline and the hydroxyproline of actinomycin I. In actinomycin V, the radioactivity was distributed in proline and ketoproline. Thus 4-hydroxy-L-proline and 4-keto-L-proline originate from L-proline. When ^{14}C-labeled L-proline and unlabeled 4-hydroxy-L-proline were added concurrently to *S. antibioticus,* the specific activity of proline in actinomycins I and V was not diluted, but that of 4-hydroxyproline in actinomycin I was. Evidently, 4-hydroxy-L-proline is directly incorporated. Confirming this, radiolabeled 4-hydroxy-L-proline was incorporated primarily into the hydroxyproline moiety of actinomycin I, with only a small amount of the radioactivity in proline and ketoproline (Katz et al., 1962).

To establish that 4-keto-L-proline was not a precursor of 4-hydroxyproline, L-[*cis*-4-^{3}H]proline was added to cultures of *S. antibioticus*. As predicted, the tritium label was completely retained in hydroxproline, but no tritium was detected in ketoproline. When L-[*trans*-4-^{3}H]proline was used, neither hydroxyproline nor ketoproline was labeled. It was concluded that conversion of L-proline to 4-hydroxy-L-proline occurs with complete retention of configuration at C-4. Only the *trans*-hydrogen at C-4 is displaced, and 4-ketoproline is not an intermediate (Salzman et al., 1965). The oxygen of the hydroxyl group of 4-hydroxy-L-proline comes from atmospheric oxygen, not water (Diegelmann et al., 1969). It is not yet clear whether proline is free or peptide-bound when it is hydroxylated. The various possibilities are summarized in Fig. 6.2.

d. *Sarcosine moiety*. The fourth position in actinomycin peptides is always occupied by sarcosine. Radioisotope experiments conducted with *S. antibioticus* and *S. chrysomallus* revealed that sarcosine is directly synthesized from glycine and methionine (Katz and Weissbach, 1962a; Ciferri et al., 1962). However, it is not clear whether free sarcosine is synthesized before incorporation into the antibiotic or whether glycine is incorporated into the antibiotic and then converted to sarcosine. The first route was favored by Katz and Weissbach (1963) on the following evidence: [N-$^{14}CH_3$]sarcosine and [1-^{14}C]sarcosine were incorporated into the antibiotic at the same rate and to the same extent. If sarcosine had been demethylated to glycine before incorporation, the $^{14}CH_3$ moiety would have been diluted with the methyl pool of the organism (Katz and Weissbach, 1963; Fukagawa et al., 1974).

Ciferri et al. (1965) proposed that glycine and sarcosine were incorporated into the antibiotic by different routes, since the incorporation of one labeled amino acid was not reduced when mixed with an unlabeled sample of the other. They assumed that there was no free sarcosine unless it was added

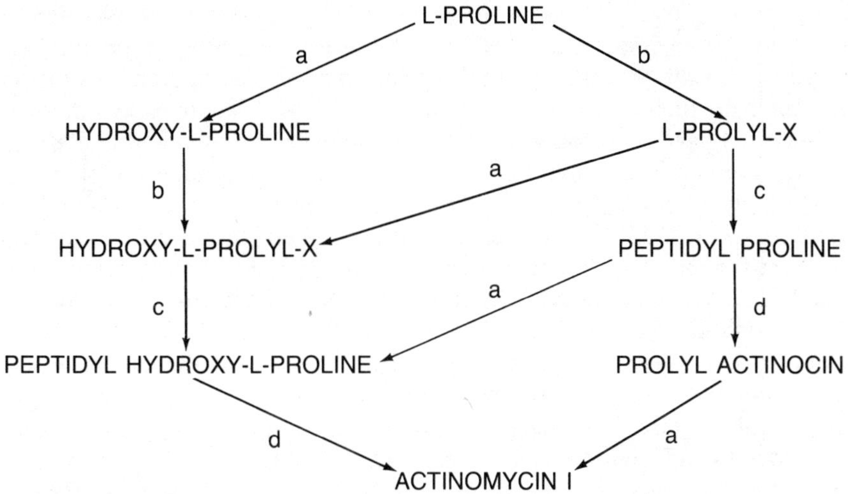

FIGURE 6.2 Possible sites of proline hydroxylation. Reactions: a = hydroxylation, b = activation, c = peptide formation, d = phenoxazinone synthesis. Source: Diegelmann et al. (1969). Reproduced with permission of Academic Press, Inc., New York.

as a supplement, and that this supplement would be incorporated into only the third position of the peptide to replace proline. In contrast, glycine would be incorporated into the fourth position, then N-methylated. The production of actinomycins II and III (containing sarcosine) in the absence of a sarcosine supplement cannot be explained by this hypothesis.

With protoplasts from *S. antibioticus,* some labeled amino acids were incorporated into actinomycin. However, sarcosine was not among these, and a supplement of sarcosine did not reduce the incorporation of labeled glycine and methionine. This suggested that in protoplasts, free sarcosine was not a precursor of peptide-bound sarcosine (Keller and Kleinkauf, 1977). Protoplasts from *S. parvulus* also failed to utilize sarcosine for actinomycin synthesis, but Hitchcock and Katz (1978) considered that this might be due to a change in the amino acid transport system.

e. *N-Methylvaline moiety.* The fifth amino acid in the actinomycin peptides is N-methyl-L-valine, or in the actinomycin Z series produced by *S. fradiae* and the Schering compounds produced by *M. floridensis,* it may be N-methyl-L-alanine (Brockmann and Stahler, 1965; Wagman et al., 1976). N-Methylvaline is synthesized from L-valine and L-methionine or glycine. The glycine may be metabolized to a C_1 unit and used to synthesize the methyl group of methionine. This may be incorporated into the methyl groups of N-methylamino acids (Katz and Weissbach, 1963).

f. *Chromophore*. Synthesis of actinomycin by *S. antibioticus* is inhibited by tryptophan analogues such as α-, 4-, 5-, or 6-methyl-DL-tryptophan. The inhibition is reversed by tryptophan or intermediates on the pathway from tryptophan to phenoxazinone, such as kynurenine, 3-hydroxykynurenine, and 3-hydroxyanthranilic acid. That the actinomycin chromophore is derived from tryptophan was supported by the selective incorporation of DL-[7-^{14}C]tryptophan. No radioactivity was detected in the peptide components (Sivak and Katz, 1962). Labeled 3-hydroxy-4-methylanthranilic acid was directly incorporated into actinomycin (Golub et al., 1969); unlabeled, both it and 3-hydroxy-4-methylkynurenine decreased the incorporation of [^{14}C]tryptophan (Weissbach et al., 1965a, b; Perlman et al., 1973). Possible metabolic pathways by which tryptophan could be metabolized to form the actinomycin chromophore are shown in Fig. 6.3.

Several enzymes associated with synthesis of the actinomycin chromophore have been isolated from, or detected in, cell extracts of actinomycin-producing organisms. Phenoxazinone synthetase, partially purified from *S. antibioticus,* catalyzes the condensation of *o*-aminophenols to the corre-

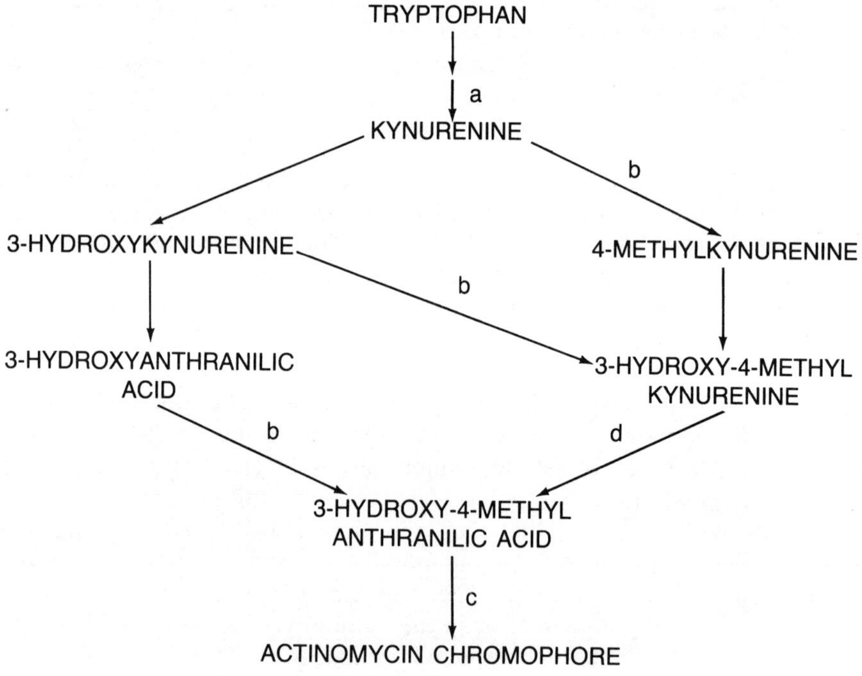

FIGURE 6.3 Possible metabolic pathways from L-tryptophan to the actinomycin chromophore. Reactions: a = kynurenine formamidase, b = methylation, c = phenoxazinone synthetase, d = activity in cell extract. Source: Perlman et al. (1973). Reproduced with permission of Japan Antibiotics Research Association, Tokyo.

sponding phenoxazinones (Katz and Weissbach, 1962b). When a peptide derivative, such as 3-hydroxy-4-methylanthranilyl-L-threonyl-D-valyl-L-prolyl-sarcosinyl-N-methyl-L-valine was used as a substrate, the actinocinyl peptide was readily formed by the enzyme (Salzman et al., 1969). A cell extract of *S. antibioticus* also catalyzes the conversion of 3-hydroxy-4-methylkynurenine to 3-hydroxy-4-methylanthranilic acid (Perlman et al., 1973). Brown et al. (1980) found two isozymes of kynurenine formamidase, the second enzyme in tryptophan metabolism, in the mycelium of *S. parvulus*. One was constitutive, and the other was inducible. The latter was induced only just before actinomycin synthesis began and was believed to be associated with that process.

2. Assembly of Precusors. Based on current information, the biosynthesis of actinomycin proceeds as shown in Fig. 6.4.

3. Controlled Biosynthesis. Exogenously added amino acids or analogs related in structure to the constituent amino acids can be incorporated into peptidolactone antibiotics in competition with endogenously synthesized

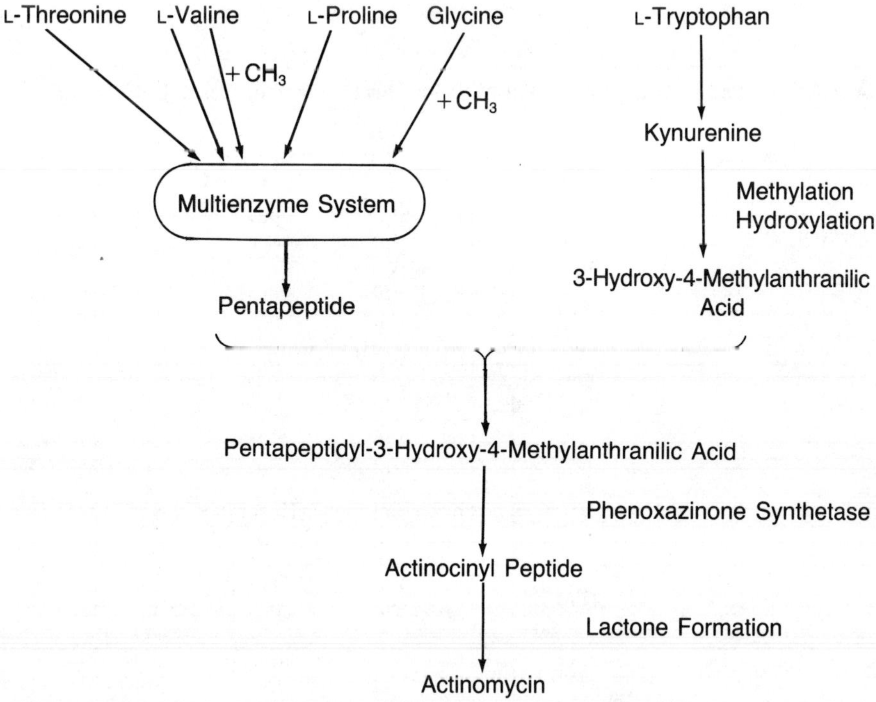

FIGURE 6.4 Actinomycin biosynthesis.

precursors. As a consequence, new analogs can be produced, or the synthesis of minor components can be increased. It is accepted in general that these replacements are due to the broad specificities of antibiotic-synthesizing systems (Lipmann et al., 1971; Fujikawa et al., 1968; Kurahashi, 1974; Laland and Zimmer, 1973; Kleinkauf et al., 1978). Actinomycins obtained by adding amino acids to the medium (controlled biosynthesis) are listed in Table 6.3.

The three sites in actinomycin where substitutions occur are (1) valine or *allo*-isoleucine, (2) proline, and (3) N-methylvaline.

a. *Valine or allo-isoleucine site*. The valine or *allo*-isoleucine site and the N-methylvaline site are affected together by addition of a branched-chain amino acid or its biochemical precursor. Schmidt-Kastner (1956) reported that DL-valine added to a culture of *S. chrysomallus* greatly increased production of actinomycin IV, which is normally quite small. When DL-isoleucine was added, two new actinomycins (E_1 and E_2) containing N-methylisoleucine were produced. Katz et al. (1961) also showed that four stereoisomers of isoleucine and two isomers of threonine (a precusors of isoleucine) elicited production of actinomycins E_1 and E_2. The order of effectiveness was D-isoleucine > D-*allo*-isoleucine > L-isoleucine > D-threonine > L-threonine. Yajima et al. (1972) clarified the mechanism by which isoleucine stereoisomers are interconverted in an actinomycin-producing organism (Fig. 6.5).

TABLE 6.3 Amino Acid Sequences of Actinomycins Obtained by Controlled Biosynthesis

Supplement	*Actinomycin*	*Site of Actinomycin Peptides* 1	2	3	4	5
Pipecolic acid	Pip_2	Thr	Val	Pip	Sar	MeVal
		Thr	Val	Pip	Sar	MeVal
	$Pip_{1\alpha}$	Thr	Val	4-KetoPip	Sar	MeVal
		Thr	Val	Pip	Sar	MeVal
	$Pip_{1\beta}$	Thr	Val	Pip	Sar	MeVal
		Thr	Val	Pro	Sar	MeVal
	$Pip_{1\gamma}$	Thr	Val	4-HyPip	Sar	MeVal
		Thr	Val	Pip	Sar	MeVal
	$Pip_{1\delta}$	Thr	Val	4-KetoPip	Sar	MeVal
		Thr	Val	Pro	Sar	MeVal
	$Pip_{1\varepsilon}$	Thr	Val	4-HyPip	Sar	MeVal
		Thr	Val	Pro	Sar	MeVal
Sarcosine	F_1	Thr	*a*Ileu	Sar	Sar	MeVal
		Thr	Val	Sar	Sar	MeVal
	F_2	Thr	*a*Ileu	Sar	Sar	MeVal
		Thr	Val	Pro	Sar	MeVal

TABLE 6.3 (Continued)

		Site of Actinomycin Peptides				
Supplement	*Actinomycin*	*1*	*2*	*3*	*4*	*5*
	F_3	Thr	*a*Ileu	Sar	Sar	MeVal
		Thr	*a*Ileu	Sar	Sar	MeVal
	F_4	Thr	*a*Ileu	Sar	Sar	MeVal
		Thr	*a*Ileu	Pro	Sar	MeVal
	F_8	Thr	Val	Sar	Sar	MeVal
		Thr	Val	Sar	Sar	MeVal
	F_9	Thr	Val	Pro	Sar	MeVal
		Thr	Val	Sar	Sar	MeVal
Isoleucine stereoisomers	E_1	Thr	*a*Ileu	Pro	Sar	MeVal
		Thr	Val Ileu	Pro	Sar	Me*a*Ileu
	E_2	Thr	*a*Ileu	Pro	Sar	Me*a*Ileu
		Thr	*a*Ileu	Pro	Sar	Me*a*Ileu
cis- or *trans*-4-Methylproline	K_{1c}	Thr	Val	Pro	Sar	MeVal
		Thr	Val	*cis*-4-MePro	Sar	MeVal
	K_{2c}	Thr	Val	*cis*-4-MePro	Sar	MeVal
		Thr	Val	*cis*-4-MePro	Sar	MeVal
	K_{1t}	Thr	Val	Pro	Sar	MeVal
		Thr	Val	*trans*-4-MePro	Sar	MeVal
	K_{2t}	Thr	Val	*trans*-4-MePro	Sar	MeVal
		Thr	Val	*trans*-4-MePro	Sar	MeVal
Azetidine-2-carboxylic acid	Azetomycin I	Thr	Val	Pro	Sar	MeVal
		Thr	Val	Azet	Sar	MeVal
	Azetomycin II	Thr	Val	Azet	Sar	MeVal
		Thr	Val	Azet	Sar	MeVal
cis- or *trans*-4-Chloroproline	*cis*-Chloro A	Thr	Val	*cis*-4-C1Pro	Sar	MeVal
		Thr	Val	*cis*-4-C1Pro	Sar	MeVal
	cis-Chloro B	Thr	Val	Pro	Sar	MeVal
		Thr	Val	*cis*-4-C1Pro	Sar	MeVal
	trans-Chloro A	Thr	Val	*trans*-4-C1Pro	Sar	MeVal
		Thr	Val	*trans*-4-C1Pro	Sar	MeVal
	trans-Chloro B	Thr	Val	Pro	Sar	MeVal
		Thr	Val	*trans*-4-C1Pro	Sar	MeVal

Abbreviations: Pip = pipecolic acid, 4-KetoPip = 4-keto-pipecolic acid, 4-HyPip = 4-hydroxy-pipecolic acid, *cis*-4-MePro = *cis*-4-methylproline, *trans*-4-MePro = *trans*-4-methylproline, Azet = azetidine-2-carboxylic acid, *cis*-4-C1Pro = *cis*-4-chloroproline, *trans*-4-C1Pro = *trans*-4-chloroproline. Other abbreviations are the same as in Table 6.2.
From Mauger and Katz (1978). Reproduced with permission of Elsevier Scientific Publishing Company, Amsterdam.

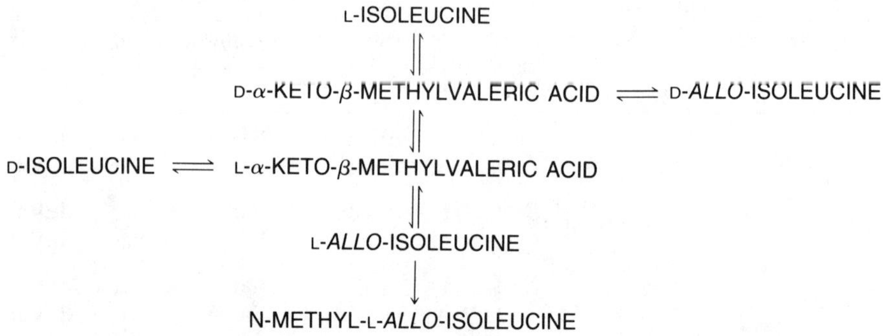

FIGURE 6.5 Possible pathway for synthesis of N-methyl-L-*allo*-isoleucine and D-isoleucine. From Yajima et al. (1972). Reproduced with permission of Academic Press, Inc., New York.

b. *Proline site*. The amino acid at the third position is the most easily replaceable. Adding sarcosine to cultures of *S. chrysomallus* gave several new actinomycins that could be distinguished from each other by their content of sarcosine and proline (Schmidt-Kastner, 1960). Adding sarcosine to *S. antibioticus* greatly increased synthesis of actinomycins II and III, which are minor products under normal fermentation conditions, while production of actinomycin IV decreased (Katz and Goss, 1959; Goss and Katz, 1960). That the proline in actinomycin IV was replaced by the exogenously supplied sarcosine was confirmed by the structure elucidation of actinomycins II and III (Johnson and Mauger, 1959). The variety of new actinomycins that has been obtained by adding proline analogs is indicated in Table 6.3. The following substances can also be incorporated: 4-hydroxy-L-proline (actinomycin I), L-thiazolidine-4-carboxylic acid (two new actinomycins), and 5-methylproline (Katz and Goss, 1958; Formica and Katz, 1973; Formica and Apple, 1976; Yoshida et al., 1966; Katz, 1967; Nishimura and Bowers, 1967).

4. Regulation of Actinomycin Biosynthesis. Actinomycin biosynthesis may be regulated by catabolite repression. Phenoxazinone synthetase, which synthesizes the actinomycin chromophore, is produced just before antibiotic appears in the culture (Marshall et al., 1968). Synthesis of the enzyme is severely repressed by glucose. Cultures of *S. antibioticus* using glucose as a carbon source grow vigorously but do not make phenoxazinone synthetase, and thus produce no actinomycin, until all of the glucose has been consumed (Gallo and Katz, 1972).

Ochi and Katz (1978) have reported the possible involvement of a plasmid in actinomycin synthesis by *S. parvulus* and *S. antibioticus*. When these organisms were treated with curing agents such as acriflavine, ethidium bromide, and novobiocin, loss of antibiotic-producing ability occurred at much higher frequencies than loss of ability to form aerial mycelium and the incidence of auxotrophic progeny. Protoplast fusion and cosynthesis studies also supported plasmid involvement.

While control mechanisms such as nitrogen metabolite regulation, phosphate effects, feedback inhibition, or the action of bioregulators may influence actinomycin synthesis, these have not yet been seriously explored. Progress in elucidating the biosynthetic pathway has not only supplied potentially useful new actinomycins and improved yields but has provided a valuable model for clarifying biosynthetic mechanisms for other antibiotics. With the background now acquired, a study of the metabolic control of actinomycin biosynthesis would undoubtedly pay large dividends.

C. Biological Properties

There have been many excellent reviews of the biological properties of actinomycin, and only brief mention of them will be made here. Actinomycins are active against gram-positive bacteria but inactive or only slightly active against gram-negative bacteria and fungi. Since actinomycins are highly toxic in animals, the antibiotics are not used clinically except in certain cancer chemotherapies. First report of actinomycin anticancer activity indicated that Ehrlich carcinoma in mice and Walker carcinoma in rats were suppressed by actinomycin C (Hackmann, 1952). Subsequently, the ability of actinomycins (mainly actinomycins C and D) to suppress various cancers was observed in many laboratories. To date, actinomycins, either alone or in combination with other drugs or other therapies, have been proven effective in Wilms's tumor in children, choriocarcinoma, and rhabdomyosarcoma (Farber, 1966; Fernbach and Martyn, 1966; Wolff et al., 1968; D'Angio, 1972; Li, 1971; Lewis, 1972; Pinkel and Pratt, 1973).

III. VIRGINIAMYCIN AND RELATED COMPOUNDS

The virginiamycin family of antibiotics is highly active against gram-positive bacteria and mycoplasmas and has very low toxicity. They are effective against experimental infections in animals, but because of their low solubility in water, clinical use in humans is restricted to topical applications. They are used as growth promoters as well as therapeutically in the veterinary field. They promote substantial improvements in feed utilization, and their lack of toxicity, poor readsorption, negligible tissue fixation, and other characteristics are ideal for this purpose. Although penicillin and tetracycline mixtures are the main growth promoters now used, the potential for development and spread of antibiotic-resistant strains makes it desirable that antibiotics for animal use be different from those in human use. Because of this, the importance of the virginiamycins will undoubtedly increase.

A. Producing Organisms

Many antibiotics of the virginiamycin family are produced by a variety of organisms (Table 6.4.). These generally synthesize two or more active compounds belonging to groups A and B.

B. Classification and Structures

According to the classification proposed by Tanaka (1975) and by Vazquez (1975), the virginiamycin family of antibiotics is divided into two major groups, A and B, which are antimicrobially synergistic. Group A antibiotics are non-peptidyl macrocyclic lactones, and group B are peptidolactones. Group A can be divided into three subgroups: A-I, A-II, and A-III. The structures are shown in Fig. 6.6.

Group B can be further classified into two subgroups, B-I and B-II, but all have part of their structure in common: the N-terminal amino acid is threonine, and its amino group is masked by 3-hydroxypicolinic acid. The hydroxyl group of threonine and the carboxyl group at the C-terminus are linked by an ester bond. The antibiotics in subgroup B-I (Fig. 6.7) contain seven moieties. Many structural variations exist. All are microbial products except patricins A and B, which are chemically synthesized. In

TABLE 6.4 Organisms Producing the Virginiamycin Family of Antibiotics

Microorganism	*Product*	*Reference*
Streptomyces virginiae	Virginiamycins (staphylomycins, No. 899)	Somer and Dijck (1955)
S. alborectus	Virginiamycins	Ogata et al. (1978)
S. loidensis	Vernamycins	Donovick et al. (1961)
S. graminofaciens	Streptogramins	Charney et al. (1953)
S. ostreogriseus	Ostreogrycins	Ball et al. (1958)
S. mitakaensis	Mikamycins	Arai et al. (1958)
S. pristinae	Pristinamycins	Benazet et al. (1962)
S. olivaceous	PA114A, B(synergistins)	Celmer and Sobin (1956)
S. conganensis	F1370A, B	Linder et al. (1960)
S. griseus }	Viridogrisein (etamycin)	Bartz et al. (1955)
S. griseoviridus }	Griseoviridin	
S. griseoviridus P8648	Neoviridogriseins Griseoviridin	Okumura et al. (1979c)
Streptomyces sp.	Etamycin	Heinemann et al. (1955)
Actinoplanes philippinensis	A-2315A, B, and C	Hamill and Stark (1975)
A. auranticolor	Plauracins	Celmer et al. (1975)
Actinoplanes sp	A17002A, B, C, and D	Martinelli et al. (1979)
Actinomadura flava	Madumycins	Brazhnikova et al. (1975)
Micromonospora sp.	Vernamycins	Liu et al. (1981)
Actinomyces daghestanicus	Antibiotic 6613 (etamycin)	Kruglyak et al. (1961)

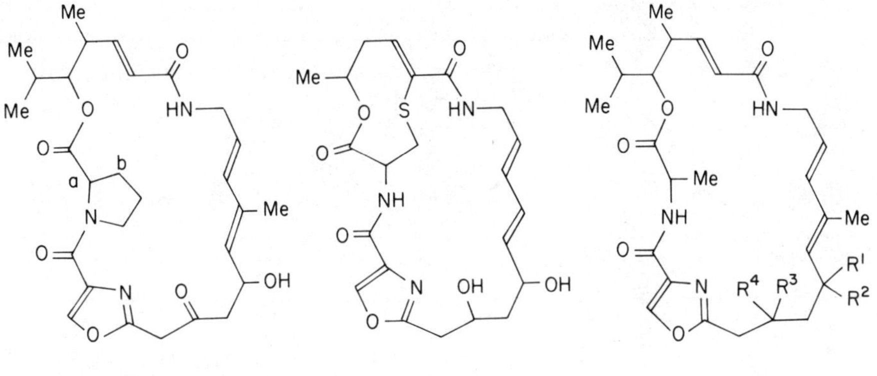

A-I A-II A-III

A-I
a) ab unsaturated: virginiamycin M (Durant et al., 1974) = vernamycin A (Donovick et al., 1961) = streptogramin A (Vazquez, 1962) = ostreogrycin A (Delpierre et al., 1966) = mikamycin A (Okabe, 1959) = pristinamycin IIA (Preud'Homme et al., 1968) = synergistin A-I (Celmer and Sobin, 1956)
b) ab saturated: virginiamycin M_2 (Durant et al., 1974) = ostreogrycin G (Kingston et al., 1966) = pristinamycin IIB (Preud'Homme et al., 1968)

A-II
griseoviridin (Fallona et al., 1964; Birnbaum and Hall, 1976)

A-III
a) R_1 = R_3 = H, R_2 = R_4 = OH: A-2315A (Chamberlin and Chen, 1977) = CP-35763 (Moppet and Whipple, 1977) = madumycin II (Brazhnikova et al., 1975) = A17002F (Martinelli et al., 1979)
b) R_1 = H, R_2 = OH, R_3R_4 = O: A-2315B (Chamberlin and Chen, 1977) = CP36926 (Moppet and Whipple, 1977) = madumycin I (Brazhnikova et al., 1975)
c) R_1 = R_2 = R_3 = H, R_4 = OH: A17002C (Martinelli et al., 1979)

FIGURE 6.6 Structures of virginiamycin group A antibiotics.

contrast, subgroup B-II contains few members (Fig. 6.8). Structural variation has arisen only through controlled biosynthesis.

C. Biochemistry

Up to the present, there have been only a few reports on the biosynthesis of the A group antibiotics. The reason for the neglect of such interesting structures may be that the "building stones" of the A group cannot easily be excised by chemical degradation. It is difficult to locate any precursors that have been incorporated into the molecule. However, there are now many kinds of ^{13}C-labeled substrates available, and ^{13}C-NMR spectroscopy can be used to determine, without chemical degradation, where label is

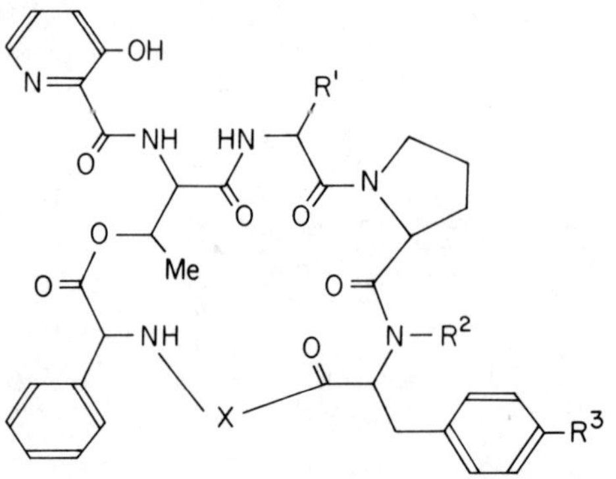

Antibiotic	R_1	R_2	R_3	X
Virginiamycin S	C_2H_5	CH_3	H	4-oxopipecolic acid
Virginiamycin S_2	C_2H_5	H	H	4-hydroxypipecolic acid
Virginiamycin S_3	C_2H_5	CH_3	H	3-hydroxy-4-oxopipecolic acid
Virginiamycin S_4	CH_3	CH_3	H	4-oxopipecolic acid
Vernamycin B_α	C_2H_5	CH_3	$N(CH_3)_2$	4-oxopipecolic acid
Vernamycin B_β	C_2H_5	CH_3	$NHCH_3$	4-oxopipecolic acid
Vernamycin B_γ	CH_3	CH_3	$N(CH_3)_2$	4-oxopipecolic acid
Vernamycin B_δ	CH_3	CH_3	$NHCH_3$	4-oxopipecolic acid
Vernamycin C	C_2H_5	CH_3	$N(CH_3)_2$	aspartic acid
Ostreogrycin B_3	C_2H_5	CH_3	$N(CH_3)_2$	3-hydroxy-4-oxopipecolic acid
Patricin A	C_2H_5	CH_3	H	proline
Patricin B	C_2H_5	CH_3	H	pipecolic acid

Virginiamycins (Vanderhaeghe and Parmentier, 1960; Vanderhaeghe et al., 1971); Vernamycin B_α (Bodanszky and Ondetti, 1964) = mikamycin B (Watanabe, 1961) = streptogramin B (Verwey et al., 1958) = pristinamycin IA (Preud'Homme et al., 1968) = ostreogrycin B (Eastwood et al., 1960); Vernamycin B_β (Bodanszky and Ondetti, 1964) = ostreogrycin B_2 (Eastwood et al., 1960) = pristinamycin IB (Preud'Homme et al., 1968); Vernamycin B_γ (Bodanszky and Ondetti, 1964) = ostreogrycin B_1 (Eastwood et al., 1960) = pristinamycin IC (Preud'Homme et al., 1968); Vernamycin B_δ (Bodanszky and Ondetti, 1964); Vernamycin C = doricin (Bodanszky and Sheehan, 1964); Ostreogrycin B (Cox et al., 1970); Patricins (Bodanszky et al., 1968).

FIGURE 6.7 Structures of virginiamycin group B-I antibiotics.

distributed in the product. Biosynthesis of the B group has been studied extensively, but details of the mechanisms remain to be clarified.

1. Biosynthesis of the A Group. Early studies with ^3H- and ^{14}C-labeled precursors showed that acetate, methionine, serine, proline, glycine, and glucose were possible precursors of virginiamycin M_1 (Roberfroid and Dumont 1967). These results were confirmed by Kingston and Kolpak (1980) using

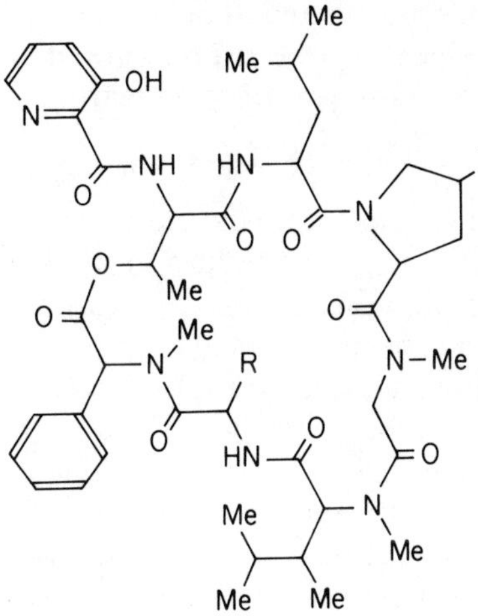

Antibiotic	R	X
Neoviridogrisein I	C_2H_5	H
Neoviridogrisein II	CH_3	H
Neoviridogrisein III	C_2H_5	OH
Viridogrisein	CH_3	OH
Neoviridogrisein-MP	CH_3	CH_3

Neoviridogriseins I and III (Okumura et al., 1979b)
Neoviridogrisein II (Okumura et al., 1979a) = etamycin B (Chopra et al., 1979)
Viridogrisein (Haskell et al., 1955) = etamycin (Sheehan et al., 1958) = 6613 (Kruglyak et al., 1961) = neoviridogrisein IV (Okumura et al., 1979c)
Neoviridogrisein-MP (Okumura et al., 1982)

FIGURE 6.8 Structures of virginiamycin group B-II antibiotics.

[2-^{13}C]acetate, [1,2-^{13}C]acetate, L-[methyl-^{13}C]methionine, and DL-[3-^{13}C]-serine. The ^{13}C-labeled antibiotic samples were analyzed by ^{13}C-NMR spectroscopy. From the enhancement of signals assigned to each carbon, the biosynthetic origin of the carbon skeleton was elucidated. The dehydroproline moiety is derived from proline, the oxazole ring from serine, and most carbon atoms of the macrocyclic ring from acetate.

The structure of griseoviridin, which is produced by *Streptomyces griseoviridus* together with viridogrisein, was proposed by Ames and Bowman (1956) and by Fallona et al. (1964). In 1976, the structure was corrected after X-ray analysis. It contains four recognizable units: D-cysteine, serine, an unsaturated ω-aminodecanoic acid, and a δ-hydroxyhexanoic acid (Birnbaum and Hall, 1976). The ^{13}C-NMR spectrum of griseoviridin has been obtained, and signals have been assigned to each of the carbon atoms of the antibiotic (Bycroft, 1977). Consequently, knowledge of the biosynthesis awaits only the preparation of ^{13}C-labeled griseoviridin from appropriate precursors. Bycroft has postulated that the group A virginiamycins are initially formed as depsipeptides from a mixture of amino acid and hydroxy acid precursors and are subsequently modified.

2. Biosynthesis of the B Group. In most cultures producing virginiamycins, the B group antibiotic is a minor product in the complex. Etamycin is an exception, and perhaps for this reason studies on the biosynthesis of the B group components have been restricted to this substance. The biosynthetic precursors of etamycin were established, by incorporation experiments with radioactive compounds, to be as follows: 3-hydroxypicolinic acid is derived from L-lysine, L-threonine from L-threonine, D-leucine from L-leucine, 4-*allo*-hydroxy-D-proline from L-proline, sarcosine from glycine and L-methionine, β,N-dimethyl-L-leucine from L-leucine and L-methionine, L-alanine from L-alanine, and L-phenylsarcosine from L-phenylalanine and L-methionine. 3-Hydroxypicolinic acid, sarcosine, and 4-hydroxy-L-proline are also directly incorporated into appropriate positions in the antibiotic. Since labeled 4-hydroxy-L-proline was incorporated more specifically and with less dilution than labeled proline into the 4-*allo*-hydroxy-D-proline moiety of etamycin, hydroxylation was postulated to be the initial step, followed by inversion of configuration at position 2 (Hook and Vining, 1973).

Using L-[^{14}C-(U)]leucine and L-[^{14}C-methyl]methionine in short-term incubations, Kamal and Katz (1976) showed that L-leucine, and not the D-enantiomer, is the direct precursor of D-leucine in the antibiotic. Since chloramphenicol did not inhibit etamycin synthesis, they concluded that the biosynthesis of etamycin is directed by a nonribosomal mechanism, like that used for other peptide antibiotics. Katz et al. (1979) detected free *trans*-L-[^{14}C-(U)]hydroxyproline in the intracellular pool and medium of *S. griseoviridus* cultures incubated for short periods with L-[^{14}C-(U)]proline under conditions where etamycin was synthesized. They also observed that unlabeled *trans*-4-hydroxy-L-proline specifically inhibited incorporation of labeled L-proline and concluded that free L-proline was hydroxylated to *trans*-4-hydroxy-L-proline. This would be converted to 4-*allo*-hydroxy-D-proline during incorporation into etamycin.

The 3-hydroxypicolinic acid in etamycin was specifically labeled by L-[^{14}C-(U)]lysine and not, as in pyridomycin (see later), by DL-[1-^{14}C]-aspartic acid and [1,3-^{14}C]glycerol (Hook and Vining, 1973). For the biosynthesis of β,N-dimethyl-L-leucine in etamycin, L-[^{14}C-(U)]leucine supplied the carbon skeleton, and L-[^{14}C-methyl]methionine provided both the N-methyl and β-methyl groups. The mechanism proposed for its biosynthesis, based on the rule of α-epimerization (Walker et al., 1970; Walker and Perlman, 1971; Bodanszky and Perlman, 1968), is shown in Fig. 6.9.

3. Controlled Biosynthesis. Since the biosynthesis of virginiamycin B group antibiotics is directed by a nonribosomal mechanism (Kamal and Katz,

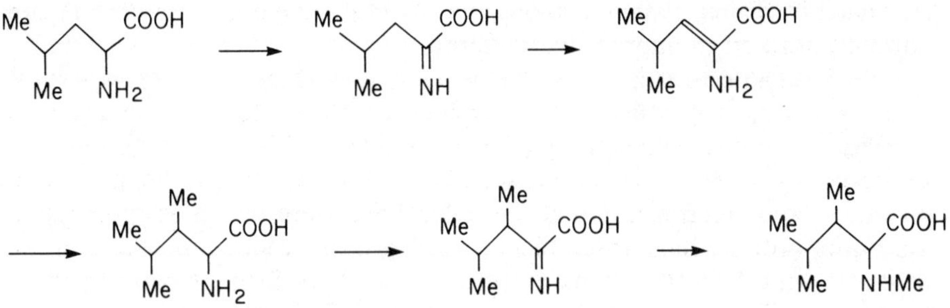

FIGURE 6.9 Possible mechanism for formation of L-β,N-dimethyl-L-leucine from L-leucine. From Walker and Perlman (1971). Reproduced with permission of John Wiley & Sons, Inc., New York.

1976), congeners should be produced by supplementing the medium with amino acids. Addition of certain amino acids to *S. griseoviridus* strain P8648, which produces viridogrisein (= etamycin) and griseoviridin, yielded such compounds. With L-α-amino-*n*-butyric acid, neoviridogriseins I and III were obtained. In these, the L-alanine moiety is replaced with L-α-amino-*n*-butyric acid (Fig. 6.8). The D-enantiomer inhibited both growth and antibiotic production and completely abolished the synthesis of neoviridogriseins I and III (Okumura et al., 1979b). Direct incorporation of the L-enantiomer was confirmed with the labeled amino acid (Y. Okumura and Y. Fukagawa, unpublished data). Replacement of alanine by α-amino-*n*-butyric acid seems to be common in virginiamycins of group B-I. Even without supplementation, *Streptomyces virginiae*, *Streptomyces ostreogriseus*, and *Streptomyces loidensis* synthesize antibiotics that contain α-amino-*n*-butyric acid instead of alanine. The reasons for this are thought to be that: (1) group B-I producers have larger intracellular pools of α-amino-*n*-butyric acid than group B-II producers, and (2) in group B-I producers, the enzyme responsible has higher affinity for α-amino-*n*-butyric acid than for alanine.

Streptomyces griseoviridus produced neoviridogrisein II when supplemented with proline. In this product, the 4-*allo*-hydroxy-D-proline of viridogrisein is replaced with D-proline. Addition of D-proline to the culture inhibited production not only of neoviridogrisein II but also of viridogrisein (Okumura et al., 1979a). [^{14}C-(U)]Proline was incorporated into both neoviridogrisein II and viridogrisein (Y. Okumura, unpublished data), indicating that D-proline in the antibiotic is derived from the L-enantiomer and not the D-enantiomer. This may seem inconsistent with earlier studies (Hook and Vining, 1973; Katz et al., 1979), which indicated that hydroxylation precedes epimerization. However, it can be explained as follows: The formation of 4-hydroxy-L-proline by the hydroxylation system is a rate-limiting step in antibiotic synthesis. In consequence, L-proline added in excess to the medium is accepted by the peptide synthetase and epimerized while being incorporated into the antibiotic. Presumably, the affinity for 4-hydroxy-

L-proline is higher than for L-proline, so that neoviridogrisein II is not normally formed in appreciable quantities.

Neoviridogrisein II is 1.7–3.5 times more active than viridogrisein against some bacteria, suggesting a strong influence of the site of this imino acid moiety on antimicrobial activity (Okumura et al., 1979a). When L-azetidine-2-carboxylic acid, L-thiazolidine-4-carboxylic acid, L-pipecolic acid, and positional and stereoisomers of methylprolines were tested as precursors, the methylprolines had most effect on viridogrisein production. The addition of 4-methylproline (D-*cis* and L-*trans* mixture) to *S. griseoviridus* strain P8648 gave a new product named neoviridogrisein-MP. This had *cis*-4-methyl-D-proline at the expected site. Since the organism uses only the L form of an amino acid, the L-*trans* form is presumably incorporated with inversion of configuration at C-2. The inversion may be catalyzed by an enzyme like hydroxyproline-2-epimerase (Okumura et al., 1982).

Another strain of *S. griseoviridus,* PD04955, produces neoviridogriseins as normal antibiotic components. Along with etamycin (= viridogrisein) were compounds containing proline instead of 4-*allo*-hydroxy-D-proline and others containing glycine. The proportion of each component in the product depended on the medium, especially on the nitrogen source (Chopra et al., 1979).

4. Regulation of Virginiamycin Biosynthesis. Yanagimoto and Terui (1971) reported that an inducer for virginiamycin synthesis was formed just before initiation of antibiotic production by *S. virginiae* and that it was not one of the precursors of the antibiotic. The purified substance induced antibiotic synthesis at a concentration of 0.02 μg/ml. It was a γ- or δ-lactone containing a hydroxybutyl residue and two methyl groups. γ-Undecalactone and γ-nonalactone also exhibited the inducing activity (Yanagimoto et al., 1979). The substance has structural features similar to those of A-factor (Khokhlov, 1980), which induces the production of streptomycin (Khokhlov and Tovarova, 1972).

D. Biological Activities

The antimicrobial activity and mode of action of virginiamycins have been the subject of excellent reviews (Tanaka, 1975; Vazquez, 1975; Cocito, 1979). The newly isolated neoviridogriseins are generally active against gram-positive bacteria only. Neoviridogrisein II is the most active compound, and the synergistic relationship between it and the virginiamycin group A antibiotics has been demonstrated by Okumura et al. (1979d).

IV. QUINOXALINE ANTIBIOTICS

Quinoxaline antibiotics contain two quinoxaline moieties in their structure. There are two groups, the quinomycins and the triostins, which differ in the sulfur bridge system. These antibiotics are highly active against gram-positive bacteria and certain cancers, but they are very toxic in animals. Consequently, none are used clinically.

A. Producing Organisms and Structures

Quinoxaline antibiotic producers and the structures of the antibiotics are listed in Table 6.5 and Fig. 6.10, respectively. Besides the compounds listed in Fig. 6.10, there have been some analogs obtained by controlled biosynthesis.

B. Biochemistry

Radioisotope experiments demonstrated that the constituents of triostin are derived from L-serine, L-valine, L-cysteine, and L-alanine. The methyl group of L-methionine supplies the N-methyl groups, and tryptophan is an efficient precursor of the quinoxaline moieties. The incorporation of ^{14}C-tryptophan into triostin was decreased by unlabeled kynurenine, kynuramine, and quinoxaline-2-carboxylic acid; the quinoxaline-2-carboxylic acid supplement increased synthesis of the antibiotic. These results indicate that tryptophan was metabolized to quinoxaline-2-carboxylic acid through kynurenine and kynuramine and then incorporated into the antibiotic (Yoshida et al., 1970). Since chloramphenicol blocked protein synthesis while enhancing incorporation of amino acid precursors into the antibiotic, the biosynthesis of triostin is undoubtedly directed by a nonribosomal template (Yoshida and Katagiri, 1969).

Adding isoleucine enantiomers to *Streptomyces* sp. 732 selectively enhanced quinomycin B formation and completely inhibited quinomycin C

TABLE 6.5 Organisms Producing Quinoxaline Antibiotics

Microorganism	*Product*	*Reference*
Streptomyces sp. 65-24	Quinomycin A	Carter et al. (1954)
S. aureus Q-11A	Quinomycin A	Ueda et al. (1954)
S. abikoensis	Quinomycin A	Ueda and Umezawa (1956)
S. echinatus	Quinomycin A	Corbaz et al. (1957a)
S. echinatus x-948	Quinomycin A	Berger et al. (1957)
S. aureus 723	Quinomycins A, B_0, C	Kuroya et al. (1961)
S. flavochromogenes	Quinomycin C	Ivanitskaya et al. (1961)
S. griseolus 3950	Quinomycins A, C	Ivanitskaya et al. (1961)
S. lavendulae	Quinomycin A	Maksimova et al. (1965)
S. griseoplanus	Quinomycin A	Maksimova et al. (1965)
S. fumosus	Quinomycin C	Maksimova et al. (1965)
S. mellinus nov. sp.	Quinomycin A	Maksimova et al. (1965)
S. aureus S-2-210	Triostins A, B_0, C	Kuroya et al. (1961)

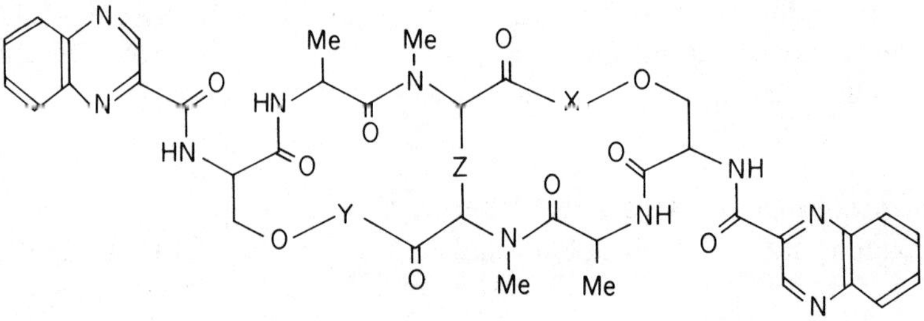

Antibiotic		X	Y	Z
Quinomycin	A	MeVal	MeVal	thioacetal
	B_0	MeVal	Mealleu	thioacetal
	C	DiMealleu	DiMealleu	thioacetal
	D	MeVal	Mealleu	thioacetal
	E	Mealleu	DiMealleu	thioacetal
Triostin	A	MeVal	MeVal	disulfide
	B_0	MeVal	DiMealleu	disulfide
	B	Mealleu	DiMealleu	disulfide
	C	DiMealleu	DiMealleu	disulfide

Abbreviations: MeVal = N-methylvaline, Mealleu = N-methyl-*allo*-isoleucine, DiMealleu = N, γ-dimethyl-*allo*-isoleucine
Sources: Otsuka and Shoji (1967), Martin et al. (1975)

FIGURE 6.10 Structures of quinoxaline antibiotics.

production; the order of effectiveness for quinomycin B synthesis was DL-isoleucine > D-isoleucine > L-isoleucine (Yoshida and Katagiri, 1967). Quinomycin B contains two N-methyl-L-*allo*-isoleucine residues, and their biosynthesis is assumed to follow the route deduced for this amino acid in actinomycin (Katz et al., 1961). In the controlled biosynthesis of quinoxaline antibiotics, replacement of the chromophore is a characteristic feature. Fig. 6.11 lists the products in which one or both chromophores are replaced with other residues. The biological activity of these compounds has been examined, but none had an improved chemotherapeutic index.

V. OTHER PEPTIDOLACTONE ANTIBIOTICS

A. From the Bacilli

The genus *Bacillus* produces many antibiotics, which are mainly small polypeptides. They can be divided into three major groups: linear peptides, cyclic peptides, and peptidolactones. Table 6.6 lists the producers of peptidolactones. It should be noted that antibiotics from the bacilli, unlike those from the actinomycetes, do not contain N-methylamino acids.

The biosynthesis of these antibiotics has not been examined.

Name	X	Y	Bridge	Reference
Quinomycin A	I	I	thioacetal	Otsuka and Shoji (1967)
Triostin A	I	I	disulfide	Kuroya et al. (1961)
NX-quinomycin A	I	III	thioacetal	Yoshida et al. (1968)
QN-quinomycin A	III	III	thioacetal	Yoshida et al. (1968)
Quinazomycin	I	V	thioacetal	Khan et al. (1969)
Biquinazomycin	V	V	thioacetal	Arif et al. (1970)
NX-triostin	I	III	disulfide	Yoshida et al. (1970)
QN-triostin	III	III	disulfide	Yoshida et al. (1970)
2QCl	II	II	thioacetal	Bojesen et al. (1981)
2TP	VI	VI	thioacetal	Bojesen et al. (1981)
2QnM	IV	IV	thioacetal	Bojesen et al. (1981)
2QnMb	IV	IV	disulfide	Bojesen et al. (1981)

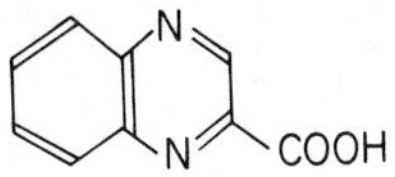
I

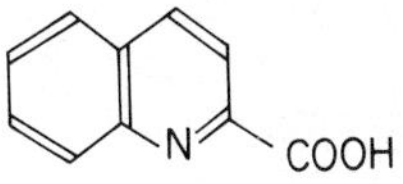
III

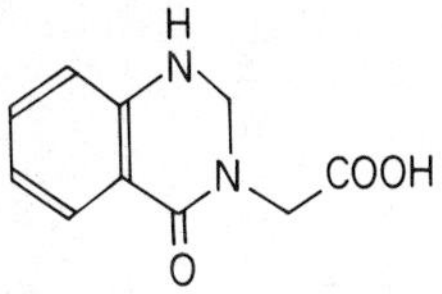
V

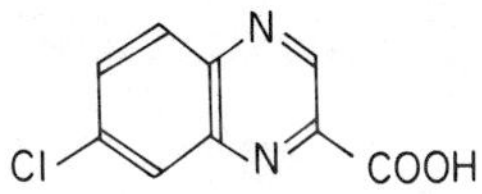
II

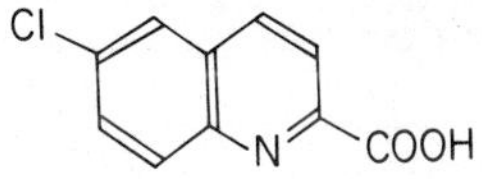
IV

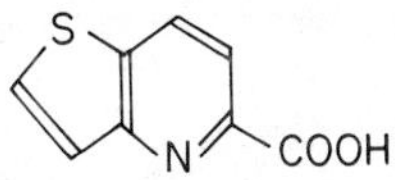
VI

FIGURE 6.11 Structures of quinoxaline antibiotics obtained by replacement of the chromophore(s). The generalized structure is:

X-PEPTIDOLACTONE-Y.

B. From the Actinomycetes

Some peptidolactone antibiotics from the actinomycetes do not fall into the groups described in previous sections. Examples are shown in Fig. 6.12, and the producers are listed in Table 6.7.

From studies on the biosynthesis of pyridomycin with radiolabeled

TABLE 6.6 Peptidolactone antibiotics from the genus *Bacillus*

Organism	*Antibiotic*	*Reference*
B. brevis	Brevistin	Shoji and Kato (1976)
B. circulans	Permetin	Takeuchi et al. (1979)
B. circulans	Polypeptin	Sogan (1976)
B. mesentericus	Esperin	Thomas and Ito (1969)
B. subtilis	Surfactin	Kakinuma et al. (1969)
B. subtilis	TL-119	Shoji et al. (1975)
B. subtilis	A-3302	Ogawa et al. (1976)

Pyridomycin

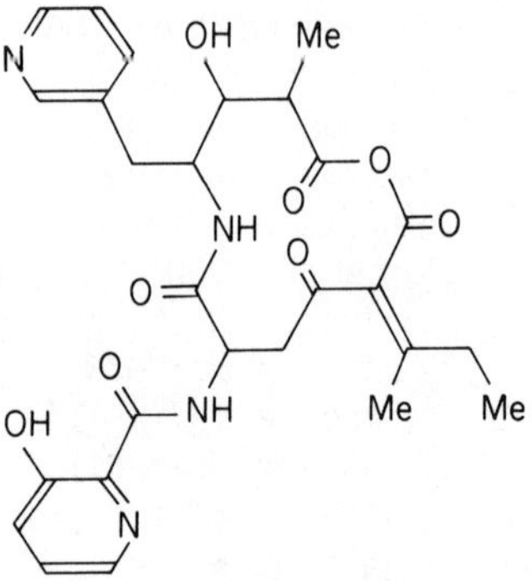

Monamycin

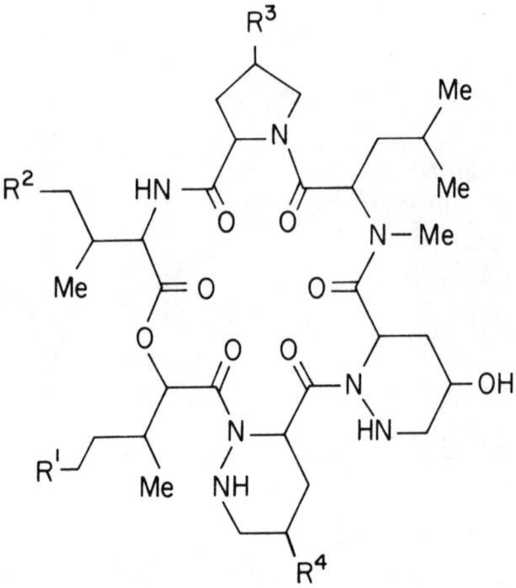

Monamycin	A	B_1	B_2	B_3	C	D_1	D_2	E	F	G_1	G_2	G_3	H_1	H_2	I
R_1	H	H	H	Me	Me	Me	H	Me	Me	H	H	Me	Me	H	Me
R_2	H	H	Me	H	H	H	Me	Me	Me	H	Me	H	H	Me	Me
R_3	Me	Me	H	H	Me	Me	Me	Me	Me	Me	H	H	Me	Me	Me
R_4	H	H	H	H	H	H	H	H	H	Cl	Cl	Cl	Cl	Cl	Cl

Globomycin (SF-1902)

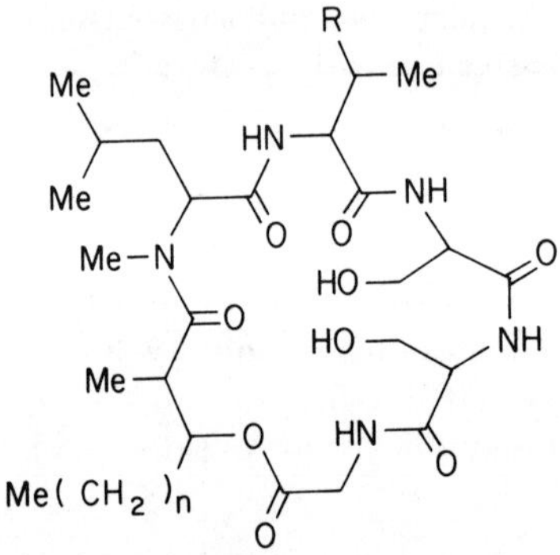

Globomycin	SF-1902A_1	SF-1902A_2	SF-1902A_3	SF-1902A_{4a}	SF-1902A_{4b}	SF-1902A_5
R	Et	Et	Me	Et	Me	Et
n	5	3	5	6	7	7

FIGURE 6.12 Structures of peptidolactone antibiotics from the actinomycetes.

TABLE 6.7 Peptidolactone antibiotics from the actinomycetes

Organism	*Antibiotic*	*Reference*
Actinoplanes awajinensis	Mycoplanecin	Arai et al. (1979)
Streptoverticillium cinnamonensis	Globomycin	Inukai et al. (1978)
Streptomyces halstedii	Globomycin	Inukai et al. (1978)
S. neohygroscopicus	Globomycin	Inukai et al. (1978)
S. hagronensis	Globomycin	Inukai et al. (1978)
S. hygroscopicus	SF-1902(F_1-F_5)	Omoto et al. (1981)
S. violaceochromogenes	Lipopeptin	Tsuda et al. (1980)
S. jamaicensis	Monamycin	Hassall and Magnus (1959)
S. pyridomyceticus	Pyridomycin	Okami et al. (1954)
Actinomadura luzonensis	BBM-928	Ohkuma et al. (1980)

compounds, Ogawara et al. (1968) concluded that the 3-hydroxypicolinic acid residue was derived from L-aspartic acid and glycerol or pyruvate. The L-threonine moiety came from L-threonine, the α-keto-β-methylvaleric acid residue from L-isoleucine, the 4-amino-3-hydroxy-2-methyl-5-(3-pyridyl)-pentanoic acid component from L-aspartic acid and glycerol (pyridine ring), propionic acid (terminal C-3 part), and pyruvate (C-4 position). 3-Hydroxy-picolinic acid was also directly incorporated into the antibiotic.

Monamycins are produced as a complex mixture by *S. jamaicensis* (Hassall and Magnus, 1959). The yield was reduced with increasing concentrations of inorganic phosphate in the fermentation medium. This inhibition was paralleled by an increase in the amount of mycelium produced (Hall and Hassall, 1970). When sodium bromide was added to the medium and chloride ions were omitted, the chlorine in the six congeners was replaced with bromine (Hall et al., 1973). The bromo compounds had the least antimicrobial activity of any monamycin component tested. Labeled amino acids were used to show that DL-glutamic acid was the precursor of piperazic acid, L-leucine was the precursor of N-methylleucine and methylproline, and L-methionine furnished the N-methyl groups (Davies et al., 1973; Arroyo et al., 1976). It is of interest that N-methylleucine was not a precursor of the N-methylleucine in the antibiotic and that 4-methylproline seemed to be synthesized from L-leucine.

Globomycin, a gram-negative specific antibiotic with spheroplast-forming activity, is produced by four different actinomycetes (Inukai et al., 1978). *Streptomyces hygroscopicus* SF-1902 produced five antibiotics, one of which was globomycin, while the others were globomycin analogs (Omoto et al., 1981). These differed from each other in the acyl and branched-chain amino acid moieties. The antimicrobial activity was sensitive to the length of the side chains, those with the longest side chains being the most active. Controlled biosynthesis should provide a means of improving the yield of the most valuable components. No information is yet available on the biochemistry of lipopeptin, mycoplanecin, and BBM-928.

VI. CONCLUDING REMARKS

Many of the studies on the biosynthesis of peptidolactone antibiotics and its regulation have been undertaken with the aim of obtaining more useful therapeutic agents or of increasing production of antibiotics in practical use. Also of interest has been the relationship between the biosynthetic mechanisms for peptidolactone synthesis and those established for the peptide antibiotics.

The physiology and the genetics of the actinomycetes, which produce most peptidolactone antibiotics, are not yet fully understood, and cell-free synthesizing systems appear to be very difficult to prepare. When further advances have been made in these areas, there will be considerable application in control of production, design of medium and fermentation conditions, and improvement of producing strains. Random selections of strains and suitable fermentation conditions require much work and time. Moreover, since random screening depends on chance, the results have no general application.

Much more work is needed before the pathway of biosynthesis and genetic regulation mechanisms are well enough understood, but recent development of genetic techniques such as protoplast fusion, transformation, and recombination should facilitate the important tasks of yield improvement and synthesis of new products with superior biological properties. In this review, I have described the existing state of our knowledge of the biosynthesis of peptidolactone antibiotics and its regulation.

ACKNOWLEDGEMENTS

I wish to thank Drs. T. Yajima (Toho University, Maebashi), E. Katz (Georgetown University, Washington, D.C.), and Y. Fukagawa and T. Ishikura (in our laboratories, Fujisawa), who kindly contributed useful information.

I am deeply indebted to Academic Press, Inc. (New York), Elsevier Scientific Publishing Co. (Amsterdam), Japan Antibiotics Research Association (Tokyo), and John Wiley & Sons, Inc. (New York) for their permission to use published materials.

REFERENCES

Albertini, A., Cassani, G., and Cifferi, O. (1964) *Biochim. Biophys. Acta 80,* 655–664.

Ames, D. E., and Bowman, R. E. (1956) *J. Chem. Soc.* 2925–2928.

Arai, M., Karasawa, S., Yonehara, H., and Umezawa, H. (1958) *J. Antibiot. 11,* 14–20.

Arai, M., Torikata, A., Enokita, R., Imai, H., and Okazaki, H. (1979) Japanese Patent 79-160,302

Arif, A. J., Singh, C., Bhaduri, A. P., Gupta, C. M., Khan, A. W., and Dhar, M. M. (1970) *Indian J. Biochem. 7,* 193.

Arroyo, V., Hall, M. J., Hassall, C. H., and Yamasaki, K. (1976) *J. Chem. Soc. Chem. Comm.* 845–846.

Ball, S., Boothroyd, B., Lees, K. A., Raper, A. H., and Smith, E. L. (1958) *Biochem. J. 68,* 24P.

Bartz, Q. R., Standiford, J., Mold, J. D., Johannessen, D. W., Ryder, A., Maretzki, A., and Haskell, T. H. (1955) *Antibiot. Annual 1954/55,* 777–783.

Benazet, F., Cosar, C., Dubost, N., Julou, L., and Mancy, D. (1962) *Semaine Therapeutique 38,* 13.

Berger, J., La Sala, E. R., Scott, W. E., Meltsner, B. R., Sternbach, L. H., Kaiser, S., Teitel, S., Mach, E., and Goldberg, M. W. (1957) *Experientia 13,* 434–436.

Birnbaum, G. I., and Hall, S. R. (1976) *J. Amer. Chem. Soc. 98,* 1926–1931.
Bodanszky, M., and Ondetti, M. A. (1964) *Antimicrob. Agents Chemother. 1963,* 360–365.
Bodanszky, M., and Perlman, D. (1968) *Nature 219,* 291–292.
Bodanszky, M., and Sheehan, J. T. (1964) *Antimicrob. Agents Chemother. 1963,* 38–40.
Bodanszky, M., Ondetti, M. A., and Sheehan, J. T. (1968) U. S. Patent 3,373,151.
Bojesen, G., Gauvrea, D., Williams, D. H., and Waring, M. J. (1981) *J. Chem. Soc. Chem. Comm.* 46–47.
Bossi, R., Hutter, R., Keller-Schierlein, W., Neipp, L., and Zahner, H. (1958) *Helv. Chim. Acta 41,* 1645–1652.
Brazhnikova, M. G., Kudinova, M. K., Potapova, N. P., Filippova, T. M., Borowski, E., Zielinski, J., and Golic, J. (1975) *Bioorg. Khim. 1,* 1383–1384.
Brockmann, H., and Grubhofer, N. (1949) *Naturwiss. 36,* 376–377.
Brockmann, H., and Stahler, E. A. (1965) *Naturwiss. 52,* 391–392.
Brown, D. D., Hitchcock, M. J. M., and Katz, E. (1980) *Arch. Biochem. Biophys. 202,* 18–22.
Bycroft, B. W. (1977) *J. Chem. Soc. Perkin Trans.* 2464–2470.
Carter, H. E., Schaffner, C. P., and Gottlieb, D. (1954) *Arch. Biochem. Biophys. 53,* 282–293.
Celmer, W. D., and Sobin, B. A. (1956) *Antibiot. Annual 1955/56,* 437–441.
Celmer, W. D., Cullen, W. P., Routien, J. B., Moppet, C. E., Shibakawa, R., and Tone, J. (1975) German Patent 2,516,020
Chamberlin, J. W., and Chen, S. (1977) *J. Antibiot. 30,* 197–201.
Charney, J., Fischer, W. P., Curran, C., Machlowitz, R. A., and Tytell, A. A. (1953) *Antibiot. Chemother. 3,* 1283–1286.
Chopra, C., Hook, D. J., and Vining, L. C. (1979) *J. Antibiot. 32,* 392–401.
Ciferri, O., Albertini, A., and Rossi, P. (1962) *Biochem. Biophys. Res. Comm. 7,* 264–267.
Ciferri, O., Albertini, A., and Cassani, G. (1965) *Biochem. J. 96,* 853–861.
Cocito, C. (1979) *Microbiol. Rev. 43,* 145–198.
Corbaz, R., Ettlinger, L., Gäumann, E., Keller-Schierlein, W., Kradolfer, F., Neipp, L., Prelog, V., Rëusser, P., and Zähner, H. (1957a) *Helv. Chim. Acta 40,* 199–204.
Corbaz, R., Ettlinger, L., Keller-Schierlein, W., and Zahner, H. (1957b) *Arch. Mikrobiol. 26,* 192.
Cox, B. R., Eastwood, F. W., Snell, B. K., and Todd, A. R. (1970) *J. Chem. Soc. Chem. Comm.* 1623–1624.
D'Angio, G. J. (1972) *Cancer 30,* 1528.
Davies, J. S., Foley, M. H., Hassall, C. H., and Arroyo, V. (1973) *J. Chem. Soc. Chem. Comm.* 782–783.
Delpierre, G. R., Eastwood, F. W., Gream, G. E., Kingston, D. G. I., Sarin, P. S., Todd, A. R., and Williams, D. H. (1966) *J. Chem. Soc.* C 1653–1669.
Diegelmann, R. F., Ondrejikova, O., and Katz, E. (1969) *Arch. Biochem. Biophys. 131,* 276–287.
Donovick, R., Dutcher, J. D., Heuser, L. J., and Pagano, J. F. (1961) U. S. Patent 2,990,325.
Durant, F., Evrard, G., Declercq, J. P., and German, G. (1974) *Cryst. Struct. Comm. 3,* 503–510.

Eastwood, F. W., Snell, B. K., and Todd, A. L. (1960) *J. Chem. Soc.* 2286–2292.
Fallona, M. C., Mayo, P. D., McMorris, T. C., Money, T., and Stossel, A. (1964) *Can. J. Chem. 42,* 371–393.
Farber, S. (1966) *J. Amer. Med. Assoc. 198,* 826.
Fernbach, D. J., and Martyn, D. T. (1966) *J. Amer. Med. Assoc. 195,* 1005.
Formica, J. V., and Apple, M. A. (1976) *Antimicrob. Agents Chemother. 9,* 214–221.
Formica, J. V., and Katz, E. (1973) *J. Biol. Chem. 248,* 2066–2071.
Fujikawa, K., Sakamoto, Y., Suzuki, T., and Kurahashi, K. (1968) *Biochim. Biophys. Acta 169,* 520–533.
Fukagawa, Y., Mason, K. T., and Katz, E. (1974) *Biochim. Biophys. Acta 338,* 198–212.
Gallo, M., and Katz, E. (1972) *J. Bacteriol. 109,* 659–667.
Golub, E. E., Ward, M. A., and Nishimura, J. S. (1969) *J. Bacteriol. 100,* 977–984.
Goss, W. A., and Katz, E. (1960) *Antibiot. Chemother. 10,* 221–229.
Hackmann, C. (1952) *Z. Krebsforsh. 58,* 607.
Hall, M. J., and Hassall, C. H. (1970) *Appl. Microbiol. 19,* 109–112.
Hall, M. J., Handford, B. O., Hassall, C. H., Phillips, D. A. S., and Rees, A. V. (1973) *Antimicrob. Agents Chemother. 3,* 380–383.
Hamill, R. L., and Stark, W. M. (1975) U. S. Patent 3,932,980.
Haskell, T. H., Maretzki, A., and Bartz, Q. R. (1955) *Antibiot. Annual 1954/1955,* 784–789.
Hassall, C. H., and Magnus, K. E. (1959) *Nature 184,* 1223–1224.
Heinemann, B., Gourevitch, A., Lein, J., Johnson, D. L., Kaplan, M. A., Vans, D., and Hooper, I. R. (1955) *Antibiot. Annual 1954/55,* 728–732.
Hitchcock, M. J., and Katz, E. (1978) *Antimicrob. Agents Chemother. 13,* 104–114.
Hook, D. J., and Vining, L. C. (1973) *Can. J. Biochem. 51,* 1630–1637.
Inukai, M., Enokita, R., Torikata, A., Nakahara, M., Iwado, S., and Arai, M. (1978) *J. Antibiot. 31,* 410–420.
Ivanitskaya, L. P., Kruglyak, E. B., Maksimova, T. S., and Preobrazhenskaya, T. P. (1961) *Antibiotiki 6,* 393.
Johnson, A. W., and Mauger, A. B. (1959) *Biochem. J. 73,* 535–538.
Kakinuma, A., Ouchida, A., Shima, T., Sugino, H., Isono, M., Tamura, G., and Arima, K. (1969) *Agr. Biol. Chem. 33,* 1669–1671.
Kamal, F., and Katz, E. (1976) *J. Antibiot. 29,* 944–949.
Katz, E. (1960) *J. Biol. Chem. 235,* 1090–1094.
Katz, E. (1967) in *Antibiotics* Vol. II (Gottlieb D., and Shaw, P. D., eds) pp. 276–341. Springer, Berlin.
Katz, E., and Goss, W. A. (1958) *Nature 182,* 1668–1669.
Katz, E., and Goss, W. A. (1959) *Biochem. J. 73,* 458–465.
Katz, E., and Weissbach, H. (1962a) *Biochem. Biophys. Res. Comm. 8,* 186–190.
Katz, E., and Weissbach, H. (1962b) *J. Biol. Chem. 237,* 882–886.
Katz, E., and Weissbach, H. (1963) *J. Biol. Chem. 238,* 666–675.
Katz, E., Waldron, C. R., and Meloni, M. L. (1961) *J. Bacteriol. 82,* 600–608.
Katz, E., Prockop, D. J., and Udenfriend, S. (1962) *J. Biol. Chem. 237,* 1585–1588.
Katz, E., Wise, M., and Weissbach, H. (1965) *J. Biol. Chem. 240,* 3071–3078.
Katz, E., Kamal, F., and Mason, K. T. (1979) *J. Biol. Chem. 254,* 6684–6690.

Keller, U., and Kleinkauf, H. (1977) *Arch. Biochem. Biophys. 184,* 111–124.
Khan, A. W., Bhaduri, A. P., Gupta, C. M., and Dhar, M. M. (1969) *Indian J. Biochem. 6,* 220.
Khokhlov, A. S. (1980) in *Frontiers of Bioorganic Chemistry and Molecular Biology* (Ananchenko, S. N., ed.) pp. 201–210, Pergamon Press, New York.
Khokhlov, A. S., and Tovarova, I. I. (1972) *Postepy Hig. Med. Dosw. 26,* 469–491.
Kingston, D. G. I., and Kolpak, M. X. (1980) *J. Amer. Chem. Soc. 102,* 5966–5967.
Kingston, D. G. I., Sarin, P. S., Todd, L., and Williams, D. H. (1966) *J. Chem. Soc.* C 1856–1860.
Kleinkauf, H., Koishwitz, H., Vater, J., Zocher, R., Keller, U., Mahmutoglu, I., Bauer, K., Altmann, M. M., Kittelberger, R., Marahiel, M., and Salnikow, J. (1978) in *Federation of European Biochemical Societies, Twelfth Meeting 1978* Vol. 55, *Regulation of Secondary Product and Plant Hormone Metabolism* (Luckner, M., and Schreiber, K., eds.) pp. 37–47, Pergamon Press, New York.
Kruglyak, E. B., Konstantinova, N. V., and Sulavko, L. A. (1961) *Antibiotiki 6,* 298–302.
Kurahashi, K. (1974) *Ann. Rev. Biochem. 43,* 445–459.
Kuroya, M., Ishida, N., Katagiri, K., Shoji, J., Yoshida, T., Mayama, M., Sato, K., Matsuura, S., Nimomi, Y., and Shiratori, O. (1961) *J. Antibiot. Ser. A, 14,* 324–329.
Laland, S. G., and Zimmer, T. L. (1973) In *Essays in Biochemistry* Vol. IX (Campbell, P. N., and Dickens, F., eds.) pp. 31–57, Academic Press, New York.
Lewis, J. L., Jr. (1972) *Cancer 30,* 1517.
Li, M. C. (1971) *Ann. Intern. Med. 74,* 102.
Linder, F., Wallhausser, K. H., and Weidenmuller, H. L. (1960) German Patent 1,072,773.
Lipmann, F., Gevers, W., Kleinkauf, H., and Roskosky, R., Jr. (1971) *Adv. Enzymol. 35,* 1–34.
Liu, W.-C., Seiner, V., Dean, L. D., Trejo, W. H., Principe, P. A., Meyers, E., and Sykes, R. B. (1981) *J. Antibiot. 34,* 1515–1516.
Maksimova, T. S., Kovsharova, I. N., and Proshlgakova, V. V. (1965) *Antibiotiki 10,* 298.
Manaker, R. A., Gregory, F. J., Vining, L. C., and Waksman, S. A. (1955) *Antibiot. Annual 1954/55,* 853–857.
Marshall, R., Redfield, B., Katz, E., and Weissbach, H. (1968) *Arch. Biochem. Biophys. 123,* 317–323.
Martin, D. G., Mizsak, S. A., Biles, C., Stewart, J. C., Baczynsky, J. L., and Meulman, P. A. (1975) *J. Antibiot. 28,* 332–336.
Martinelli, E., Zerilli, F., Volpe, G., Pagani, H., and Cavalleri, B. (1979) *J. Antibiot. 32,* 108–114.
Mason, K. T., Shaw, G. J., and Katz, E. (1977) *Arch. Biochem. Biophys. 180,* 509–513.
Mauger, A. B., and Katz, E. (1978) in *Antibiotics: Isolation, Separation and Purification* (Weinstein, M. J., and Wagman, G. H., eds.) pp. 1–38, Elsevier Scientific Publishing Co., Amsterdam.
Meienhofer, J., and Atherton, E. (1973) *Adv. Appl. Microbiol. 16,* 203–300.
Moppet, C. E., and Whipple, E. B. (1977) Abstracts of the Seventeenth Interscientific Conference on Antimicrobial Agents and Chemotherapy, p. 243.

Nishimura, J. S., and Bowers, W. F. (1967) *Biochem. Biophys. Res. Comm. 28,* 665–670.

Ochi, K., and Katz, E. (1978) *J. Antibiot. 31,* 1143–1148.

Ogata, K., Matsuura, M., Irie, H., Ueno, T., Tani, Y., and Yamada, H. (1978) *J. Antibiot. 31,* 1313–1315.

Ogawa, Y., Mori, H., Ichihashi, M., Ueno, T., Nakashima, T., Fukami, H., Nakajima, R., and Ida, H. (1976) Abstracts of the Fourteenth Professional Symposium on Peptide Chemistry, pp. 123–126.

Ogawara, H., Maeda, K., and Umezawa, H. (1968) *Biochemistry 7,* 3296–3302.

Ohkuma, H., Sakai, F., Nishiyama, Y., Ohbayashi, M., Imanishi, H., Konishi, M., Miyaki, T., Koshiyama, H., and Kawaguchi, H. (1980) *J. Antibiot. 33,* 1087–1097.

Okabe, K. (1959) *J. Antibiot. Ser. A, 12,* 86–89.

Okami, Y., Maeada, K., and Umezawa, H. (1954) *J. Antibiot. Ser. A, 7,* 55–56.

Okumura, Y., Takei, T., Sakamoto, M., Ishikura, T., and Fukagawa, Y. (1979a) *J. Antibiot. 32,* 584–592.

Okumura, Y., Takei, T., Sakamoto, M., Ishikura, T., and Fukagawa, Y. (1979b) *J. Antibiot. 32,* 1002–1010.

Okumura, Y., Okamura, K., Takei, T., Kouno, K., Lein, J., Ishikura, T., and Fukagawa, Y. (1979c) *J. Antibiot. 32,* 575–583.

Okumura, Y., Sakamoto, M., Takei, T., Ishikura, T., and Fukagawa, Y. (1979d) *J. Antibiot. 32,* 1130–1136.

Okumura, Y., Fukagawa, Y., Okamoto, R., and Ishikura, T. (1982) *Agr. Biol. Chem. 46,* 731–737.

Omoto, S., Ogino, H., and Inouye, S. (1981) *J. Antibiot. 34,* 1416–1423.

Otsuka, H., and Shoji, J. (1967) *Tetrahedron 23,* 1535.

Perlman, D., Otani, S., Perlman, K. L., and Walker, J. E. (1973) *J. Antibiot. 26,* 289–296.

Pinkel, D., and Pratt, C. (1973) in *Cancer Medicine* (Holland, J. F., and Frei, E., eds.) pp. 1900–1907, Lea & Febiger, Philadelphia.

Preud'Homme, J., Tarridec, P., and Belloc, A. (1968) *Bull. Soc. Chim. France* 585–591.

Rao, K. V., and Renn, D. W. (1960) *Antimicrob. Agents Chemother. 1959,* 490.

Roberfroid, M., and Dumont, P. (1967) *Ind. Chim. Belg. 32,* 307–309.

Salzman, L. A., Katz, E., and Weissbach, H. (1964) *J. Biol. Chem. 239,* 1864–1866.

Salzman, L. A., Weissbach, H., and Katz, E. (1965) *Proc. Nat. Acad. Sci. U.S.A. 54,* 542–546.

Salzman, L. A., Weissbach, H., and Katz, E. (1969) *Arch. Biochem. Biophys. 130,* 536–546.

Schmidt-Kastner, G. (1956) *Medizin und Chemie 5,* 463.

Schmidt-Kastner, G. (1960) *Ann. N.Y. Acad. Sci. 89,* 299–303.

Sheehan, J. C., Zachan, H. G., and Lawson, W. B. (1958) *J. Amer. Chem. Soc. 80,* 3349–3355.

Shoji, J., and Kato, T. (1976) *J. Antibiot. 29,* 380–389.

Shoji, J., Hinoo, H., Wakisaka, Y., Koizumi, K., and Mayama, M. (1975) *J. Antibiot. 28,* 126–128.

Sivak, A., and Katz, E. (1962) *Biochim. Biophys. Acta 62,* 80–90.

Sivak, A., Meloni, M. L., Nobili, F., and Katz, E. (1962) *Biochim. Biophys. Acta 57,* 283–289.

Sogan, J. A. (1976) *J. Med. Chem. 19,* 1228–1231.
Somer, P. D., and Dijck, P. V. (1955) *Antibiot. Chemother. 5,* 632–639.
Takeuchi, Y., Murai, A., Takahara, Y., and Kainosho, M. (1979) *J. Antibiot. 32,* 121–129.
Tanaka, N. (1975) in *Antibiotics* Vol. III (Corcoran, J. W., and Hahn, F. E., eds.) pp. 487–497, Springer, Berlin.
Thomas, D. W., and Ito, T. (1969) *Tetrahedron 25,* 1985–1990.
Tsuda, K., Kihara, T., Nishii, M., Nakamura, G., Isono, K., and Suzuki, S. (1980) *J. Antibiot. 33,* 247–248.
Ueda, M., and Umezawa, H. (1956) *J. Antibiot. Ser. A, 9,* 86–87.
Ueda, M., Tanigawa, Y., Okami, Y., and Umezawa, H. (1954) *J. Antibiot. Ser. A, 7,* 125–126.
Vanderhaeghe, H., and Parmentier, G. (1960) *J. Amer. Chem. Soc. 82,* 4414–4422.
Vanderhaeghe, H., Janssen, G., and Compernolle, F. (1971) *Tetrahedron Lett.* 2687–2688.
Vazquez, D. (1962) *Biochim. Biophys. Acta 61,* 849–851.
Vazquez, D. (1975) in *Antibiotics* Vol. III (Corcoran, J. W., and Hahn, F. E., eds.) pp. 521–534, Springer, Berlin.
Verwey, W. F., West, M. K., and Miller, A. K. (1958) *Antibiot. Chemother. 8,* 500–505.
Wagman, G. H., Marquez, J. A., Watkins, P. D., Gentile, F., Murawski, A., Patel, M., and Weinstein, M. J. (1976) *Antimicrob. Agents Chemother. 9,* 465–469.
Waksman, S. A., and Gregory, F. J. (1954) *Antibiot. Chemother. 4,* 1050–1056.
Waksman, S. A., and Woodruff, H. B. (1940) *Proc. Soc. Exper. Biol. Med. 45,* 609–614.
Walker, J. E., and Perlman, D. (1971) *Biotech. Bioeng. 13,* 371–379.
Walker, J. E., Bodanszky, M., and Perlman, D. (1970) *J. Antibiot. 23,* 255–256.
Watanabe, K. (1961) *J. Antibiot. Ser. A, 14,* 1–13.
Weissbach, H., Redfield, B., Beaven, V., and Katz, E. (1965a) *Biochem. Biophys. Res. Comm. 19,* 523–530.
Weissbach, H., Redfield, B., Beaven, V., and Katz, E. (1965b) *J. Biol. Chem. 240,* 4377–4381.
Wolff, J. A., Krivitt, W., Newton, W. A., and D'Angio, G. J. (1968) *N. Engl. J. Med. 279,* 290.
Yajima, T., Grigg, M. A., and Katz, E. (1972) *Arch. Biochem. Biophys. 151,* 565–575.
Yanagimoto, M., and Terui, G. (1971) *J. Ferm. Technol. 49,* 611–618.
Yanagimoto, M., Yamada, Y., and Terui, G. (1979) *J. Ferm. Technol. 57,* 6–14.
Yoshida, T., and Katagiri, K. (1967) *J. Bacteriol. 93,* 1327–1331.
Yoshida, T., and Katagiri, K. (1969) *Biochemistry 8,* 2645–2651.
Yoshida, T., Mauger, A., Witkop, B., and Katz, E. (1966) *Biochem. Biophys. Res. Comm. 25,* 66–72.
Yoshida, T., Kimura, Y., and Katagiri, K. (1968) *J. Antibiot. 21,* 465–467.
Yoshida, T., Kimura, Y., and Katagiri, K. (1970) in *Progress in Antimicrobial Anticancer Chemotherapy* Vol. II pp. 1160–1165, University of Tokyo Press, Tokyo.

CHAPTER

7

Macrolides

Satoshi Ōmura
Yoshitake Tanaka

I. INTRODUCTION

Macrolide antibiotics are a class of compounds containing a 12-, 14-, or 16-membered macrocyclic lactone ring to which one or more aminosugars and/or deoxysugars is attached (Woodward, 1957). Recently, "atypical" or "fused" macrolides such as avermectins (Burg et al., 1979), milbemycins (Takiguchi et al., 1980), and nargenicins (Celmer et al., 1980) have been reported. New criteria may be needed to classify or to redefine macrolide antibiotics. We will limit our discussion to "typical" macrolides, which do not possess bicyclic structures and are antibacterial agents. The antifungal polyene macrolide antibiotics are discussed in Chapter 8.

Current trends in industrial production of antibiotics in Japan are characterized by a sharp increase in β-lactams and a corresponding decrease in tetracyclines and chloramphenicol (Okachi and Nara, 1980). Production of macrolides and aminoglycosides has been virtually unchanged for the past five years. A worldwide survey would probably show a similar tendency. In the calendar year 1978, 800 tons of antibiotics worth 638×10^9 yen were marketed in Japan; this accounted for 22.8% of the whole market for pharmaceutical drugs (2794×10^9 yen). Macrolides (30×10^9 yen, 120 tons) represented about 5% in value of the antibiotic market (Okachi and Nara, 1980).

Recently, macrolides have attracted increasing attention in the search for new antibiotics by screening soil microorganisms and by chemical and microbial modification. In this chapter, we will survey the biosynthesis of macrolide antibiotics and present knowledge of the metabolic regulation of macrolide biosynthesis. Recent attempts to obtain new macrolides by biochemical methods will also be included.

II. USE OF MACROLIDE ANTIBIOTICS IN THERAPY

Pikromycin was discovered in 1950 and the structure was characterized by Brockmann and Oster (1957). Before the discovery of pikromycin, proactinomycin had been found by Gardner and Chain (1942). Its physicochemical properties show a close resemblance to those of the macrolides, but the structure of proactinomycin has never been determined. To date, more than 80 macrolide antibiotics have been structurally characterized, among which the 14-membered erythromycin and oleandomycin (Fig. 7.1) and the 16-membered leucomycin, josamycin, spiramycin, and midecamycin (Fig. 7.2) are currently used in antibiotic therapy. Josamycin is identical with leucomycin A_3 (Ōmura et al., 1970). Leucomycin and spiramycin, as well as another 16-membered macrolide, tylosin, have veterinary and fishery applications.

Despite their structural diversity, macrolide antibiotics exhibit a similar antimicrobial spectrum. Macrolides are active against gram-positive bacteria, gram-negative cocci such as *Streptococcus* and *Neisseria,* and *Haemophilus.* In addition, they are active against *Mycoplasma.* The latter activity is of particular importance, since *Mycoplasma* cause serious diseases of the respiratory organs and are insensitive to β-lactams and aminoglycosides. Thus macrolides are the drug of choice against such infections.

Studies on the mode of action of erythromycin established that the antibiotic inhibits protein synthesis in bacteria (Schmid, 1971). Erythromycin binds specifically and in a 1:1 ratio to 50S ribosomes of *Escherichia coli* and inhibits peptidyl transfer. It does not primarily affect the binding of peptidyl-tRNA or amino acyl-tRNA to the ribosomes. A similar mechanism was suggested for 16-membered macrolides (Haupt et al., 1976; Siegrist et al., 1981; Tago and Nagano, 1970; Vazquez, 1967).

The ineffectiveness of most macrolide antibiotics against gram-negative rods such as *E. coli* is mainly due to their inability to penetrate the bacterial outer membrane. This was shown indirectly by the efficient binding of [^{14}C]-erythromycin to *E. coli* ribosomes and by interference with this binding by nonlabeled derivatives of erythromycin and leucomycin (Ōmura et al., 1977a). By using this binding assay technique, it is now possible to examine whether

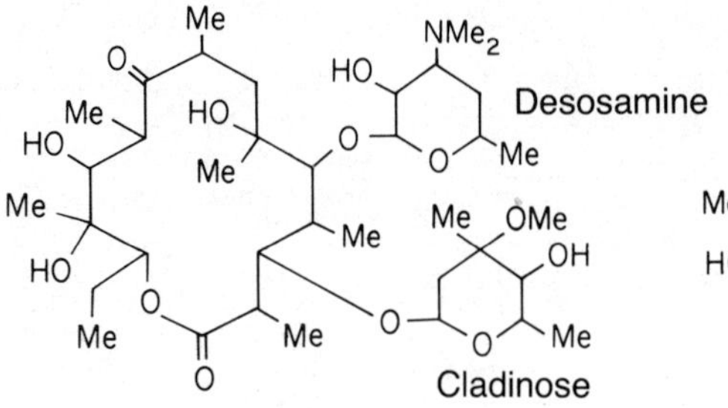

Erythromycin A
Erythromycins
Megalomicins
Lankacidins
Kujimycin A
XK-41-B2
23672RP
(15)

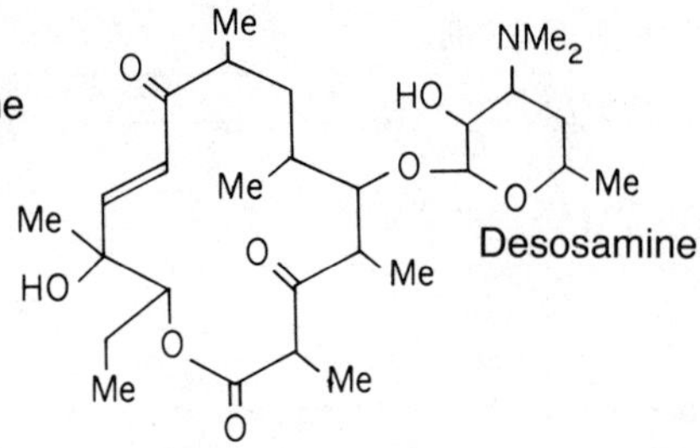

Pikromycin
Narbomycin
(2)

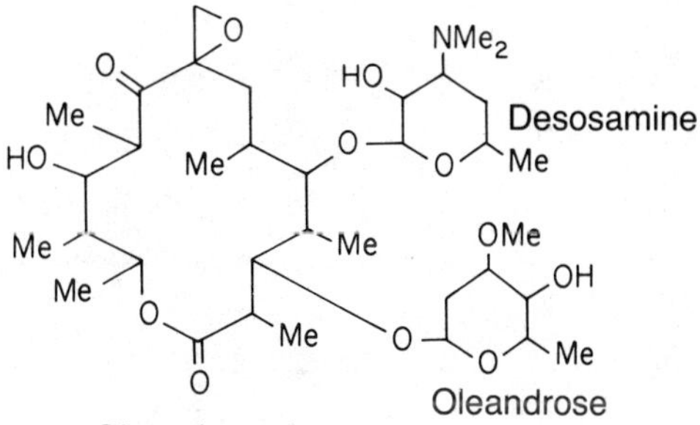

Oleandomycin
Oleandomycins
(2)

FIGURE 7.1 Structures of representative 14-membered macrolide antibiotics. The name of the compound shown is italicized. Other antibiotics with the same aglycone carbon skeleton and the total number of such compounds (in parentheses) are listed below this name.

the altered antibacterial spectrum of a chemically modified macrolide antibiotic is due to altered ability to bind to ribosomes or to penetrate cells.

Macrolide antibiotics show low toxicity, and only limited side effects (nephrotoxicity or kidney disorder) have been noted in patients treated with high doses of erythromycin estolate over a period of more than two weeks (Robinson, 1962; McCormack et al., 1977). Macrolide antibiotics show no cross-resistance to β-lactams and aminoglycosides. However, several lines of evidence indicate that macrolide-resistant staphylococci and streptococci of clinical origin are now increasing.

As in other classes of antibiotics, there is a need for novel macrolides and more potent derivatives with activity against these resistant strains and against gram-negative rods that occasionally cause opportunistic infections.

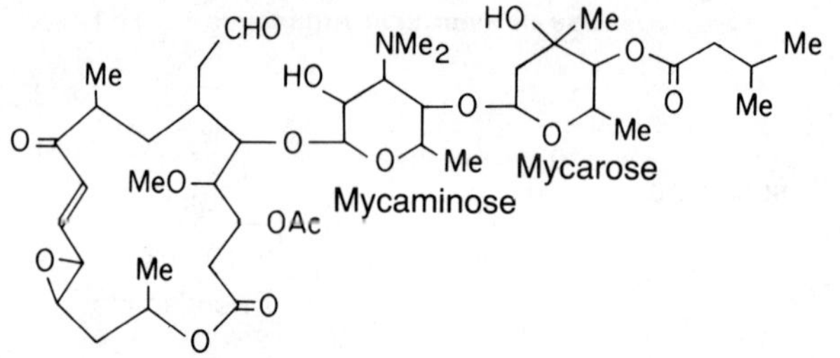

Magnamycin A

Magnamycins	Maridomycins
(Carbomycins)	Espinomycins
Leucomycins	Midecamycins
Spiramycins	Platenomycins
Turimycins	Deltamycins
Niddamycins	
(42)	

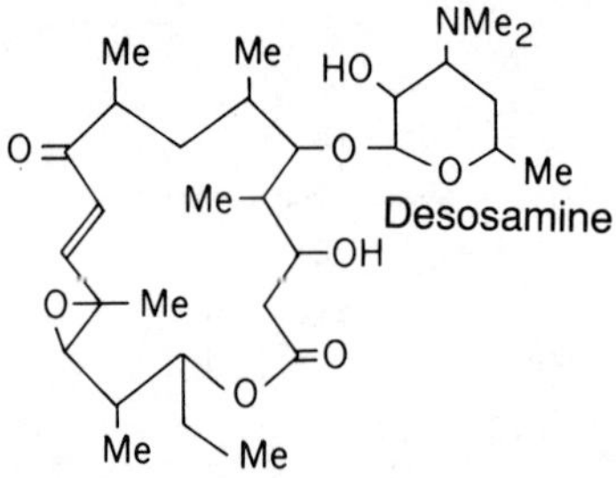

Juvenimicin A_2
(1)

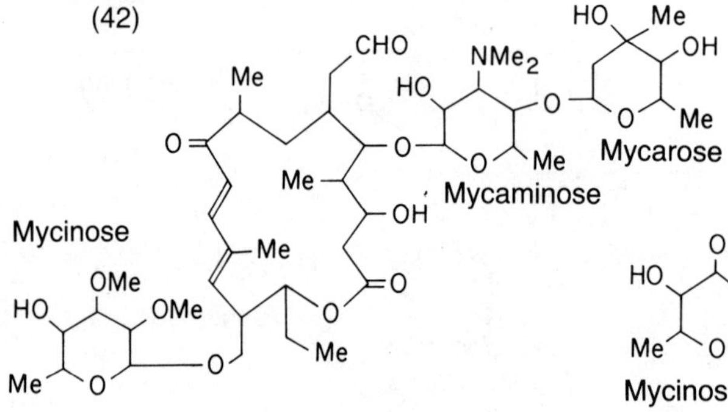

Tylosin

Cirramycins	Relomycin
Rosaramicins	Macrocin
Juvenimicins	B-58941
Angolamycin	M-4365
(Shincomycin)	Staphycoccomycin
(16)	A-6888

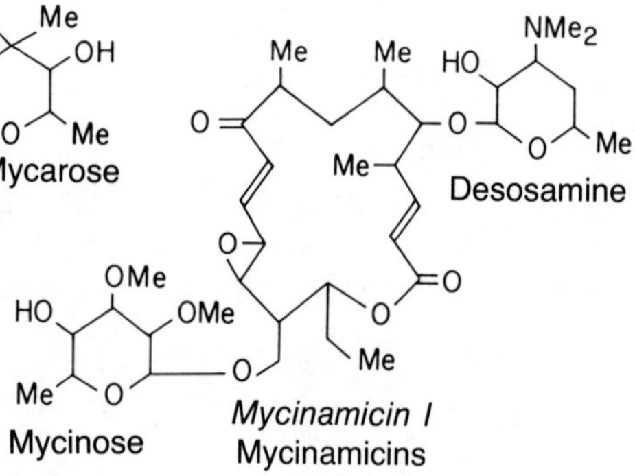

Mycinamicin I
Mycinamicins
(5)

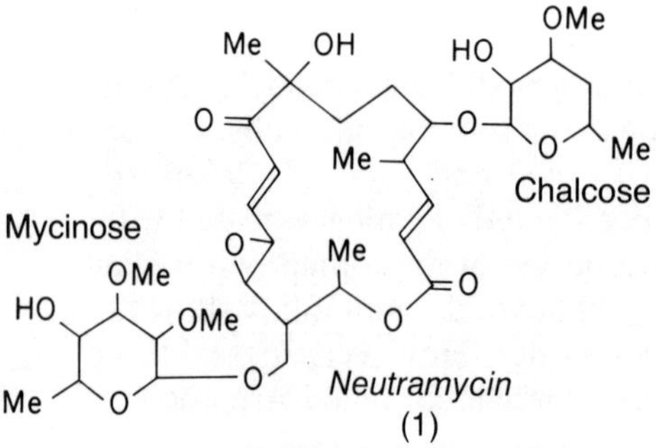

Neutramycin
(1)

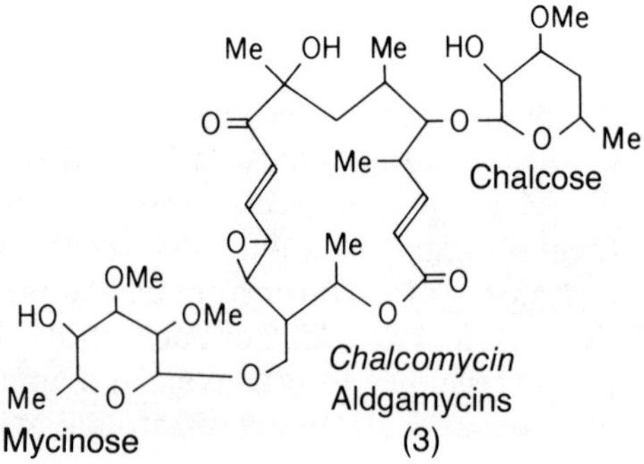

Chalcomycin
Aldgamycins
(3)

FIGURE 7.2 Structures of representative 16-membered macrolide antibiotics. The name of the compound shown is italicized. Other antibiotics with the same aglycone carbon skeleton and the total number of such compounds (in parentheses) are listed below this name.

III. BIOSYNTHESIS OF 14-MEMBERED MACROLIDE ANTIBIOTICS

A. Aglycone Formation

From early studies on erythromycin, it was assumed that the mechanism of lactone formation was similar to that established for saturated fatty acid biosynthesis. Addition of an appropriate amount of propionate or propanol to culture media stimulated erythromycin production by *Streptomyces erythraeus* (Raczyńska-Bojanowska et al., 1970). Gerzon et al. (1956) proposed that, of 21 carbons in the lactone ring of the erythromycins, the seven methyl groups originated from propionate. Later studies by Kaneda et al. (1962) supported this. Subsequent experiments using [^{14}C]methylmalonate showed that one of the seven propionates was a "primer," incorporated into carbons 13, 14, and 15. The rest were "chain-extending units," incorporated into the remaining carbons of the lactone ring (Friedman et al., 1964). Thus it was demonstrated that condensation of the propionates involves a mechanism similar to that of fatty acid synthesis (Fig. 7.3).

Ōmura ct al. (1976) utilized ^{13}C-labeled precursors and ^{13}C-NMR spectroscopy to study the biosynthesis of pikromycin in *Streptomyces flavochromogenes* (Fig. 7.3). One acetate and six propionates were incorporated

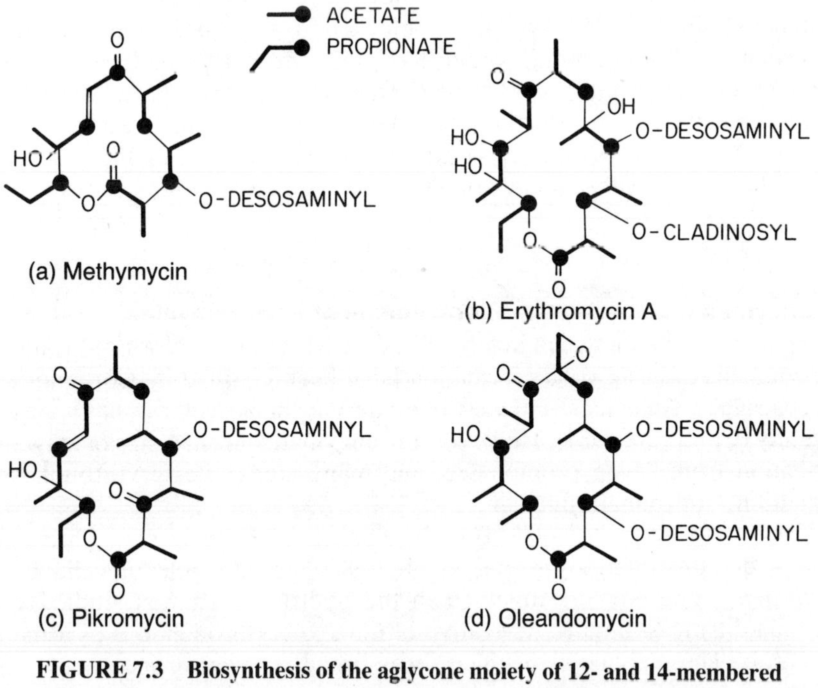

FIGURE 7.3 Biosynthesis of the aglycone moiety of 12- and 14-membered macrolide antibiotics. Sources: (a) Birch et al. (1964), (b) Kaneda et al. (1962), (c) Ōmura et al. (1976), (d) Grisebach and Hofheinz (1964).

intact into the aglycone. Oleandomycin, another clinically important (as its triacetate) 14-membered macrolide produced by *Streptomyces antibioticus*, has an epoxide at the C-8 position. The biosynthetic scheme shown in Fig. 7.3 was proposed by Grisebach and Hofheinz (1964).

B. The Pathway after Aglycone Formation

Studies on the course of erythromycin biosynthesis after aglycone formation were markedly accelerated by the discovery of erythronolide B in a culture broth of the wild-type strain of *S. erythraeus* (Tardrew and Nyman, 1964). Hung et al. (1965) demonstrated that erythronolide B was an obligatory intermediate in erythromycin biosynthesis by showing that [^{14}C]erythronolide B was synthesized de novo from [^{14}C]propionate and that [^{14}C]erythronolide B was transformed to [^{14}C]erythromycins. Shortly afterwards, J. R. Martin and Rosenbrook (1967) isolated various blocked mutants that accumulated biosynthetic intermediates such as 6-deoxyerythronolide B and 3-*O*-mycarosylerythronolide B as well as erythronolide B. Some mutants were able to transform the intermediates to erythromycins. Martin and Rosenbrook examined biosynthetic relationships between the intermediates by the method that McCormick et al. (1960) called "cosynthesis." 6-Deoxyerythronolide and 3-*O*-mycarosylerythronolide B, produced by mutant strains 9EI 262 and 8EI 57, respectively, were transformed to erythromycins by a third mutant, strain 2NU 153, which was blocked at a site before erythronolide B formation. Using mutant strain 2NU 153, the biosynthetic relationship between erythromycins A, B, and C was also clarified (J. R. Martin and Goldstein, 1970; J. R. Martin et al., 1966). The combined results lead to the pathway for erythromycin biosynthesis summarized in Fig. 7.4. It is supported by the subsequent finding of erythromycin D in cultures (Majer et al., 1977).

C. Enzymes Involved in the Biosynthesis of Erythromycins

The nature of the enzymes involved in at least three steps of erythromycin biosynthesis after aglycone formation have been clarified by Corcoran and his colleagues. An extensive review of erythromycin biosynthesis has appeared recently (Corcoran, 1981). However, the enzymatic mechanisms of aglycone formation from seven propionates, and many steps after erythronolide B formation, still remain unclear.

That 6-deoxyerythronolide B is the immediate precursor or erythronolide B was demonstrated by Hung et al. (1965) using whole mycelia of *S. erythraeus*. The enzyme involved in the hydroxylation was studied with the 6,6a-ditritio-9-dihydro derivative of 6-deoxyerythronolide B as a model substrate. The enzyme's specificity was broad enough for it to accept the 9-dihydro derivative as a substrate, permitting a sensitive radioisotope assay. A crude cell-free extract from the mycelia of *S. erythraeus* was fractionated

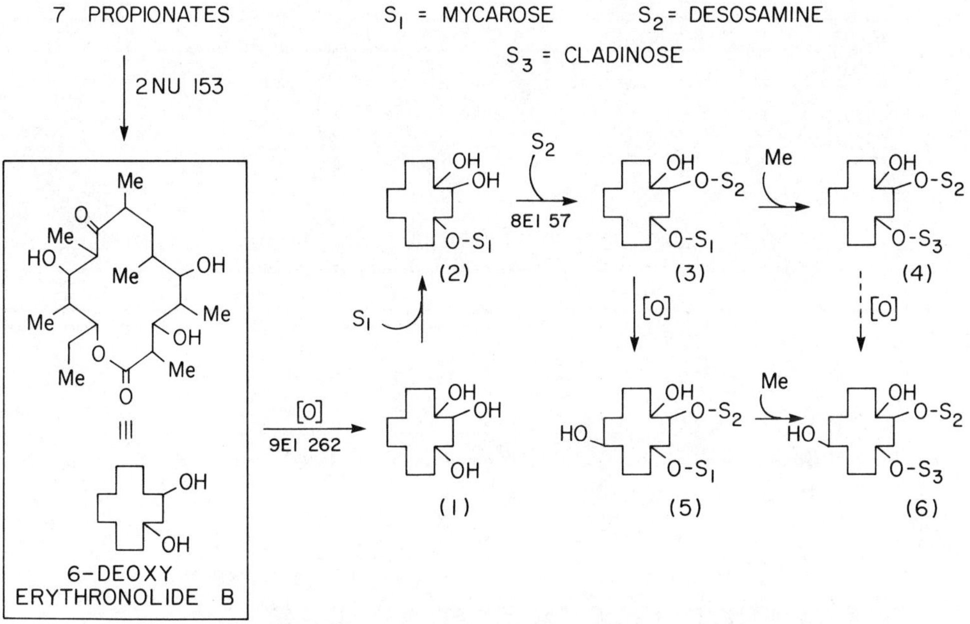

FIGURE 7.4 Biosynthetic pathways to erythromycin A in *Streptomyces erythraeus*. Possible blocked steps in mutants 2NU 153, 9EI 262 and 8EI 57 are indicated. The compounds are: (1) erythronolide B, (2) 3-*O*-mycarosylerythronolide B, (3) erythromycin D, (4) erythromycin B, (5) erythromycin C, (6) erythromycin A.

by ammonium sulfate precipitation and column chromatography into three protein components. According to Corcoran, the first component was a heat-stable heme protein containing cytochrome P-448 as an essential moiety; the second was a flavoprotein (FAD-specific) of the cytochrome P-450 reductase type; and the third component was a heat-labile non-heme iron protein of the ferredoxin type. All three components were required for hydroxylation, and full activity could be reconstituted in the presence of an NADPH-regenerating system. Scheme 7.1 shows the mechanism proposed for the overall hydroxylation reaction (Nishikiori et al., 1978).

The C-12 hydroxylase activity required for the transformation of erythromycin D to C was detected in a cell extract from *Streptomyces erythraeus* (Corcoran and Vygantas, 1977). When 9-dihydroerythromycin D was incubated with the crude enzyme preparation, corresponding 9-dihydro derivatives of erythromycin C and erythromycin A were formed. Formation of the latter product suggested the presence of erythromycin C *O*-methyl transferase activity in the same crude enzyme preparation. This was later confirmed. The *O*-methyl transferase requires S-adenosylmethionine as a methyl donor and has a relatively narrow substrate specificity. However, erythromycin C and erythromycin D were methylated equally well, yielding erythromycins

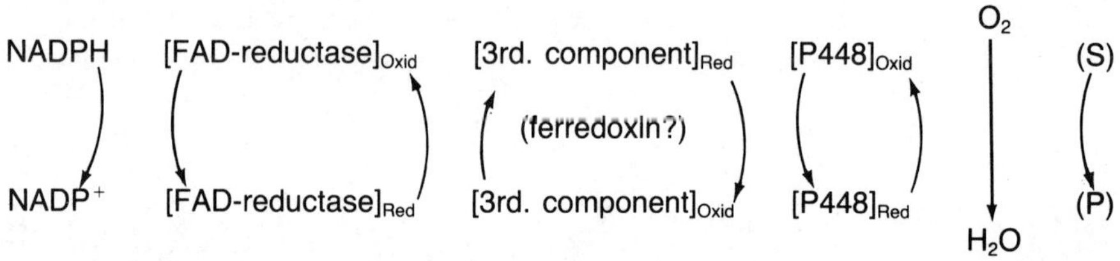

SCHEME 1 Mechanism proposed for the hydroxylation of 9-dihydro-6-deoxyerythronolide B (S) to form 9-dihydroerythronolide B (P).

A and B, respectively. On this evidence, erythromycin B may be an intermediate in a branch path to erythromycin A (Corcoran, 1981).

IV. BIOSYNTHESIS OF 16-MEMBERED MACROLIDE ANTIBIOTICS

A. Aglycone Formation

The biosynthesis of 16-membered macrolide antibiotics has been reviewed (Ōmura and Kitao, 1979; Ōmura and Nakagawa, 1975, 1981).

The biosynthetic origin of the carbon skeleton of the lactone ring in magnamycin A has been extensively studied by ^{14}C-precursor experiments. Grisebach and Achenbach (1962) and Corcoran and Chick (1966) suggested that C-7 to C-16 in the lactone ring of magnamycin arose from four acetates and one propionate but that the other carbons (C-1 to C-6 and a formylmethyl group at C-6) were derived from precursors other than acetate. Later, Ohno et al. (1974) and Gersch et al. (1977) suggested incorporation of butyrate into the aglycone moieties of maridomycin and turimycin (the latter is identical with leucomycin, maridomycin, or platenomycin; Haupt et al., 1976).

^{13}C-Nuclear magnetic resonance (^{13}C-NMR) spectroscopy was applied by Ōmura et al. (1975a, 1977b) to study the biosynthesis of 16-membered macrolide antibiotics. This technique brought about remarkable progress. The biosynthesis of the aglycone moiety of leucomycin has been clarified (Fig. 7.5); only the origin of the carbons at C-3 and C-4 remains unknown (Ōmura et al., 1977b). In a similar way, biosynthesis of the aglycone moieties of tylosin (Ōmura et al., 1975b, 1977b), rosaramicin (formerly rosamicin) (Ganguly et al., 1976), and mycinamicin (Puar et al., 1981) has been studied. ^{13}C-NMR spectroscopy was also useful for analyzing the metabolism of butyrate to propionate in *Streptomyces fradiae* (Ōmura et al., 1977b).

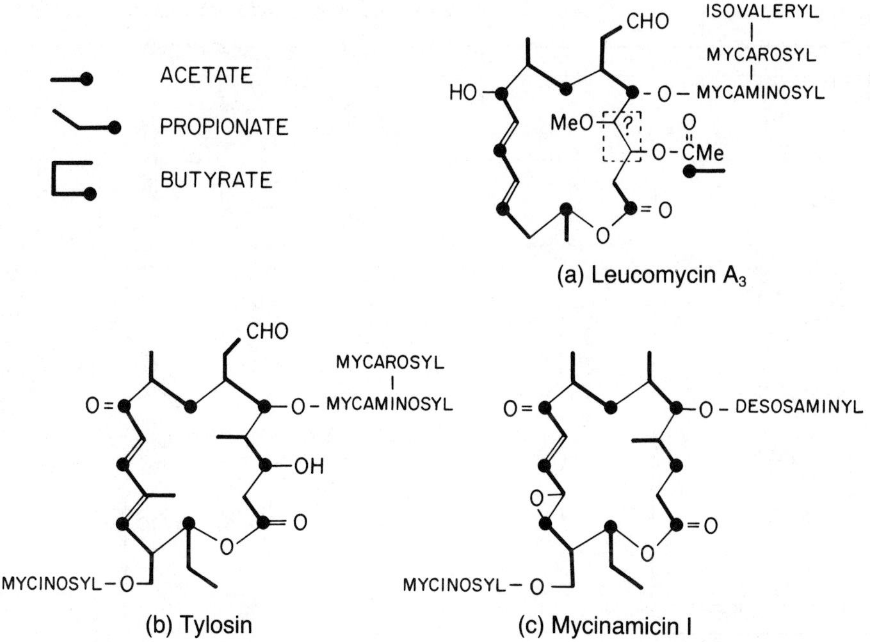

(a) Leucomycin A_3

(b) Tylosin

(c) Mycinamicin I

FIGURE 7.5 Biosynthesis of the aglycone moieties of 16-membered macrolide antibiotics. Sources: (a, b) Ōmura et al. (1975b, 1977b), (c) Puar et al. (1981).

B. Biosynthesis of Macrolide Sugars

Deoxysugars and aminosugars in macrolide antibiotics are biosynthesized from D-glucose. L-Sugars are attached at C-3 by an α-linkage, and D-sugars at C-5 by a β-linkage. Celmer (1971) proposed a 4,5-dehydrosugar as a common intermediate for macrolide sugars. Pape and Brillinger (1973) studied cell-free biosynthesis of mycarose in a tylosin-producing strain of *Streptomyces rimosus* and proposed a sequence *via* TDP-4-keto-6-deoxy-D-glucose.

C. The Pathway after Aglycone Formation

In the formation of macrolide antibiotics, the lactone ring made initially is oxidized and reduced; then sugars are attached to it. The glycosylated intermediate is again oxidized, reduced, and acylated sequentially, leading to the final product, as illustrated below.

1. Biosynthesis of platenomycin—Studies with blocked mutants. Furumai et al. (1973, 1975) used "cosynthesis" to study biosynthesis of the 16-membered macrolide, platenomycin. They obtained many platenomycin nonproducing mutants from *Streptomyces platensis*. Antibiotic formation by pairs of

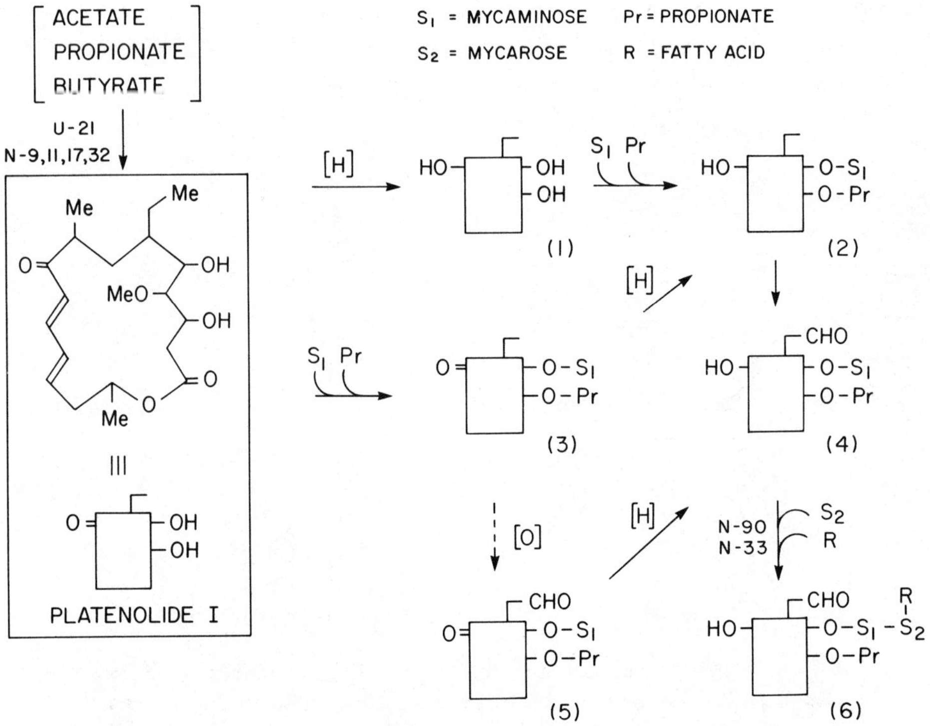

FIGURE 7.6 Biosynthetic pathways to platenomycins in *Streptomyces platensis*. Possible blocked steps in mutants N-9, -11, -17, -32, -33 and -90 and U-21 are indicated. The compounds are: (1) platenolide II, (2) 5-*O*-mycaminosyl-3-*O*-propionylplatenolide II, (3) 5-*O*-mycaminosyl-3-*O*-propionylplatenolide I, (4) demycarosylplatenomycin, (5) 9-dehydro-demycarosylplatenomycin, (6) platenomycins A_1 (R = isovaleryl), B_1 (R = propionyl) and C_2 (R = acetyl).

"secretor" and "convertor" mutants made it possible to detect the producers of biosynthetic intermediates. Several platenomycin intermediates were isolated from the fermentation broths of the mutants and placed in biosynthetic sequence (Fig. 7.6).

2. Biosynthesis of tylosin and spiramycin—Studies using the enzyme inhibitor cerulenin. Instead of using blocked mutants, Ōmura et al. (1978) used cerulenin to specifically inhibit the "condensing enzyme" involved in the condensation of acyl-ACP with malonyl-ACP in the biosynthesis of fatty acids (Greenspan et al., 1977; Nomura et al., 1972; Ōmura et al., 1976). Cerulenin also inhibits the biosynthesis of polyketide-derived metabolites such as leucomycin, tetracycline, 6-methylsalicylic acid (Ōmura, 1976), daunorubicin (McGuire et al., 1979), and candicidin (J. F. Martin, 1976) without affecting cell growth. Thus the antibiotic can supply mycelia in which de novo synthesis of aglycone is inhibited but other paths (sugar metabolism) are unaffected. Such cerulenin-grown mycelia do not produce macrolide antibiotics but are capable of transforming added lactonic compounds to the final product. Consequently, it is possible to determine whether or not a given compound is an intermediate in the biosynthetic pathway.

Possible intermediates in the formation of tylosin were obtained chemically from tylosin and were incubated with cultures of *S. fradiae* in the presence of cerulenin. Analysis of the products showed the pathway: mycaminosyltylonolide > demycarosyltylosin > tylosin (Ōmura et al., 1978). Thus sequential glycosylation of the aglycone moiety by mycaminose, mycinose, and mycarose was suggested.

Recently, protylonolide was isolated from a blocked mutant (a mycaminose idiotroph) of *S. fradiae* KA-427 (Ōmura et al., 1980b). Protylonolide is a 16-membered lactone with the basic carbon skeleton of tylosin aglycone; it is a precursor of tylosin, since it is efficiently converted to tylosin by the parent strain (Ōmura et al., 1980b). To further clarify the steps in tylosin biosynthesis, many derivatives of protylonolide and tylosin-related compounds were synthesized chemically and biotransformed using *S. fradiae* KA-427 grown in the presence of cerulenin. The results suggested that the pathway to tylosin after protylonolide proceeds as shown in Fig. 7.7 (Ōmura et al., 1982). The proposed pathway is consistent with the previous results of Ōmura et al., (1978), and with the results that Seno et al., (1977) obtained with *S. fradiae*.

The same technique was used to determine the pathway of spiramycin biosynthesis in *Streptomyces ambofaciens* (Ōmura et al., 1979c). While forocidin I was converted into neospiramycin I, spiramycin I, and spiramycin II (and III), it was not transformed into leucomycin V, indicating that the linkage of forosamine to C-9 in the aglycone precedes the attachment of mycarose to C-4′ of mycaminose (Fig. 7.8). Spiramycin I (C-3 hydroxyl) was rapidly converted into spiramycin II, but neospiramycin II (3-*O*-acetyl) was not. These results suggested that acylation of the lactone at C-3 occurs in the final step of spiramycin biosynthesis. The C-3 acylating enzyme was isolated and characterized (Ōmura et al., 1979a). As will be described later, its synthesis is induced by glucose and suppressed by butyrate (Kitao et al., 1979a).

3. Biosynthesis of acyl moieties. Grisebach and Achenbach (1963), Furumai et al. (1975), and Ohno et al. (1974) suggested that the 4″-O-isovaleryl and 4″-O-propionyl groups on mycarose were derived from L-leucine and L-isoleucine, respectively. Maridomycin production (Miyagawa et al., 1979) and leucomycin production (Vezina et al., 1979) were enhanced in valine- and leucine-analog-resistant strains derived from the wild-type organisms.

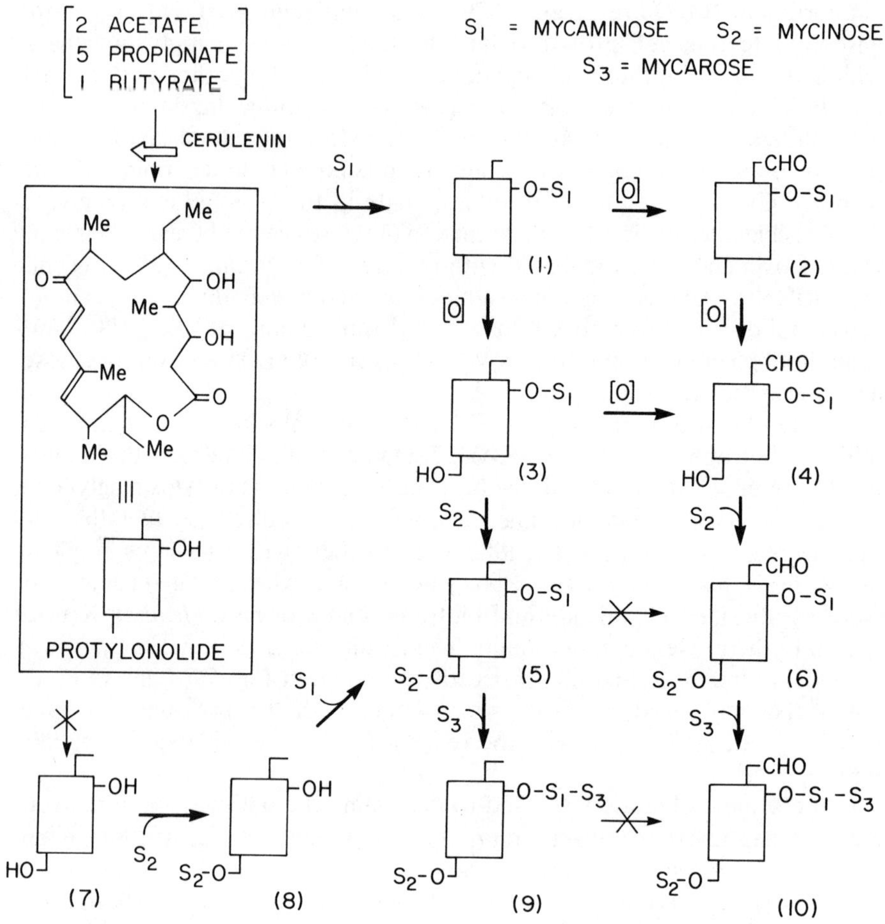

FIGURE 7.7 Postulated pathways for the biosynthesis of tylosin in *Streptomyces fradiae*. The compounds are: (1) 5-*O*-mycaminosylprotylonolide, (2) deepoxycirramycin A, (3) 20-deoxy-5-*O*-mycaminosylrelonolide (4) 5-*O*-mycaminosyltylonolide, (5) 20-deoxydemycarosylrelomycin (6) demycarosyltylosin, (7) 23-hydroxyprotylonolide, (8) 20-deoxy-23-*O*-mycinosylrelonolide, (9) 20-deoxyrelomycin, (10) tylosin. Source: Ōmura et al., (1982).

V. CLASSIFICATION OF MACROLIDE ANTIBIOTICS BASED ON THE LACTONE RING SKELETON

Macrolide antibiotics are usually classified according to the size of the ring system. On the basis of structural and biosynthetic characteristics of the aglycone moiety, the 14-membered macrolides can be subdivided into three groups: These are the erythromycin, pikromycin, and oleandomycin groups

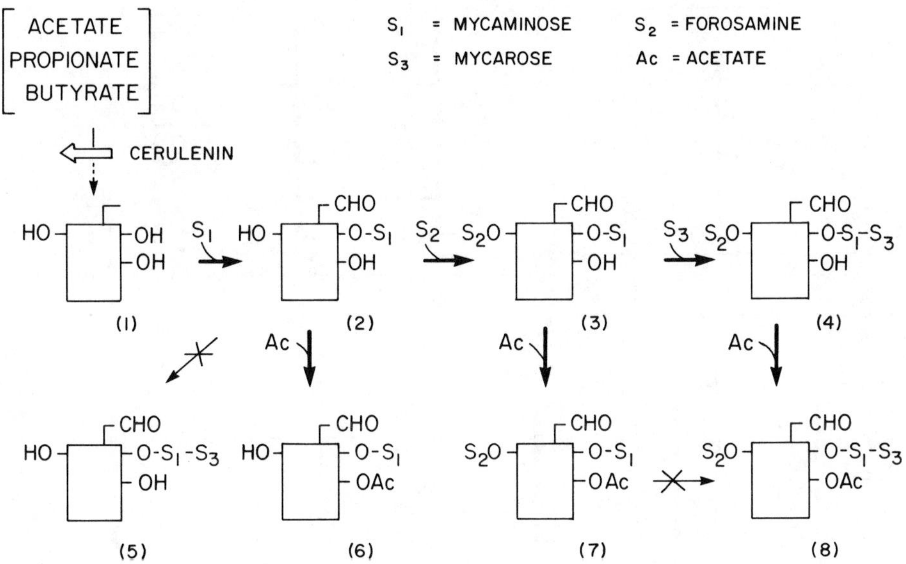

FIGURE 7.8 Biosynthetic pathways to spiramycin II in *Streptomyces ambofaciens*. The compounds are: (1) platenolide II, (2) forocidin I, (3) neospiramycin I, (4) spiramycin I, (5) leucomycin V, (6) forocidin II, (7) neospiramycin II, (8) spiramycin II.

(Figs. 7.1 and 7.9). The 16-membered macrolides were formerly classified into four (Ōmura and Kitao, 1979) but are now classified into six groups (Figs. 7.2 and 7.9).

These groupings are based on the sequence of intermediates in aglycone biosynthesis. Fig. 7.9 shows the Fischer projections of representative 12-, 14-, and 16-membered macrolide antibiotics. Biosynthetic derivations of these antibiotics have been shown (Figs. 7.3 and 7.5). Condensation proceeds from the bottom to the top. It is obvious that, within each group, differences in the carbon skeleton simply depend on the sequence of condensation of the two to four different kinds of precursors.

In 14-membered macrolides, the first (P_1) and third (P_3) precursors are acetate or propionate, while the rest are all propionate. Thus at least four species of 14-membered aglycone can exist. Three have been discovered. The remaining one should be of the acetate (P_1)-propionate (P_2)-acetate (P_3)-(propionate)$_4$(P_4–P_7) type. On the same basis, at least 72 kinds of 16-membered aglycone are possible, provided that only acetate, propionate, and butyrate are the precursors. To date, six of the 72 species have been identified. Not all the possible candidates will be detected, but the range of possible variation is significant; we should still be able to find many new macrolide antibiotics with novel types of aglycones.

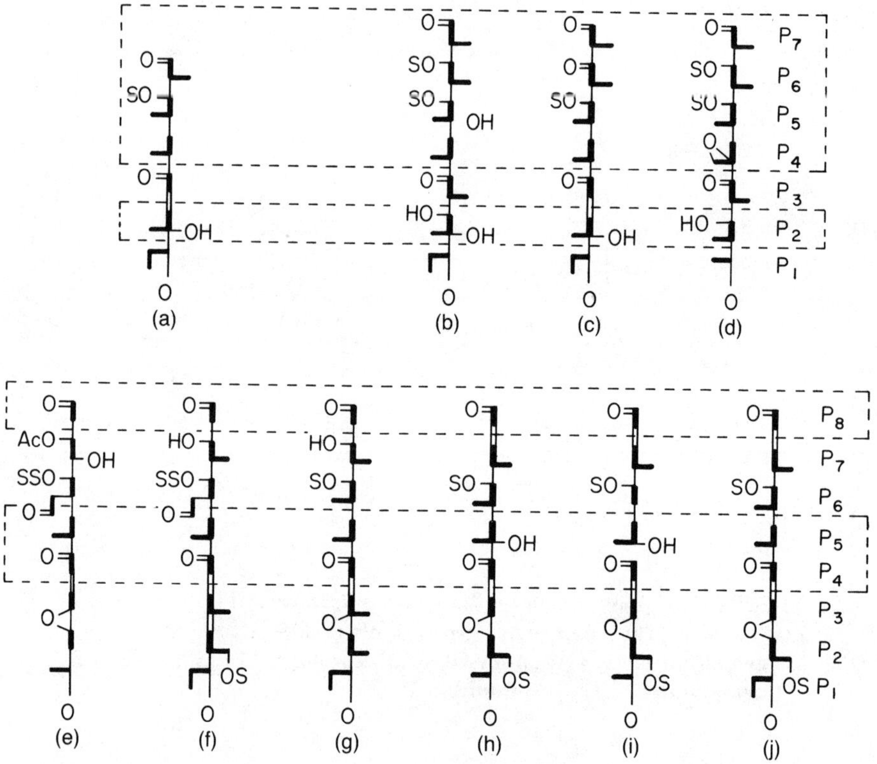

FIGURE 7.9 Fischer projections of 12-, 14- and 16-membered macrolide antibiotics and their classification according to the carbon skeleton of the lactone ring. Absolute configurations are based on experimental data, where available, or on the Celmer model (Celmer, 1971). S denotes a sugar moiety. Pn indicates precursor. Aglycone synthesis proceeds by condensation of precursors P_1, P_2 . . . in this order. Positions occupied by common precursors are enclosed in boxes.

The macrolides depicted above are: (a) methymycin, (b) erythromycin A, (c) pikromycin, (d) oleandomycin, (e) magnamycin A, (f) tylosin, (g) juvenimicin A_2, (h) chalcomycin, (i) neutramycin, and (j) mycinamicin I.

VI. REGULATION OF BIOSYNTHESIS

Secondary metabolism is generally subject to common regulatory processes, like carbon catabolite repression and phosphate regulation, which are not usually encountered in primary biosynthetic metabolism. Much needs to be clarified before we understand the regulatory mechanisms involved and information on the regulation of macrolide biosynthesis is quite limited. Therefore the following discussion is presented mainly in the hope of stimulating further experimental work.

A. Regulatory Aspects of Aglycone Formation

Interest has been focused on the nature of the aglycone-forming enzyme (polyketide-condensing enzyme). However, there is no report so far of the detection of in vitro activity or isolation and characterization of such an enzyme. The identity or nonidentity of polyketide- and fatty acid–condensing enzymes still remains undetermined. Apparent resemblance between the two has been observed in several cases (Corcoran, 1981; Hutchinson et al., 1981). Nonidentity has also been suggested by several workers. 6-Methylsalicyclic acid synthetase from *Penicillum urticae* was purified sufficiently to free it from fatty acid–condensing enzyme activity (Scott et al., 1974). Other evidence for their nonidentity has been suggested from studies on the mode of action of cerulenin. The antibiotic inhibits the synthesis of 6-methylsalicylic acid, tetracycline, and several macrolide antibiotics at levels that exert little effect on the growth of the microorganism employed, suggesting that there is little inhibition of their fatty acid biosynthesis (Ōmura, 1976). Examination of the sequence profile of the precursors of the macrolide antibiotic aglycones (Fig. 7.9) suggests that the species condensed is strictly controlled at each step. Consequently, it is unlikely that aglycone formation is catalyzed by a fatty acid synthetase with the substrates for fatty acid synthesis merely replaced by the substrates for polyketide synthesis.

Mutants with genetic blocks at sites before lactone formation can be used to investigate biochemical mechanisms. Strain NP-10 of *S. fradiae,* which is unable to produce tylosin but able to transform added protylonolide to tylosin (Ōmura et al., 1980b), an analogous mutant strain 2NU 153 of *S. erythraeus* (J. R. Martin et al., 1966) as well as mutants N-32 and U-21 of *S. platensis* (Furumai et al., 1975) are probably defective in lactone formation and should prove useful in biosynthetic studies. Recently, Ghisalba and Nüesch (1981) reported isolating an intermediate of rifamycin B biosynthesis that consisted of a C_7N starter unit and the first three (of a total of ten) fatty acid components. This intermediate must be formed at a very early stage in the biosynthesis of the ansa chain. Since there are structural and biosynthetic similarities between the ansa moiety of the ansamycins and the aglycone of macrolide antibiotics (Brufani et al., 1973), such intermediates may also be found in mutants of macrolide producers.

B. Carbon Catabolite Regulation

Suppression of secondary metabolite production by glucose was first described by Soltero and Johnson (1953), who observed it during penicillin production by *Penicillium chrysogenum*. Similar effects of rapidly metabolized carbon sources on the production of antibiotics such as streptomycin, actinomycin, tetracycline, chloramphencol, and cephalosporin were later reported (Hostalek, 1980; J. F. Martin and Demain, 1980). Demain called this phenomenon carbon catabolite regulation of secondary metabolism. That its effect was exerted at the enzyme level was demonstrated for puromycin biosynthesis (*O*-demethylpuromycin *O*-methyltransferase) (Sankaran and Pogell, 1975) and actinomycin biosynthesis (phenoxazinone synthetase) (Gallo and Katz, 1972). Synthesis of the 16-membered macrolide tylosin in *S. fradiae* is severely depressed by glucose (Mardry et al., 1979). 2-Deoxyglucose which, as a glucose analog, was probably transported into the mycelia and phosphorylated but not metabolized further, also caused inhibition. Another glucose analog, α-methylglucoside, which might not be phosphorylated, had no suppressing effect. Glucose and 2-deoxyglucose also interfered with oxidation of methyl oleate, a precursor-donating compound.

Interesting examples of inducible synthesis of an enzyme were reported by Kitao et al. (1979b). In the course of studying precursors of aglycone synthesis, they found that addition of 0.1% *n*-butyrate increased leucomycin production by *Streptomyces kitasatoensis,* as estimated by a microbiological assay. The increased leucomycin titer was accompanied by an increase in the ratio of leucomycin A_1 to A_3, brought about as follows: synthesis of comycin A_1 (C-3 hydroxyl) to leucomycin A_3 (C-3 acetyl), is induced by glucose, and the induction is inhibited by *n*-butyrate. The same mechanism affects spiramycin synthesis in *S. ambofaciens* (Kitao et al., 1979a). An important application of this phenomenon is the directed production of the most potent antibiotic components, leucomycin A_1 and spiramycin I, by addition of *n*-butyrate to fermentation media. This is the first example in which glucose induces synthesis of an enzyme involved in secondary metabolism. However, Abdollahi and Buchanan (1981) have recently suggested that glucose induces an enzyme involved in aflatoxin biosynthesis.

Relief from regulation is of practical importance for antibiotic production. It has been achieved chiefly by environmental manipulation. Repressing carbon sources are replaced by nonrepressing ones; these are usually slowly metabolized compounds (J. F. Martin and Demain, 1980). Another way of relieving microorganisms from carbon catabolite regulation is to grow cultures in a chemostat with low levels of glucose in the medium. Gray and Bhuwapathanapun (1980) have carried out experiments of this kind. High specific production of tylosin and relomycin were achieved at glucose concentrations below 1%. The optimal amount was 0.5%.

C. Phosphate Regulation

Synthesis of macrolide antibiotics, like that of numerous other secondary metabolites, is suppressed if too much inorganic phosphate is included in the medium (J. F. Martin, 1977; Weinberg 1974). An increase in the initial phosphate level from 5 mM to 10 mM led to 80% reduction of tylosin production by a streptomycete without any effect on growth (Mardry et al., 1979). Similar phosphate effects were described for tylosin synthesis by *S. fradiae* (Vu-Trong et al., 1981) and for the synthesis of turimycin by *Streptomyces hygroscopicus* (Gersch, 1980), In another leucomycin producer, *S. kitasatoensis,* increasing the initial phosphate level from 5 mM to 125 mM caused only 50% reduction in antibiotic production (Ōmura et al., 1980c). Spiramycin production by *S. ambofaciens* is also only weakly sensitive to phosphate (S. Ōmura, unpublished). The reduction of tylosin synthesis was ascribed partly to a lowered level (60–70% of the control value) of methylmalonyl-CoA carboxytransferase (Vu-Trong et al., 1981). The effect of phosphate is most severe just before antibiotic synthesis begins and almost disappears once the cells are committed (Gersch, 1980; Vu-Trong et al., 1981). Although repression and inhibition could be involved in phosphate regulation of macrolide biosynthesis, only repression has been postulated.

Inorganic phosphate appears to show its repressive effect through control of the level of an intracellular mediator. Three kinds of mediator have been postulated. In tylosin biosynthesis by *S. fradiae,* intracellular adenylates (ADP, AMP, and ATP, in order of decreasing concentration) reached a maximum level and rapidly decreased before the onset of tylosin synthesis. With increasing initial phosphate levels, the individual adenylate levels increased and remained high throughout the fermentation with concomitant suppression of tylosin synthesis. Adenylate energy charge values were constant (0.4–0.5) and identical in the media with various phosphate concentrations (Vu-Trong et al., 1981). Glucose showed similar effects on tylosin production, adenylate levels, and adenylate energy charge level (Vu-Trong et al., 1980). Thus it was suggested that the level of the adenylate pool rather than the adenylate energy charge plays a significant role in mediating the phosphate (and carbon) regulation. In another phosphate-susceptible tylosin producer, glucose-6-phosphate was postulated to be the mediator (Mardry et al., 1979). It has been proposed that cyclic AMP is involved in the phosphate regulation of turimycin synthesis by *S. hygroscopicus* (Gersch, 1980).

Arsenite has an intriguing effect on the synthesis of narbomycin, a 14-membered macrolide of the pikromycin group. This substance is well-known as a respiratory inhibitor and is also a phosphate analog in some secondary metabolic processes, such as erythromycin biosynthesis (Vanek and Majer, 1967). Addition of 0.3 mM of sodium arsenite to *S. venezuelae* decreased narbomycin synthesis but increased accumulation of the aglycone narbonolide (Maezawa et al., 1974). Arsenite was assumed to affect biosynthesis of the sugar moiety.

Release of macrolide synthesis from phosphate regulation is usually achieved by controlling the initial concentration in the medium (that is, environmental manipulation). In chemostat cultures of *S. fradiae,* a low phosphate level (3.4 mM) increased the rate of tylosin production (Gray and Bhuwapathanapun, 1980). Relomycin production was not affected. Suppression of turimycin synthesis by phosphate was reversed by cyclic AMP or sodium fluoride, an activator of adenylate cyclase, in *S. hygroscopicus* (Gersch, 1980). J. F. Martin et al. (1979) have isolated mutants of *Streptomyces griseus* deregulated in phosphate control of the polyene macrolide, candicidin (see Chapter 8). The technique seems applicable to phosphate-sensitive producers of other antibiotics but no reports of its wider use are yet available.

D. Nitrogen Regulation

The effect of nitrogen sources on antibiotic production was described first by Rake and Donovick (1946). Later, Dulaney (1948) encountered it along with a phosphate effect. Both observations related to streptomycin production by *S. griseus*. Like carbon sources and inorganic phosphate, ammonium ions and nitrogen-containing organic compounds that are rapidly metabolized suppress the synthesis of several antibiotics (Aharonowitz, 1980; Drew and Demain, 1977; Nakayama, 1977). A direct relationship between levels of nitrogen source and onset of secondary metabolism has been shown in patulin biosynthesis by the fungus *P. urticae* (Grootwassink and Gaucher, 1980). Here, synthesis of *m*-hydroxybenzyl alcohol dehydrogenase, a secondary metabolism-specific enzyme, began only after exhaustion of nitrate from the medium. It should be noted that, in this example of nitrogen regulation, the antibiotic does not contain nitrogen atoms. A close relationship has also been shown to exist between ammonium ion levels in β-lactam antibiotic-producing cultures and the activity of enzymes, such as glutamate dehydrogenase and glutamine synthetase, involved in nitrogen metabolism (Aharonowitz, 1979; Streicher and Tyler, 1981).

Nitrogen regulation of macrolide biosynthesis has been reported occasionally. Addition of soybean meal to *S. erythraeus* cultures at idiophase caused a marked reduction in the rate of antibiotic synthesis and incorporation of labeled propionate (Smith et al., 1962). An increase in ammonium sulfate concentration stimulated mycelial growth but decreased oleandomycin production by *S. antibioticus* (Surikova et al., 1976). With *S. kitasatoensis* in a chemically defined medium, 3 mM ammonia reduced leucomycin production by 50% or more (Ōmura et al., 1980c). The effect was most noticeable when ammonium ions were added in the middle-to-late logarithmic phase of growth. The kinetics were similar to those for phosphate regulation of macrolide biosynthesis described above and to those for nitrogen regulation of cephamycin biosynthesis in *Streptomyces clavuligerus* (Aharonowitz and Demain, 1979). Thus it appears that ammonium ions supressed leucomycin production by inhibiting the synthesis of enzyme rather than the activity of preformed enzymes.

Relief of macrolide-producing microorganisms from nitrogen regulation has relied so far on environmental manipulation. Suppressive nitrogen sources such as ammonia have been replaced by nonsuppressive ones such as amino acids. Recently, Ōmura et al. (1980c) observed a threefold to eight-fold increase in leucomycin production by *S. kitasatoensis* when supplements of magnesium phosphate were added to complex media or ammonia-containing defined media. Mycelial growth doubled at most. The ammonia concentration decreased markedly in the supernatant fluids and increased in the sediment

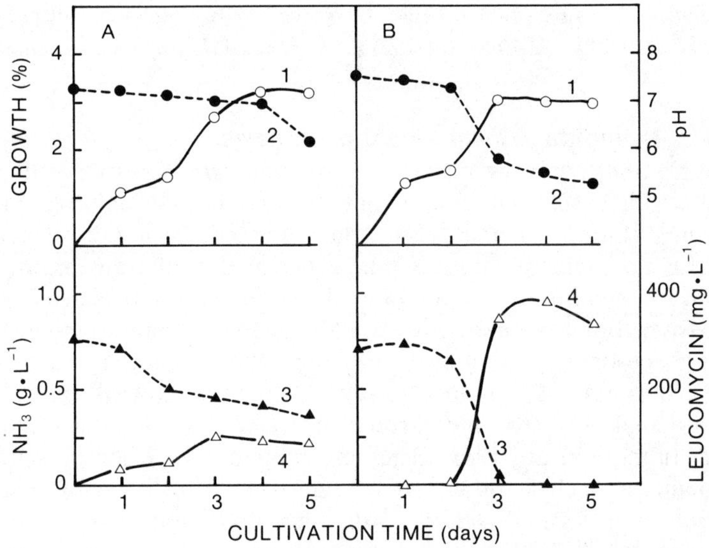

FIGURE 7.10 Growth (1), pH changes (2), nitrogen utilization (3) and leucomycin production (4) in a chemically defined medium: (A) without and (B) with added (1%) magnesium phosphate. The medium for *S. kitasatoensis* KA-429 contained: 3% glycerol, 0.5% glucose, 1% NH_4 lactate, 0.5% $CaCO_3$, 0.05% K_2HPO_4, 0.05% $MgSO_4{\cdot}7H_2O$, trace metals (Fe, Mn, Zn, Co, Cu) at 1 mg/L, pH at 7.5.

fractions (Fig. 7.10). It was concluded that magnesium phosphate trapped ammonium ions from the medium, creating an ammonium ion–limited condition and relief from nitrogen regulation. Production of spiramycin was also enhanced (Ōmura et al., 1980d).

VII. PLASMIDS

Until recently, the role of genetics in studies of antibiotic biosynthesis has centered on blocking pathways and altering regulatory properties to enhance antibiotic productivity. Since the report that plasmids might be involved in aureothricin biosynthesis by *Streptomyces kasugaensis* (Okanishi et al., 1970), biosynthetic studies have included such genetic aspects as mapping and plasmid manipulation. Most of the basic information and useful techniques have come from studies on *Streptomyces venezuelae* (Okanishi et al., 1974) and *Streptomyces coelicolor* A3(2) (Hopwood, 1967; Sermonti and Spada-Sermonti, 1955). A connection between plasmid functions and the biosynthesis of methylenomycin has been demonstrated (Hopwood, 1978). The production of other antibiotics, the resistance of producers to their own antibiotics, and regulation of antibiotic excretion have also been associated with plamids—sometimes without strong supporting evidence (Hopwood, Chapter 2).

A. Role of Plasmids in Macrolide Biosynthesis

Spiramycin nonproducers appear in *S. ambofaciens* KA-1028 (ISP-5053) at a frequency of 1–10%, either spontaneously or after curing treatments with acriflavine. Strains S-1 and AF-30, thus isolated, harbored no detectable plasmid in their cleared lysates but were capable of transforming added forocidin I into spiramycin (Ōmura et al., 1979b). From these early studies, it was assumed that plasmid genes would function in spiramycin biosynthesis at some stage before forocidin I formation. Later, using a new method of detecting plasmids, it was possible with a small (0.5 ml) volume of culture broth to detect both free and chromosome-membrane complex–associated plasmids (Ōmura et al., 1981). The new technique revealed that the above two mutants as well as others (AF-11 and QN-25) of the same phenotype still harbored a plasmid (pSA1, 53Md) that was identical with that borne by the parent KA-1028. In these mutants and in the parent, a substantial amount of the plasmid appeared to be associated with the chromosome-membrane complex. The bound fraction in the mutants was slightly greater than in the parent.

A $pSA1^-$ mutant (strain U-1717) subsequently obtained was a strict nonproducer. Upon regeneration from protoplasts, strain U-1717 was found to recover partial antibiotic productivity, on a solid medium only, at a frequency of more than 99%. This suggested that the spiramycin production (*spp*)-negative mutation was not linked to pSA1. On receiving pSA1 DNA by mating from the producing parent strain KA-1028 or the non-producing QN-25, regenerated strains (U-1717R-24, for instance) were fully capable of producing spiramycin, on both solid and liquid media (Ikeda et al., 1982). Thus, although the locus has not yet been mapped, *spp* should be located on the chromosome; the role of pSA1, if any, may lie in enhancing excretion of spiramycin.

The above results are consistent with other studies on the distribution of plasmids among macrolide-producing streptomycetes (Table 7.1) (Ōmura et al., 1981). Only six of 22 macrolide producers examined contained plasmids, a finding suggesting that macrolide biosynthesis need not be plasmid-specified. Even when the biosynthesis of a macrolide is related to a plasmid function, the relationship seems strain-specific. No plasmid was detected in four independently isolated strains of *S. kitasatoensis* producing leucomycin (Ōmura et al., 1981), while in *S. hygroscopicus,* production of turimycin depended on the presence of a plasmid (Kähler and Noack, 1974). In another leucomycin producer, *Streptoverticillium reticuli,* a plasmid of 48 Md was detected. This disappeared after curing treatments with concomitant loss of the capacity to form leucomycin, melanin, and aerial mycelium (Schrempf and Goebel, 1978).

TABLE 7.1 Distribution of Plasmids Among 12-, 14-, and 16-membered Macrolide-producing Actinomycetes

Strain	*Products*	*Ring size*	*Size of Plasmid (megadaltons)*
Streptoverticillium reticuli	Leucomycin	16	48[a]
Streptomyces kitasatoensis 66-14-38	Leucomycin	16	—[b]
S. fradiae C-373	Tylosin & relomycin	16	—
S. hygroscopicus KCC S-4039	Tylosin & relomycin	16	56
S. sp. AM-4440	Cirramycin	16	—
S. cirratus JTB-3	Cirramycin	16	115
S. sp. KA-430	Angolamycin	16	—
S. sp. AM-4939	Angolamycin	16	57
S. ambofaciens KA-1028	Spiramycin	16	53
S. bikiniensis KA-421	Chalcomycin	16	19 and 38
S. erythraeus ISP-5517	Erythromycin	14	—
S. erythraeus NRRL 2338[c]	Erythromycin	14	7 (and 5 others)[c]
S. brunneogriseus MCRL-0129	Albocycline	14	—
S. spinichromogenes var. *kujimyceticus* A-121	Kujimycin	14	—
Streptomyces sp. AM-4900	Pikromycin	14	—
Streptomyces sp. AM-7762	Methymycin & narbomycin	12, 14	—

a. From Schrempf and Goebel (1978).
b. —, Not detected.
c. S. Ōmura, unpublished.

B. Plasmids and Macrolide Resistance in Macrolide-Producing Streptomycetes

Resistance to their own antibiotic (self-resistance) in producers of erythromycin and two 16-membered macrolides has been investigated by Yang-Graham and Weisblum (1979). They isolated ^{3}H-labeled 23S rRNA's from the macrolide producers and analyzed them for N^6,N^6-dimethyladenine. This base, which had previously been detected in erythromycin-resistant staphylococci (Lai and Weisblum, 1971), was present, in the erythromycin producer, *S. erythraeus*. Moreover, threefold to fivefold more N^6-monomethyladenine was found in producers of tylosin (*S. fradiae*) and of cirramycin (*Streptomyces cirratus*) than in control streptomycetes producing streptomycin or oxytetracycline. This accounts for the low affinity of erythromycin for ribosomes from erythromycin producers (Teraoka and Tanaka, 1974). Alteration of the target site by plasmid-determined N^6,N^6-dimethylation is the principal mechanism of macrolide resistance in pathogenic bacteria (Gryczan et al., 1980; Horinouchi and Weisblum, 1980; Shimizu et al., 1970). It is intriguing that an analogous mechanism appears to be involved in self-resistance in macrolide-producing microorganisms. However, definitive evidence for the involvement of plasmids in self-resistance is still lacking.

VIII. HYBRID BIOSYNTHESIS—A NEW APPROACH TO NEW MACROLIDE ANTIBIOTICS

Penicillin G, a biologically synthesized analog of the naturally occurring penicillin N, is obtained by adding phenylacetic acid to cultures of prototrophic strains of *P. chrysogenum*. The technique is called directed biosynthesis (Werner and Demain, 1981). It is only rarely applicable to the synthesis of analogs in other classes of antibiotics. Addition of a precursor (acetate in macrolide biosynthesis, for example) usually does not increase production of the desired antibiotic. This may be due to the strict substrate specificity of a relevant enzyme or to regulatory effects such as feedback (or feedforward) inhibition and repression caused by the precursors and their analogs. Some of these difficulties can be overcome.

Mutational biosynthesis (Shier et al., 1969) is a technique of practical importance devised for the biosynthesis of new antibiotic derivatives using analogs of biosynthetic precursors and mutants, called idiotrophs by Nagaoka and Demain (1975), blocked in the biosynthesis of the precursors. Incorporation of analogs is more efficient because of the absence of competing natural intermediates. Four new aminoglycosides (hybrimycins) were produced in this way (Shier et al., 1969). The technique has been extended to other classes of antibiotics. A new semisynthetic macrolide, 3-*O*-oleandrosyl-5-*O*-desosaminylerythronolide A oxime was obtained by incubating erythronolide A oxime with an aglycone idiotroph of the oleandomycin producer, *S. antibioticus* (LeMathieu et al., 1976). Unfortunately, the compound was biologically inactive.

To eliminate the tedious task of isolating blocked mutants for mutational biosynthesis, a new method was devised by Ōmura et al. (1980a). This uses the enzyme inhibitor, cerulenin, as illustrated in Fig. 7.11. The biosynthesis of a precursor B of an antibiotic AB is blocked with an enzyme inhibitor, leaving the other pathways intact; then B′, an analog of B, is added to yield AB′, a hybrid antibiotic. B′ may be a chemically modified B, or a natural analog synthesized by a different microorganism. In the latter case, AB′ is synthesized interspecifically by two microorganisms. The term "hybrid biosynthesis" was thus proposed.

A hybrid product 5-*O*-desosaminylprotylonolide (Fig. 7.12) was obtained from protylonolide, the 16-membered macrolide aglycone moiety of tylosin, by incubating it in the presence of cerulenin with a producer of pikromycin, a 14-membered macrolide with desosamine as its sugar moiety (Ōmura et al., 1980a). The compound was identified as antibiotic M-4365 G_1, a natural compound discovered by Kinumaki et al. (1977). Nonetheless, the result promises a route to novel hybrid macrolides in future trials. By using various inhibitors and precursors, it should be possible to assemble versatile combinations of synthetic and microbially produced intermediates without mutation. It is necessary only that product formation follow reasonable biosynthetic routes.

IX. CONCLUDING REMARKS

This article summarizes present knowledge of the biosynthesis of macrolides and its regulation. It provides basic information for future advances in their genetics, production by fermentation, and structure determination and for the elaboration of new derivatives.

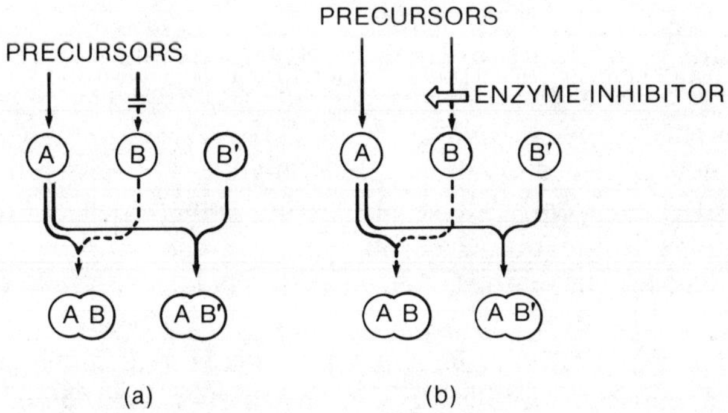

FIGURE 7.11 The basic concepts in (a) mutational biosynthesis using an idiotroph and (b) hybrid biosynthesis using an enzyme inhibitor. A and B are intermediates; B′ is an analogue of B; AB and AB′ are antibiotic products.

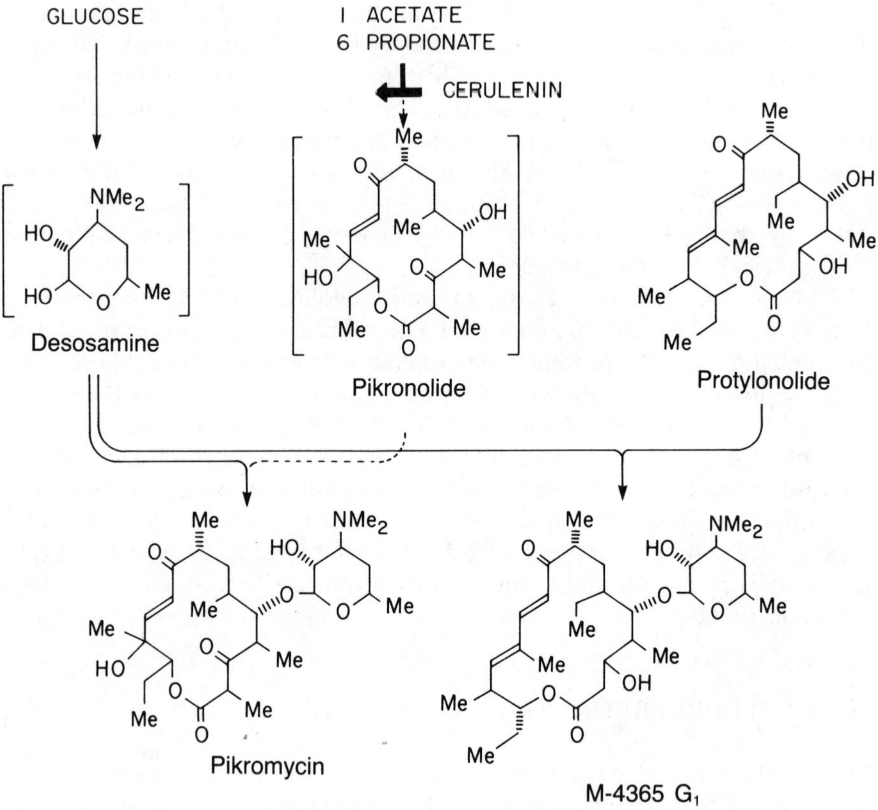

FIGURE 7.12 Hybrid biosynthesis of β-D-desosaminylprotylonolide (= M-4365 G_1) in *Streptomyces* sp. AM-4900. Protylonolide was prepared by fermentation using *S. fradiae* strain KA-427, No. 261.

There are many routes to new macrolide antibiotics. Total synthesis and its potential applications should be available in the near future. Chemical modification has supplied many useful derivatives for modern antibiotic therapy. A new potent derivative of leucomycin, 3″-*O*-propionyleucomycin A_5 (TMS-19Q), has recently been reported (Sakakibara et al., 1981). Screening of soil samples is the classical, but still the most fruitful, way to new macrolides. As has been suggested in this article, many macrolides await discovery. Although most currently known macrolide producers are restricted to *Streptomyces* and *Micromonospora*, there is no reason why these genera alone have acquired the ability to synthesize such antibiotics.

Biochemical modifications have been briefly discussed. No doubt, recombinant DNA techniques will be used to great advantage for the elaboration of new derivatives when suitable host–vector systems are available. They are now under active development in many laboratories. However, before we impatiently rush into this sort of work, further basic biochemical information must be obtained. Even some key steps in macrolide biosynthesis are still far from being fully understood. When the biosynthetic pathways, enzyme properties, gene structure, and mechanisms of gene expression, with many of which this chapter has been concerned, are adequately known, more progress will be possible through gene manipulation in the synthesis of novel macrolides.

REFERENCES

Abdollahi, A., and Buchanan, R. L. (1981) *J. Food Sci. 46,* 143–146.

Aharonowitz, Y. (1979) in *Genetics of Industrial Microorganisms* (Sebek, O. K., and Laskin, A. I., eds.) pp. 210–217, American Society for Microbiology, Washington, D.C.

Aharonowitz, Y. (1980) *Ann. Rev. Microbiol. 34,* 209–233.

Aharonowitz, Y., and Demain, A. L. (1979) *Can. J. Microbiol. 25,* 61–67.

Birch, A. J., Djerassi, C., Dutcher, J. D., Majer, J., Perlan, D., Pride, E., Richards, R. W., and Thomson, P. J. (1964) *J. Chem. Soc.* 5274–5278.

Brockmann, H., and Oster, R. (1957) *Chem. Ber. 90,* 605–617.

Brufani, M., Kluepfel, D., Lancini, G. C., Leitich, J., Mesentsev, A. S., Prelog, V., Schmook, F. P., and Sensi, P. (1973) *Helv. Chim. Acta 56,* 2315–2323.

Burg, R. W., Miller, B. M., Baker, E. E., Birnbaum, J., Currie, S. A., Hartman, R., Kong, T., Monaghan, R. L., Olson, G., Putter, I., Tunac, J. B., Wallick, H., Stapley, E. O., Ōiwa, R., Ōmura, S. (1979) *Antimicrob. Agents Chemother. 15,* 361–367.

Celmer, W. D. (1971) *Pure Appl. Chem. 28,* 413–453.

Celmer, W. D., Chmurny, G. N., Moppett, C. E., Ware, R. S., Watts, P. C., and Whipple, E. B. (1980) *J. Amer. Chem. Soc. 102,* 4203–4209.

Corcoran, J. W. (1981) in *Antibiotics* Vol. IV, *Biosynthesis* (Corcoran, J. W., ed.) pp. 132–174, Springer, Berlin.

Corcoran, J. W., and Chick, M. (1966) in *Biosynthesis of Antibiotics,* (Snell, J. F., ed.) pp. 149–201, Academic Press, New York.

Corcoran, J. W., and Vygantas, A. M. (1977) *Fed. Proc. 36,* 663.

Drew, S. W., and Demain, A. L. (1977) *Ann. Rev. Microbiol. 31,* 342–356.

Dulaney, E. L. (1948) *J. Bacteriol. 56,* 305–313.

Friedman, S. M., Kaneda, T., and Corcoran, J. W. (1964) *J. Biol. Chem. 239,* 2386–2391.

Furumai, T., Seki, Y., Takeda, K., Kinumaki, A., and Suzuki, M. (1973) *J. Antibiot. 26,* 708–710.

Furumai, T., Takeda, K., and Suzuki, M. (1975) *J. Antibiot. 28,* 789–797.

Gallo, M., and Katz, E. (1972) *J. Bacteriol. 109,* 659–667.

Ganguly, A. K., Lee, B. K,, Brambilla, R., Condon, R., and Sorre, O. (1976) *J. Antibiot. 29,* 976–977.

Gardner, A. D., and Chain, E. (1942) *Brit. J. Exp. Pathol. 23,* 123.

Gersch, D. (1980) *Proc. Biochem. 15,* 21–25.

Gersch, D., Bocker, H., and Thrum, H. (1977) *J. Antibiot. 30,* 488–493.

Gerzon, K., Flynn, E. H., Sigal, M. V., Jr., Wiley, F. P., Monahan, R., and Quarck, U. C. (1956) *J. Amer. Chem. Soc. 78,* 6396–6408.

Ghisalba, O., and Nüesch, J. (1981) *J. Antibiot. 34,* 64–71.
Gray, P. P., and Bhuwapathanapun, S. (1980) *Biotechnol. Bioeng. 22,* 1785–1804.
Greenspan, M. D., Macknow, R. C., and Ōmura, S. (1977) *Lipid 12,* 729–731.
Grisebach, H., and Achenbach, H. (1962) *Z. Naturforsch. 17b,* 6–8.
Grisebach, H., and Achenbach, H. (1963) *Experientia 19,* 6–7.
Grisebach, H., and Hofheinz, W. (1964) *J. Roy. Inst. Chem. 88,* 332–340.
Grootwassink, J. W. D., and Gaucher, G. M. (1980) *J. Bacteriol. 141,* 443–455.
Gryczan, T. J., Grandi, G., Hahn, J., and Dubnau, D. (1980) *Nucl. Acid Res. 8,* 6081–6097.
Haupt, I., Fricke, H., Cerna, J., and Rychlik, I. (1976) *J. Antibiot. 29,* 1314–1319.
Hopwood, D. A. (1967) *Bacteriol. Rev. 31,* 373–403.
Hopwood, D. A. (1978) *Ann. Rev. Microbiol. 32,* 373–392.
Horinouchi, S., and Weisblum, B. (1980) *Proc. Nat. Acad. Sci. U.S.A. 77,* 7079–7083.
Hostalek, Z. (1980) *Folia Microbiol. 25,* 445–450.
Hung, P. P., Marks, C. L., and Tardrew, P. L. (1965) *J. Biol. Chem. 240,* 1322–1326.
Hutchinson, C. R., Kurobane, I., Cane, D. E., Hasler, H., and McInnes, A. G. (1981) *J. Amer. Chem. Soc. 103,* 2477–2480.
Ikeda, H., Tanaka, H., and Ōmura, S. (1982) *J. Antibiot. 35,* 497–516.
Kähler, R., and Noack, D. (1974) *Z. Allg. Microbiol. 14,* 529–533.
Kaneda, T., Butte, J. C., Taubman, S. B., and Corcoran, J. W. (1962) *J. Biol. Chem. 237,* 322–328.
Kinumaki, A., Harada, K., Suzuki, T., Suzuki, M., and Okuda, T. (1977) *J. Antibiot. 30,* 450–454.
Kitao, C., Ikeda, H., Hamada, H., and Ōmura, S. (1979a) *J. Antibiot. 32,* 593–599.
Kitao, C., Miyazawa, J., and Ōmura, S. (1979b) *Agr. Biol. Chem. 43,* 833–839.
Lai, C. J., and Weisblum, B. (1971) *Proc. Nat. Acad. Sci. U.S.A. 68,* 856–860.
LeMathieu, R. A., Ax, H. A., Blount, J. F., Carson, M., Despreaux, C. W., Pruss, D. L., Scannell, J. P., Weiss, F., and Kierstead, R. W. (1976) *J. Antibiot. 29,* 728–734.
Maezawa, I., Hori, T., and Suzuki, M. (1974) *Agr. Biol. Chem. 38,* 91–96, 539–542.
Majer, J., Martin, J. R., Egan, R. S., and Corcoran, J. W. (1977) *J. Amer. Chem. Soc. 99,* 1620–1622.
Mardry, N., Sprinkmeyer, R., and Pape, H. (1979) *Eur. J. Appl. Microbiol. Biotechnol. 7,* 365–370.
Martín, J. F. (1976) *Dev. Ind. Microbiol. 17,* 223–231.
Martín, J. F. (1977) *Adv. Biochem. Eng. 6,* 105–127.
Martín, J. F., and Demain, A. L. (1980) *Microbiol. Rev. 44,* 230–251.
Martín, J. F., Naharro, G., Liras, P., and Villanueva, J. R. (1979) *J. Antibiot. 32,* 600–606.
Martin, J. R., and Rosenbrook, W. (1967) *Biochemistry 6,* 435–440.
Martin, J. R., and Goldstein, A. W. (1970) in *Progress in Antimicrobial and Antitumor Chemotherapy* Vol. II, pp. 1112–1116, University of Tokyo Press, Tokyo.
Martin, J. R., Perun, T. J., and Girolami, R. L. (1966) *Biochemistry 5,* 2852–2856.
McCormack, W. M., George, H., Donner, A., Kodgis, L. F., Alpert, S., Lowe, E. W., and Kass, E. H. (1977) *Antimicrob. Agents Chemother. 12,* 630–635.

McCormick, J. R. D., Hirsch, U., Sjolander, N. O., and Ooerschuk, A. P. (1960) *J. Amer. Chem. Soc. 82,* 5006–5007.
McGuire, J. C., Hamilton, B. K., and White, R. J. (1979) *Proc. Biochem. 14,* 2–5.
Miyagawa, K., Suzuki, M., and Uchida M. (1979) *Agr. Biol. Chem. 43,* 1111–1116.
Nagaoka, K., and Demain, A. L. (1975) *J. Antibiot. 28,* 627–635.
Nakayama, K. (1977) *Hakko To Taisha 36,* 1–33.
Nishikiori, T., Wiegand, R. J., Martens, M. G., and Corcoran, J. W. (1978) *Fed. Proc. 37,* 1721.
Nomura, S., Horiuchi, T., Hata, T., and Ōmura, S. (1972) *J. Antibiot. 25,* 365–368.
Ohno, H., Harada, S., and Kishi, T. (1974) *J. Antibiot. 27,* 442–448.
Okachi, R., and Nara, T. (1980) *Biotechnol. Bioeng. 22* (S1), 65–81.
Okanishi, M., Ohta, T., and Umezawa, H. (1970) *J. Antibiot. 23,* 45–47.
Okanishi, M., Suzuki, K., and Umezawa, H. (1974) *J. Gen. Microbiol. 80,* 389–400.
Ōmura, S. (1976) *Bacteriol. Rev. 40,* 681–697.
Ōmura, S., and Kitao, C. (1979) *Hakko To Kogyo 37,* 749–764.
Ōmura, S., and Nakagawa, A. (1975) *J. Antibiot. 28,* 401–433.
Omura, S., and Nakagawa, A. (1981) in *Antibiotics* Vol. IV, *Biosynthesis* (Corcoran, J. W., ed.) pp. 175–192, Springer, Berlin.
Ōmura, S., Hironaka, Y., and Hata, T. (1970) *J. Antibiot. 23,* 511–513.
Ōmura, S., Nakagawa, A., Neszmelyi, A., Gero, S. D., Sepulchre, A.-M., Piriou, F., and Lukacs, G. (1975a) *J. Amer. Chem. Soc. 97,* 4001–4009.
Ōmura, S., Nakagawa, A., Takeshima, H., Miyazawa, J., and Kitao, C. (1975b) *Tetrahedron Lett.* 4503–4506.
Omura, S., Takeshima, H., Nakagawa, A., and Miyazawa, J. (1976) *J. Antibiot. 29,* 316–317.
Ōmura, S., Nakagawa, A., Sakakibara, H., Okekawa, O., Brandsch, R., and Pestka, S. (1977a) *J. Med. Chem. 20,* 732–736.
Ōmura, S., Takeshima, H., Nakagawa, A., Miyazawa, J., Piriou, F., and Lukacs, G. (1977b) *Biochemistry 16,* 2860–2866.
Ōmura, S., Kitao, C., Miyazawa, J., Imai, H., and Takeshima, H. (1978) *J. Antibiot. 31,* 254–256.
Ōmura, S., Ikeda, H., and Kitao, C. (1979a) *J. Biochem. 86,* 1753–1758.
Ōmura, S., Ikeda, H., and Kitao, C. (1979b) *J. Antibiot. 32,* 1058–1060.
Ōmura, S., Kitao, C., Hamada, H., and Ikeda, H. (1979c) *Chem. Pharm. Bull. 27,* 176–182.
Ōmura, S., Ikeda, H., Matsubara, H., and Sadakane, N. (1980a) *J. Antibiot. 33,* 1570–1572.
Ōmura, S., Kitao, C., and Matsubara, H. (1980b) *Chem. Pharm. Bull. 29,* 1963–1965.
Ōmura, S., Tanaka, Y., Kitao, C., Tanaka, H., and Iwai, Y. (1980c) *Antimicrob. Agents Chemother. 18,* 691–695.
Ōmura, S., Tanaka, Y., Tanaka, H., Takahashi, Y., and Iwai, Y. (1980d) *J. Antibiot. 33,* 1568–1569.
Ōmura, S., Ikeda, H., and Tanaka, H. (1981) *J. Antibiot. 34,* 480–481.
Ōmura, S., Sadakane, H., and Matsubara, H. (1982) *Chem. Pharm. Bull. 30,* 223–229.

Pape, H., and Brillinger, G. U. (1973) *Arch. Microbiol. 88,* 25–35.
Puar, M. S., Lee, B. K., Munayyer, H., Brambilla, R., and Weitz, J. A. (1981) *J. Antibiot. 34,* 619–620.
Raczyńska-Bojanowska, K., Rafalski, A., and Ostrowka-Krysiak, B. (1970) *Acta Biochim. Pol. 17,* 331–339.
Rake, G., and Donovick, R. (1946) *J. Bacteriol. 52,* 222–226.
Robinson, M. M. (1962) *Antimicrob. Agents Chemother. 1961,* 394–400.
Sakakibara, H., Okekawa, O., Fujiwara, T., Aizawa, M., and Ōmura, S. (1981) *J. Antibiot. 34,* 1011–1018.
Sankaran, L., and Pogell, B. M. (1975) *Antimicrob. Agents Chemother. 8,* 721–732.
Schmid, E. E. (1971) *Antibiot. Chemother. 17,* 52–66.
Schrempf, H., and Goebel, W. (1978) Abstracts of the Third International Symposium on the Genetics of Industrial Microorganisms, p. 40.
Scott, A. I., Beedling, L. C., Georgopapadakou, N. H., and Subbarayan, C. R. (1974) *Bioorg. Chem. 3,* 238–248.
Seno, E. T., Pieper, R. L., and Huber, F. M. (1977) *Antimicrob. Agents Chemother. 11,* 455–461.
Sermonti, G., and Spada-Sermonti, I. (1955) *Nature 176,* 121.
Shier, W. T., Reinhart, K. L., Jr., and Gottlieb., D. (1969) *Proc. Nat. Acad. Sci. U.S.A. 63,* 198–204.
Shimizu, M., Saito, T., and Mitsuhashi, S. (1970) *J. Antibiot. 23,* 467–468.
Siegrist, S., Lagouardat, J., Moreau, N., and LeGoffic, F. (1981) *Eur. J. Biochem. 115,* 323–327.
Smith, R. L., Bungay, H. R., and Pittenger, R. C. (1962) *Appl. Microbiol. 10,* 293–296.
Soltero, F. V., and Johnson, M. J. (1953) *Appl. Microbiol. 1,* 52–57.
Streicher, S. L., and Tyler, B. (1981) *Proc. Nat. Acad. Sci. U.S.A. 78,* 229–233.
Surikova, E. I., Novikova, L. M., and Gerasimova, T. M. (1976) *Antibiotiki* (Moscow) *21,* 195–200.
Tago, K., and Nagano, M. (1970) in *Progress in Antimicrobial and Anticancer Chemotherapy,* Vol. I, University of Tokyo Press, Tokyo, p. 199.
Takiguchi, Y., Mishima, H., Okuda, M., Terao, M., Aoki, A., and Fukuda, R. (1980) *J. Antibiot. 33,* 1114–1119.
Tardrew, P. L., and Nyman, M. A. (1964) US Patent 3,127,315.
Teraoka, H., and Tanaka, K. (1974) *J. Bacteriol. 120,* 316–321.
Vazquez, D. (1967) *Life Sci. 6,* 845–853.
Vanek, Z., and Majer, J. (1967) in *Antibiotics* Vol. II, (Gottlieb, D., and Shaw, P. D., eds.) pp. 154–188, Springer, Berlin.
Vézina, C., Bolduc, C., Kudelski, A., and Audet, P. (1979) *Antimicrob. Agents Chemother. 15,* 738–746.
Vu-Trong, K., Bhuwapathanapun, S., and Gray, P. P. (1980) *Antimicrob. Agents Chemother. 17,* 519–525.
Vu-Trong, K., Bhuwapathanapun, S., and Gray, P. P. (1981) *Antimicrob. Agents Chemother. 19,* 209–212.
Weinberg, E. D., (1974) *Dev. Ind. Microbiol. 15,* 70–81.
Werner, R. G., and Demain, A. L. (1981) *J. Antibiot. 34,* 551–554.
Woodward, R. B. (1957) *Angew. Chem. 69,* 50–58.
Yang-Graham, M., and Weisblum, B. (1979) *J. Bacteriol. 137,* 1464–1467.

CHAPTER

8

Polyenes

Juan-Francisco Martín

I. INTRODUCTION

The polyenes are a subgroup of the macrolide class of antibiotics discussed in the preceding chapter. They contain a lactone ring of 26–38 atoms, a polyene chromophore consisting of a series of four to seven alternating double bonds that form part of the ring, and usually one aminosugar moiety, either mycosamine (3-amino-3,6-dideoxy-D-mannose) or perosamine (4-amino, 4,6-dideoxy-D-mannose). Neutral sugars like digitoxose (2,6-dideoxy-L-ribohexopyranose), although frequent in nonpolyene macrolides, are rare in polyenes. The polyene macrolides have been traditionally divided into tetraenes, pentaenes, hexaenes, and heptaenes (Table 8.1) by the number of alternating double bonds in the chromophore (Mechlinski, 1973; Martín, 1977).

II. BIOSYNTHESIS OF POLYENES

We have to distinguish between the biosynthesis of the macrolide ring, the origin of the aromatic moiety, and the formation of the aminosugar moiety. The aromatic moiety appears to be the starter for the polymerization process, so we will describe its biosynthesis first.

TABLE 8.1 Some of the More Representative Polyene Macrolide Antibiotics

Name	*Producer Strain*	*Chromophore Subgroup*	*Aminosugar Moiety*	*Aromatic Moiety*
Etruscomycin (lucensomycin)	*Streptomyces lucensis*	Tetraene	Mycosamine[1]	None
Pimaricin (tennecetin)	*S. natalensis* *S. chattanoogensis* *S. gilveosporus*	Tetraene	Mycosamine	None
Tetrin A, B	*Streptomyces* sp.	Tetraene	Mycosamine	None
Rimocidin	*S. rimosus*	Tetraene	Mycosamine	None
Nystatin (polifungin)	*S. noursei* *S. albulus*	Tetraene	Mycosamine	None
Chainin	*Chainia* sp.	Methylpentaene	None	None
Filipin (durhamycin)	*S. filipensis*	Methylpentaene	None	None
Fungichromin (moldcidin B) (pentamycin) (lagosin)	*S. cinnamoneus* *S. roseoluteus* *S. cellulosae* *S. pentaticus*	Methylpentaene	None	None
Eurocidin A, B	*S. eurocidicus* *S. albireticuli*	Pentaene	Mycosamine	None
Flavofungin (mycoticin A)	*S. ruber* *S. flavofungini*	Carbonylpentaene	None	None
Mycoticin B	*S. flavofungini*	Carbonylpentaene	None	None
Roflamycoin (flavomycoin)	*S. roseoflavus*	Carbonyl and methylpentaene	None	None
Candihexin A, B	*S. viridoflavus*	Hexaene	Mycosamine	None
Candihexin E, F	*S. viridoflavus*	Hexaene	None	None
Flavacid	*S. flavus*	Hexaene	None	None
Dermostatin	*S. viridogriseus*	Carbonylhexaene	None	None

TABLE 8.1 (Continued)

Name	*Producer Strain*	*Chromophore Subgroup*	*Aminosugar Moiety*	*Aromatic Moiety*
Ascosin	*S. canescus*	Aromatic heptaene	Mycosamine	*p*-Aminoacetophenone
Ayfactin (aureofacin)	*S. aureofaciens* *S. viridofaciens*	Aromatic heptaene	Mycosamine	*p*-Aminoacetophenone
Aureofungin	*S. cinnamoneus*	Aromatic heptaene	Mycosamine	*p*-Aminoacetophenone
Candicidin	*S. griseus*	Aromatic heptaene	Mycosamine	*p*-Aminoacetophenone
Hamycin	*S. pimprina*	Aromatic heptaene	Mycosamine	*p*-Aminoacetophenone
Heptafungin A	*S. longisporolavendulae*	Aromatic heptaene	Mycosamine	*p*-Aminoacetophenone
Levorin	*S. levoris*	Aromatic heptaene	Mycosamine	*p*-Aminoacetophenone
Trichomycin	*S. hachijoensis*	Aromatic heptaene	Mycosamine	*p*-Aminoacetophenone
DJ-400 B_2	*S. surinam*	Aromatic heptaene	Mycosamine	*p*-Aminoacetophenone
DJ-400 B_1	*S. surinam*	Aromatic heptaene	Mycosamine	N-methyl-*p*-aminoacetophenone
Perimycin (fungimycin)	*S. coelicolor*	Aromatic heptaene	Perosamine[2]	N-methyl-*p*-aminoacetophenone
Amphotericin B	*S. nodosus*	Nonaromatic heptaene	Mycosamine	None
Candidin	*S. viridoflavus* *S. netropsis*	Nonaromatic heptaene	Mycosamine	None
Mycoheptin	*Streptoverticillium mycoheptinicum*	Nonaromatic heptaene	Mycosamine	None
Antibiotic A_2	*Streptomyces* sp.	Nonaromatic heptaene	Not determined	None

1. 3-Amino-3,6-dideoxy-D-mannopyranose.
2. 4-Amino-4,6-dideoxy-D-mannopyranose.
Updated from Martín (1977)

A. Biosynthesis of the Aromatic Moiety

Some polyenes in the heptaene subgroup (for example, ascosin, aureofungin, ayfactin, azacolutin, candicidin, hamycin, heptamycin, levorin, trichomycin, and antibiotics Sch 16656 and DJ-400B2) have an aromatic *p*-aminoacetophenone moiety. Other heptaenes (for example, candimycin, DJ-400B1, and perimycin) have an aromatic N-methyl-*p*-aminoacetophenone moiety.

1. Precursors of the Aromatic Moiety. Most aromatic moieties of secondary metabolites produced by *Streptomyces* are synthetized by the aromatic amino acid biosynthetic pathway (Martín and Liras, 1982). Polyketide-derived aromatic rings are more frequent among fungal products (Turner, 1971). Predictably, therefore, the *p*-aminoacetophenone and N-methyl-*p*-aminoacetophenone moieties of polyenes are synthesized from glucose via the aromatic pathway. 4-Aminobenzoic acid (PABA) has been identified as the immediate precursor of *p*-aminoacetophenone in candicidin. That the ring and the carboxyl group are inserted intact was concluded from the high incorporation of [ring UL-^{14}C]PABA and [7-^{14}C]PABA. The latter labeled the carboxyl group (Liu et al., 1972a,b). High specific incorporation of PABA into candicidin was observed in phosphate-limited resting cells of *Streptomyces griseus* (Martín and Liras, 1976), and 98% of the radioactivity was located in the aromatic moiety after hydrolysis. Similar results were obtained in a study of the biosynthesis of the N-methyl-*p*-aminoacetophenone moiety of perimycin (fungimycin) (Liu et al., 1972b). The N-methyl group was derived from the methyl group of L-methionine.

The aromatic amino acid pathway starts with the condensation of phosphoenolpyruvate (PEP) and erythrose-4-phosphate (E_4P) by the action of the enzyme 3-deoxy-D-arabino-heptulosonate-7-phosphate (DAHP) synthetase. This enzyme has been characterized in the candicidin producer *S. griseus* (see below). Early intermediates of the aromatic amino acid pathway (for example, shikimic acid) are very poorly incorporated into candicidin; radiotracer experiments have shown that shikimate is not taken up by the cells (J. A. Gil and J.-F. Martín, unpublished). The branching point intermediate, chorismic acid, is also not taken up by cells in vivo, but it is converted into PABA by cell extracts of *S. griseus*. (Gil et al., 1981a, b).

Neither tryptophan, phenylalanine, nor tyrosine labels candicidin under conditions in which ^{14}C-labeled PABA is efficiently incorporated. Similarly, anthranilic acid is not incorporated into candicidin, although it is incorporated into the tryptophan in proteins of *S. griseus* (J. A. Gil and J.-F. Martín, unpublished). There is no interconversion of anthranilic acid and PABA by cell extracts. These results implicate chorismic acid itself as the intermediate giving rise to PABA (Fig. 8.1).

Studies with blocked mutants have indicated that conversion of chorismate to PABA requires glutamine (Gibson et al., 1964). Genetic mapping studies of PABA-requiring mutants of *Neurospora crassa* and *Escherichia coli*

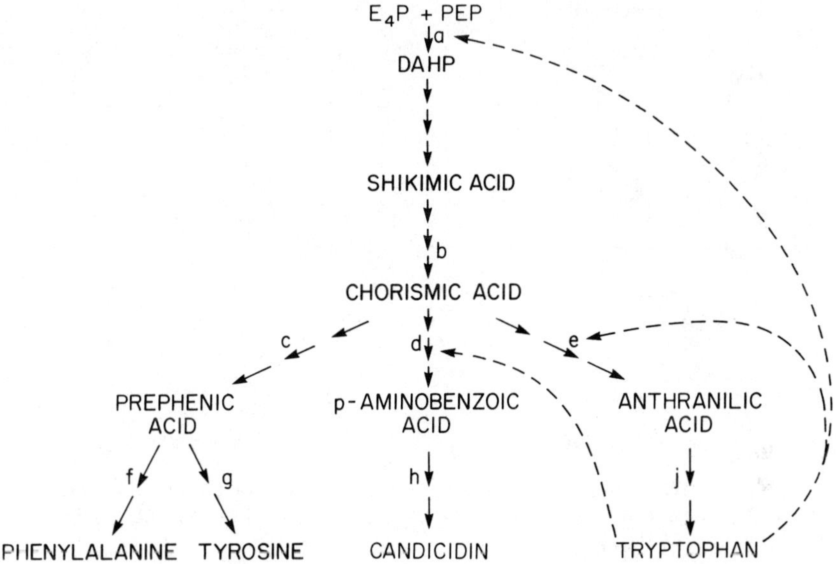

FIGURE 8.1 Biosynthetic pathway leading to aromatic amino acids and the aromatic moiety of candicidin. (a) DAHP synthetase. (b) chorismate synthetase, (c) chorismate mutase, (d) PABA synthetase, (e) anthranilate synthetase, (f) prephenate dehydratase and phenylalanine transaminase, (g) prephenate dehydrogenase and tyrosine transaminase, (h) candicidin synthetase, (j) anthranilate phosphoribosyl transferase, phosphoribosyl-anthranilate isomerase, indoleglycerol phosphate synthetase, and tryptophan synthetase. (Modified from Martín, 1977).

suggest that at least two reactions are involved (Hendler and Srinivasan, 1967; Huang and Pittard, 1967). Biochemical evidence obtained with cell extracts of yeasts supports the two-step pathway, although no intermediate between chorismic acid and PABA has ever been isolated.

2. PABA synthetase and anthranilate synthetase of polyene-producing strains. *p*-Aminobenzoic acid synthetase is an amidotransferase converting chorismic acid into PABA. The enzyme has been purified from extracts of *S. griseus* IMRU 3570. Amidotransferases may use glutamine or ammonium as the amino group donors (Buchanan, 1973). In vitro assays of PABA synthetase show that the enzyme has fivefold to tenfold higher activity when glutamine is used as the amino donor (amidotransferase activity) than when ammonium is used (aminase activity) (Fig. 8.2). Both activities of the enzyme are optimal at pH 8.5–9 and at 37°C. It is noteworthy that the PABA synthetase of *Bacillus subtilis* also has mainly amidotransferase activity (Kane and O'Brien, 1975).

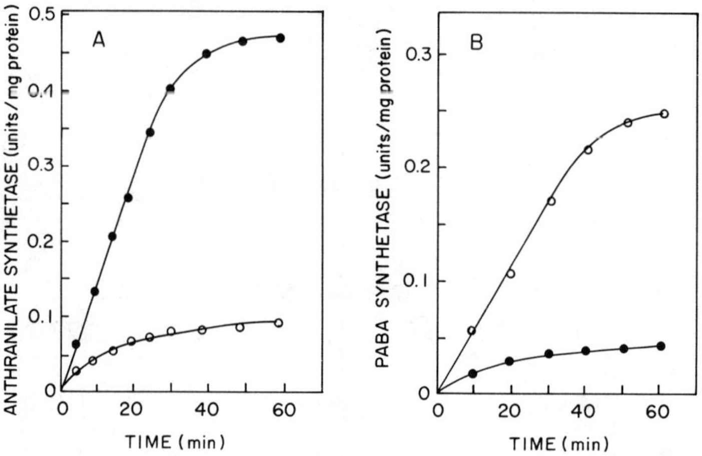

FIGURE 8.2 Aminase and amidotransferase activities of (A) anthranilate synthetase and (B) PABA synthetase. Open circles, amidotransferase activity; closed circles, aminase activity.

Anthranilate synthetase of *S. griseus* has characteristics similar to those of PABA synthetase (optimal pH: 8.5; optimal temperature: 30°C), but as Fig. 8.2 shows, it differs in having mainly aminase rather than amidotransferase activity. The anthranilate synthetase of a chloramphenicol-producing strain of *Streptomyces venezuelae* possesses both amidotransferase and aminase activity (Francis et al., 1978).

The relationship between PABA synthetase and anthranilate synthetase of *S. griseus* is intriguing. PABA synthetases of several microorganisms are composed of nonidentical subunits (Altendorf et al., 1969, 1971; Huang and Gibson, 1970; Sawula and Crawford, 1973). In *B. subtilis*, PABA synthetase is an aggregate of two nonidentical subunits. The crude enzyme will use either glutamine or ammonia as substrates. Subunit A has aminase but not amidotransferase activity. The second subunit, X, is a component of the anthranilate synthase complex. It has no PABA synthetase activity itself but complexes with subunit A to give an aggregate with amidotransferase activity (Kane and O'Brien, 1975) (Fig. 8.3).

Our results suggest that a similar situation may exist in *S. griseus*. The molecular weight of PABA synthetase estimated by gel filtration is 50,000, while anthranilate synthetase shows a molecular weight of 80,000. The higher molecular weight of anthranilate synthetase suggests that an additional subunit is present. The likelihood of common subunits in these two enzymes is supported by the fact that tryptophan, a well-known effector for the formation and activity of anthranilate synthetase, feedback inhibits and strongly represses PABA synthetase as well as anthranilate synthetase in *S. griseus* (J. A. Gil, G. Naharro, and J.-F. Martín, unpublished). The competing affinity of the two enzymes, PABA synthetase and anthranilate

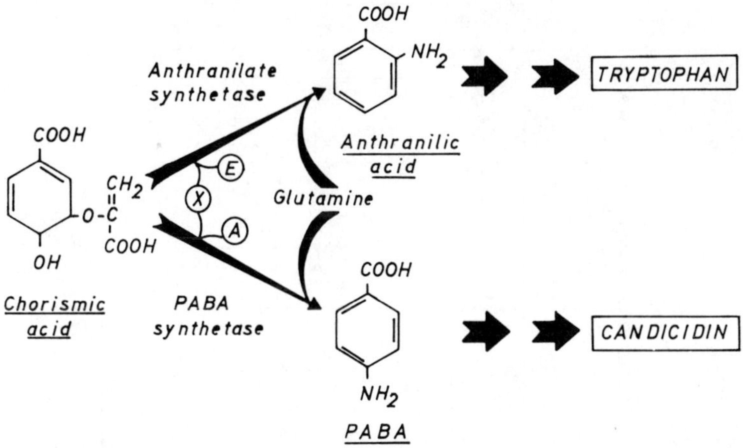

FIGURE 8.3 Competition between PABA synthetase and anthranilate synthetase for their common substrate chorismic acid. The subunit composition of PABA synthetase (X, A) and anthranilate synthetase (X, E) in *Bacillus subtilis* is indicated (from Martín et al., 1981).

synthetase, for chorismate may direct the flow of aromatic intermediates toward tryptophan (primary metabolism) if aminase activity predominates or a high intracellular pool of ammonium is available, but may favor the pathway toward candicidin (secondary metabolism) if amidotransferase activity or the pool of glutamine is higher.

3. PABA synthetase levels and polyene biosynthesis. PABA synthetase activity is not found in the phosphate-rich medium in which inoculum is grown but appears in production medium following a shift-down in phosphate concentration. It reaches a peak at about 24 h and decreases slowly thereafter, while candicidin is still being formed. This pattern of enzyme formation and disappearance is like that of many other enzymes involved in secondary metabolism (for a review, see Martín and Demain, 1980). Several mutant strains blocked in candicidin biosynthesis lack PABA synthetase activity (Gil et al., 1981b; Martín et al., 1983). Moreover, culture conditions that inhibit PABA synthetase activity to different extents cause proportionate decreases in candicidin production (Fig. 8.4). It is apparent that PABA synthetase is involved in candicidin biosynthesis.

So far, PABA synthetase activity has been found in only two strains of actinomycetes: *S. griseus* IMRU 3570 and *Streptomyces coelicolor* var. *aminophylum* IMRU 3865, which produces fungimycin, a polyene containing N-methyl-*p*-aminoacetophenone. It is not present in *S. griseus* NRRL 3851, a producer of cephamycins. Thus this enzyme may be specific for secondary metabolism, existing only in *Streptomyces* strains forming aromatic polyene

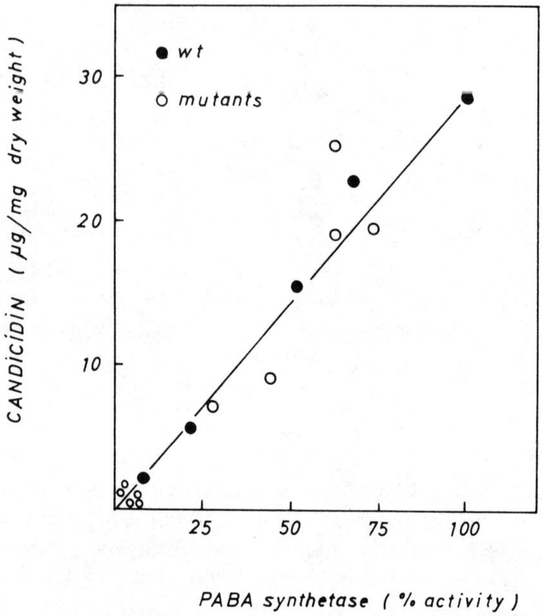

FIGURE 8.4 Correlation between PABA synthetase activity and candicidin production by the wild type under different nutritional conditions (closed circles) and by different mutants of *S. griseus* (open circles). Note the cluster of non-producing mutants which lack PABA synthetase activity (from Gil et al., 1981).

macrolide antibiotics. The existence of a different enzyme involved in PABA formation for primary metabolism cannot be excluded, although we have not detected it when purifying the enzyme. PABA synthetase minus mutants of *S. griseus* do not require PABA for growth in a minimal medium, a finding suggesting either that these strains possess a different amidotransferase that synthesizes the small amounts of PABA required as a vitamin or that they still have PABA synthetase activity below the detection level of the assay.

4. The Aromatic Moiety as the Starter Unit for Polyketide Synthesis. In the aromatic heptaenes (Bohlman et al., 1970; Dutcher, 1968), the aromatic moiety is linked to the carbon atom of the macrolide ring that bears the hydroxyl function forming the lactone bond. This suggests that cyclization occurs after a polyketide chain, started by the aromatic ring, has achieved adequate length. Evidence from biosynthetic studies on polyketide-derived products with aromatic moieties indicates that aromatic rings are indeed used to prime polyketide formation. Thus cinnamoyl starter units are used in the biosynthesis of compounds in the chalcone or stilbene series along with malonyl-CoA units (Light, 1970). Another example of an aromatic moiety that acts as a starter in the synthesis of a polyketide-derived product is the ansamycin antibiotic rifamycin (Chapter 9). Cerulenin, a specific inhibitor of polyketide synthesis, rapidly inhibits incorporation of PABA into candicidin (Martín and McDaniel, 1975d). This supports a starter function for the aromatic moiety, since inhibition of PABA incorporation would not be expected if PABA were attached after the macrolide ring was formed.

B. Regulation of Biosynthesis of the Aromatic Moiety

1. Feedback Regulation of Candicidin Biosynthesis by Aromatic Amino Acids. The aromatic biosynthetic pathway is common to all organisms, but various patterns of regulation have evolved. In many microorganisms, feedback inhibition is exerted on DAHP synthetase, the first enzyme specific for the pathway (Gibson and Pittard, 1968). The inhibition may be of different types. Examples include isoenzyme feedback inhibition in *E. coli,* sequential feedback inhibition in *Bacillus,* concerted feedback inhibition in *Rhodomicrobium,* and cumulative feedback inhibition in *Hydrogenomonas*. In addition, the first enzyme of each branch leading to a specific aromatic amino acid is usually subject to feedback inhibition by its end product. Thus anthranilate synthetase is inhibited by tryptophan; prephenate dehydrogenase and prephenate dehydratase are inhibited by tyrosine and phenylalanine, respectively. Modulation of these branches can, however, be more complicated, with some aromatic amino acids acting as activators or inhibitors of enzymes of other branches (Gibson and Pittard, 1968; Hagino and Nakayama, 1975).

The biosynthesis of candicidin is inhibited by tryptophan, phenylalanine, and, to a lesser extent, tyrosine (Líu et al. 1972a; Gil et al., 1980). A mixture of the three inhibits to a greater extent than each amino acid alone. Tryptophan strongly inhibits incorporation of labeled propionate or PABA. Incorporation of propionate is partially reversed by exogenous PABA, a result suggesting regulation at the PABA synthetase level. Results discussed below indicate that the regulatory effect of aromatic amino acids is exerted at both the PABA synthetase and DAHP synthetase levels.

2. Feedback Regulation of PABA Synthetase and Anthranilate Synthetase by Aromatic Amino Acids. PABA synthetase activity is repressed but not inhibited by tryptophan. Growth of *S. griseus* for 24 h in a medium supplemented with 10 mM tryptophan brought about a 17% decrease in PABA synthetase activity. The same concentration of phenylalanine produced a

10% repression, and tyrosine had little effect. A mixture of the three aromatic amino acids (each at 10 mM) caused a 50% repression (J. A. Gil and J.-F. Martín, unpublished), in good agreement with the effect observed in candicidin biosynthesis. A stronger effect was observed in a resting cell culture of *S. griseus*. PABA synthetase formation under these conditions was repressed 60% by 10 mM tyrosine, 88% by 10 mM phenylalanine, and 84% by 10 mM tryptophan. The difference is probably due to utilization of the amino acids for growth in the fermentation system (Martín and McDaniel, 1976).

In vitro experiments indicate that the aromatic amino acids cause little, if any, inhibition of PABA synthetase activity. Thus their main feedback effect is through repression of enzyme synthesis. Anthranilic acid and PABA also repress formation of PABA synthetase in vivo but do not inhibit the in vitro activity of the enzyme (J. A. Gil and J.-F. Martín, unpublished).

Anthranilate synthetase of *S. griseus* is strongly repressed by tryptophan but is not affected by either phenylalanine or tyrosine (G. Naharro and J.-F. Martín, unpublished). This is an important difference from PABA synthetase which is repressed by all three aromatic amino acids. Moreover, activity of anthranilate synthetase activity was 70% inhibited in vitro by 10 mM tryptophan but was not inhibited by the other aromatic amino acids. The inhibition was competitive with respect to chorismic acid. This result also contrasts with the lack of in vitro inhibition of PABA synthetase by the aromatic amino acids.

In summary, anthranilate synthetase of *S. griseus* is specifically inhibited and repressed by tryptophan, while PABA synthetase is subject to repression but not inhibition by tryptophan, phenylalanine, and tyrosine. Feedback regulation by tryptophan of anthranilate synthetase has also been reported in a chloramphenicol-producing *Streptomyces* species (Francis et al., 1978) and is probably common in members of the genus *Streptomyces*.

3. Feedback Regulation of DAHP Synthetase by Aromatic Amino Acids. In one survey, the synthetases of seven *Streptomyces* species proved to be insensitive to tryptophan (Jensen and Rebello, 1970), but more recent studies indicate that the enzyme in *Streptomyces aureofaciens* (Lingens, 1976) and *Streptomyces antibioticus* (Murphy and Katz, 1980) is feedback-regulated by tryptophan. In *S. griseus* during candicidin biosynthesis, tyrosine (10 mM) exerted a 90% repression of DAHP synthetase formation, while phenylalanine and tyrosine at the same concentration gave 40% and 50% repression, respectively. PABA had no repression effect under the same experimental conditions.

Only tryptophan inhibits DAHP synthetase activity in vitro. Kinetic studies indicate that tryptophan acts as a competitive inhibitor of erythrose-4-phosphate but is not competitive with respect to the other substrate, phosphoenolpyruvate (G. Naharro and J.-F. Martín, unpublished).

C. Biosynthesis of the Macrolide Ring

The biosynthesis of the macrolide ring includes (1) formation of acetyl-CoA precursors from glucose, (2) carboxylation of acetyl-CoA and propionyl-CoA to malonyl-CoA and methylmalonyl-CoA, respectively, (3) condensation of the precursors and cyclization of the ring, and (4) late transformations of the macrolide ring. Little new evidence has been published in recent years on these subjects. A summary of present knowledge is included here; more details can be found in previous reviews (Martín, 1977; Martín and McDaniel, 1977; Martín, 1979a).

1. Precursors. The macrolide ring of polyene antibiotics is formed from two- and three-carbon units derived from acetate and propionate. Birch et al. (1964) showed that 1-, 2-, and 3-^{14}C-labeled propionate and [1-^{14}C] acetate were incorporated into the nystatin aglycone (nystatinolide), whereas the terpenoid precursor [2-^{14}C] mevalonic acid, and the C_1 precursor [methyl-^{14}C] methionine, did not label nystatin. Degradation of nystatin from these experiments suggested that acetate was the primer unit from which the chain is extended. Similar precursor incorporations were observed in the biosynthesis of lucensomycin (Manwaring et al., 1969), amphotericin B (Linke et al., 1974; Perlman and Semar, 1965), candicidin (Liu et al., 1972a; Martín and Liras, 1976) levorin (Belousova et al., 1971), and fungimycin (Liu et al., 1972b). The failure of methyl donors to label the macrolide ring was also described by Perlman and Semar (1965) for amphotericin B and by Liu et al. (1972a) for candicidin. Side-chain methyl groups were derived instead from the methyl group of propionate units.

Phosphate-limited resting cells of *S. griseus* incorporated [^{14}C] propionate into candicidin to a much greater extent than [^{14}C] acetate. Also, [2-^{14}C] propionate yielded candicidin of higher specific activity than did [U-^{14}C] propionate or [1-^{14}C] propionate (Martín and Liras, 1976). Similar results have been reported with amphotericin B (Linke et al., 1974) and are probably due to loss of radioactivity from C-1 of propionate during the condensation process. The labeling of levorin with a mixture of [^{14}C] acetate and [^{14}C] propionate was higher than that obtained with each precursor alone (Belousova et al., 1971). This indicates that both acetate and propionate participate in the polymerization process. Straight side chains of four to six carbon atoms exist in the polyenes filipin, chainin, and etruscomycin, which might represent direct introduction of preformed 6- to 8-carbon fatty acids. Butyrate stimulates the synthesis of candicidin by resting cells of *S. griseus* (Martín and McDaniel, 1976), but there is no firm evidence for or against the direct utilization of 4-carbon precursors such as butyrate or succinate in the biosynthesis of polyenes.

2. Formation from Activated Precursors. Two types of lipid products, fatty acids and polyketides, are derived from acetate and malonate. Fatty acids are synthesized by a specific pathway in which a single acetate (or propionate) starter unit and a small number of malonate units (usually seven or eight) are joined together. Polyketides are formed by nonspecific pathways in which one starter unit (acetate or some other compound) and a highly variable number of malonate or methylmalonate units are attached together. In fatty acid synthesis, the β-keto group introduced at each condensing step is modified by reduction, dehydration, and hydrogenation prior to the next condensation step (Volpe and Vagelos, 1973). In polyketide synthesis, the β-keto group is most commonly retained. Successive condensation steps give a polyketide chain of the general type:

$$R—COCH_2COCH_2COCH_2COCH_2COSCoA$$

where R, the starter unit, varies.

The pools of acetyl-CoA and propionyl-CoA are important in the synthesis of polyenes. The acyl-CoA pool is up to 30-fold higher in a high-polifungin producing strain than in a low producer. Two carboxylating systems have been reported in the producing organism: a nonspecific ATP- and magnesium-dependent acetyl-CoA carboxylase (acetyl-CoA:carbon dioxide ligase) that utilizes carbon dioxide for carboxylation, and a carboxyltransferase (methylmalonyl-CoA:pyruvate carboxyltransferase) which uses oxalacetate as a carboxyl group donor. In addition, a third carboxylating enzyme, phosphoenolpyruvate carboxylase, also plays a role in the biosynthesis of polyene macrolides. Carboxylation of acetyl-CoA and propionyl-CoA has been correlated with the synthesis of polifungin (Roszkowski et al., 1972) and turimycin (Gräfe et al., 1975).

That precursor units in polyene macrolide biosynthesis polymerize in the same way as acetyl-CoA polymerizes during fatty acid biosynthesis was demonstrated by the use of cerulenin, an inhibitor of fatty acid synthesis (Goldberg et al., 1973; Ōmura, 1976). Cerulenin is a noncompetitive inhibitor of both the multienzyme fatty acid synthase complex of yeast and the nonassociated fatty acid synthase of *Escherichia coli* (Vance et al., 1972). It specifically inhibits the β-ketoacyl-acyl carrier protein synthase subunit, which catalyzes the condensation of fatty acyl-acyl carrier protein with malonyl-acyl carrier protein. Cerulenin inhibits the synthesis of candicidin without affecting protein synthesis or precursor uptake (Martín and McDaniel, 1975d; Martín, 1976a). The cerulenin concentration required for 50% inhibition of candicidin synthesis is similar to that required for 50% inhibition of fatty acid synthesis (Martín and McDaniel, 1975d).

"Candicidin-synthesizing" activity has been measured recently in cell extracts of *S. griseus* by determining the incorporation of [1-^{14}C] acetyl-CoA and [2-^{14}C] malonyl-CoA into candicidin. Labeled candicidin is extracted with butanol after the reaction, mixed with unlabeled candicidin, and chromatographed on silica gel G plates. In addition to the radioactive precursor, the reaction mixture includes methylmalonyl-CoA, NADPH, phosphate buffer, and cell extract. No synthesis occurs when the enzyme extract is inactivated by heating or in the absence of enzyme. Little in vitro synthesis of candicidin occurs in the absence of methylmalonyl-CoA or in reaction mixtures without NADPH. A nonproducer mutant, which was isolated in a screening of mutants blocked in candicidin biosynthesis, shows essentially no in vitro biosynthetic activity (G. Naharro and J.-F. Martín, unpublished). The requirement for NADPH is of interest, since it has been proposed that the relative formation of fatty acids and polyketide compounds depends upon the cellular pool of NADPH at different times in the fermentation (Martín, 1976b; Dimroth et al., 1970). However, no detailed study has been made of the formation of fatty acids and polyenes by cell extracts of polyene-producing strains at different concentrations of NADPH.

D. Biosynthesis of the Sugar Moieties

Two unusual aminosugars, mycosamine and perosamine, are components of many polyene antibiotics (Table 8.1). Mycosamine (Dutcher et al., 1963; Saltza et al., 1963) is the sugar component of most polyenes. It is attached by a β-glycosidic bond to the aglycone. Perosamine, the aminosugar of perimycin, is an isomer of mycosamine with the amino group at C-4 (Lee and Schaffner, 1966; Stevens et al., 1968). A new tetraene antifungal antibiotic, barodamycin, contains an aminosugar different from mycosamine (Ganju et al., 1976), but it is not known whether the sugar is perosamine. There are also several polyenes that lack an aminosugar. Most of these belong to the methylpentaene and carbonylpentaene subgroups (Martín and McDaniel, 1977). Mycosamine and perosamine are closely related to the aminosugars of nonpolyene macrolides. Similar aminosugars exist in a variety of anthracycline antibiotics and in the lipopolysaccharides of *E. coli, Salmonella, Vibrio,* and other enterobacteria (Martín, 1977).

The neutral sugar digitoxose (2,6-dideoxyribohexopyranose) has been reported recently as a component of nystatin A_3, candidinin, and polyfungin B. Digitoxose does not seem to be a direct precursor of either mycosamine or perosamine if, as in the biosynthesis of the aminosugars of nonpolyene macrolides and aminoglycoside antibiotics, TDP-4-keto-6-deoxy-D-glucose is an intermediate (Matern et al., 1973; Pape and Brillinger, 1973; Martín, 1977, 1979b).

Glucose labeled in different carbon atoms is incorporated into the macrolide sugars without extensive randomization of label (Corcoran and Chick, 1966; Vanek and Majer, 1967). Incorporation studies using resting cells of *S. griseus* showed that glucose, mannose, and galactose were less efficiently incorporated into candicidin than propionate, but the percentage of label incorporated into the mycosamine moiety was higher (Martín and Gil, 1979). More than 7.5% of the radioactivity in candicidin labeled with [1-^{14}C]glucose was located in the aminosugar moiety. [1-^{14}C]Mannose was even more efficiently incorporated into mycosamine (17.2% of the total incorporation); although this is of interest, since the sugar skeleton of mycosamine is that of a mannopyranose, a similar result was obtained when [1-^{14}C]galactose was used as the precursor. [U-^{14}C]Glucosamine was incorporated into candicidin with very low efficiency, but all of the radioactivity was located in the mycosamine moiety. Further studies are required to clarify the exact sequence of sugar activation and incorporation into the antibiotic. The attachment of the aminosugar to the macrolide ring appears to be a late step in the biosynthetic pathway, since the aglycones of macrolides have been identified as fermentation products (Hung et al., 1965). Mutants of *Streptomyces noursei* can synthesize the aglycone of polifungin but not the aminosugar moieties (Roszkowski et al., 1972). The aglycone is antibiotically inactive.

Candihexin, produced by *Streptomyces viridoflavus* strain $18A_2$, can be resolved into a mixture of chemically related components (Martín and McDaniel, 1975a). Two of the components lack the mycosamine moiety and are also inactive against yeasts or filamentous fungi (Martín and McDaniel, 1975b). Candihexin that accumulates intracellularly consists exclusively of nonglycosylated inactive components, whereas the excreted antibiotic contains a mixture of glycosylated and nonglycosylated forms (Martín and McDaniel, 1975c). The absence of intracellular glycosylated components suggests that glycosylation takes place during secretion. This is consistent with the view that the macrolide ring acts as a lipid-soluble carrier of the aminosugar during the secretion process.

E. Regulation of Biosynthesis

The biosynthesis of antibiotics is controlled by several general regulatory mechanisms in addition to specific mechanisms of feedback regulation, attenuation, or induction (Martín and Demain, 1980). In the case of polyene macrolide antibiotics, only phosphate regulation has been studied in detail, although carbon catabolite and nitrogen regulation may also exist.

Glucose, although a well-known repressor, does not greatly affect the biosynthesis of polyene antibiotics. Polyene-producing strains can use a variety of carbon sources for antibiotic biosynthesis (Martín and McDaniel, 1977), and glucose is usually one of the best. Phosphate-limited resting cells of *S. griseus* use glucose as a source of intermediates for candicidin biosynthesis. When low levels of glucose (10 mg/ml) were used, the initial rate of candicidin biosynthesis was similar to that in a medium with high glucose (50 mg/ml), but antibiotic synthesis ceased at 24 h because of glucose depletion. When the slowly utilizable disaccharide sucrose was used, the initial rate of antibiotic biosynthesis was slightly higher, suggesting a low degree of carbon catabolite regulation, but the final yield of antibiotic was similar. Lactose was poorly used by *S. griseus* as a carbon source for growth and for candicidin biosynthesis.

Carbon catabolite regulation seems to occur in the biosynthesis of candidin by *S. viridoflavus* and of candihexin by a mutant of the same strain, since slow feeding of glucose to batch cultures increased production of both antibiotics (Martín and McDaniel, 1974).

1. Phosphate Regulation. Inorganic phosphate is the crucial growth-limiting nutrient for polyene production. Phosphate controls the biosynthesis of antibiotics formed by diverse biosynthetic pathways. Moreover, its effect is a rather general one, controlling not only antibiotic biosynthesis but also differentiation in *Streptomyces* and fungi (Martín, 1976b; Pazaout et al., 1982). However, the biosynthesis of several polyenes is extremely sensitive to phosphate (Martín, 1977). Only candicidin has been studied in detail. Onset of its biosynthesis is triggered by phosphate depletion (Liras et al., 1977). When 10 mM phosphate was added at the beginning of the fermentation, this nutrient remained in excess throughout, DNA and RNA synthesis extended during the idiophase, and no candicidin was produced. Phosphate regulation is mainly a repression effect on enzymes involved in the polyene pathway, but phosphate also inhibited candicidin biosynthesis when protein synthesis was arrested with chloramphenicol (Martín and Demain, 1976).

An important question that remains to be answered unequivocally is whether phosphate is the ultimate effector in vivo or whether it simply regulates the level of a phosphate-derived intracellular effector (Co-R in the model shown in Fig. 8.5) that, in turn, controls expression of antibiotic biosynthesis. The intracellular level of ATP in *S. griseus* increases rapidly

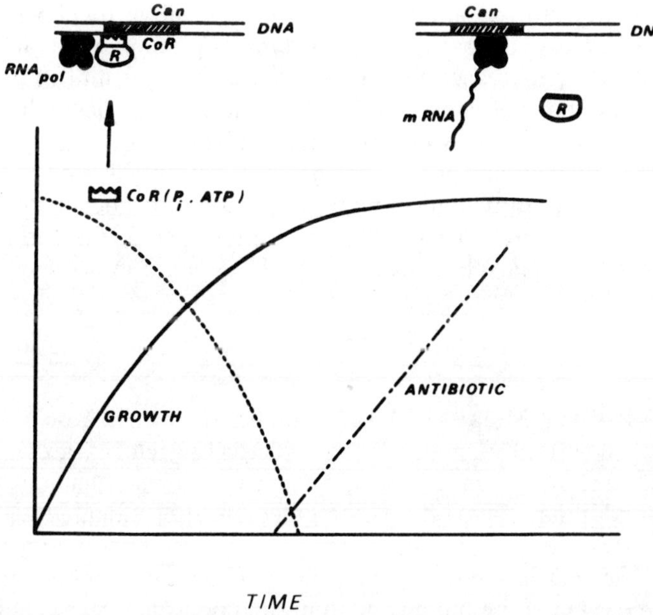

FIGURE 8.5 Model proposed to explain phosphate control of PABA synthetase and antibiotic biosynthesis at the transcriptional level. RNA pol, RNA polymerase; R, repressor; Co-R, corepressor (from Martín et al. 1979).

after the addition of 10 mM phosphate to antibiotic-producing cells, doubling within 5 min of phosphate addition (Martín and Demain, 1976). Thus ATP could have a regulatory role. This would be consistent with data from time-course studies on ATP concentration during antibiotic fermentations, which show a rapid decrease in intracellular ATP just before antibiotic synthesis begins (Liras et al., 1977). Exogenous ribonucleotides, including cyclic AMP, affect candicidin biosynthesis (Martín and Demain, 1977a), but this seems to be due to phosphate released during uptake (Martín and Demain, 1977b).

Considerable progress has been made recently in elucidating the molecular mechanism of candicidin biosynthesis through the isolation of phosphate-deregulated mutants and the characterization of key enzymes of primary and secondary metabolism.

2. Phosphate-Deregulated Mutants in Candicidin Biosynthesis Phosphate-deregulated (PD) mutants have been obtained by two different procedures. In the first, they were isolated from *S. griseus* in large numbers by plating spores on phosphate-supplemented agar medium after UV mutagenesis (Martín et al., 1979). One of them (PD-110) was studied in detail. Candicidin biosynthesis by resting cells of the mutant, measured by the incorporation of [^{14}C]-propionate into the antibiotic, was less sensitive to phosphate than was antibiotic biosynthesis by the wild type. In complex medium, yields of candicidin were increased rather than repressed by 1 mM of phosphate. Higher concentrations of phosphate had less effect on candicidin biosynthesis in the mutant than in the wild type but were still repressing. Mutant PD-110 showed normal uptake of [^{32}P]phosphate. Levels of alkaline phosphatase activity during the fermentation were as sensitive to phosphate as in the wild type (G. Naharro and J.-F. Martín, unpublished). This contrasts with the insensitivity to phosphate of alkaline phosphatase in arsenate-resistant mutants (see below).

3. Arsenate-Resistant Mutants. A large number of arsenate-resistant mutants were obtained after UV mutagenesis by selecting clones growing in the presence of 100 mM sodium arsenate. All arsenate-resistant mutants produced more candicidin than the wild type in a medium supplemented with 10 mM phosphate (Fig. 8.6). One of them showed reduced permeability to [^{32}P]phosphate, but in the others phosphate uptake was identical to that of the wild type. One of the mutants with normal phosphate permeability (AS-28) was much less sensitive than the wild type phosphate inhibition of candicidin biosynthesis and differed from the wild type in that its alkaline phosphatase was not repressed by phosphate. The same result has been

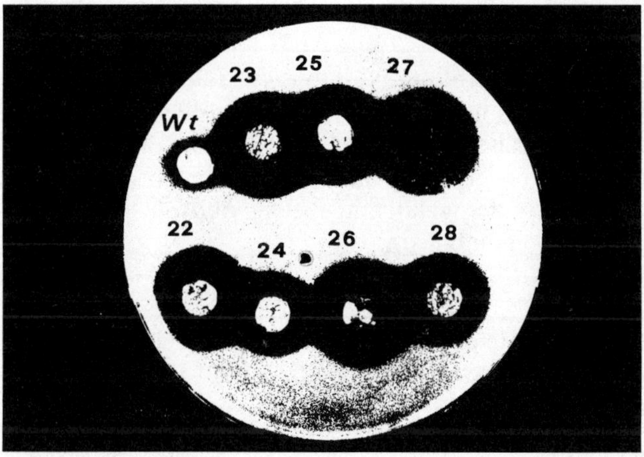

FIGURE 8.6 Production of candicidin by the wild-type strain of *S. griseus* (Wt) and arsenate-resistant mutants (numbers 22 to 28) in solid SPG medium (soya peptone 10 g·L^{-1}; glucose 10 g·L^{-1}) supplemented with 10 mM sodium phosphate (from Naharro et al. 1981).

reported for arsenate-resistant mutants of *E. coli* (Yagil and Beeri, 1977). Possibly, phosphate repression of candicidin biosynthesis is exerted by the same effector and acts by the same mechanism as phosphate repression of alkaline phosphatase.

4. Effect of Phosphate on Specific Enzymes of Primary Metabolism. Phosphorylation of fructose-6-phosphate by phosphofructokinase and ATP in the presence of Mg^{++} ions is a key enzymatic step in glycolysis. The specific activity of phosphofructokinase in wild-type cells grown in the presence of phosphate was double that of cells grown in its absence. Differential plots of phosphofructokinase-specific activity versus protein suggested induction of the enzyme by phosphate (or a phosphate-derived effector). However, simple activation by phosphate-derived metabolites cannot be excluded, since the enzyme was not purified and extracts might have contained an activator. The same level of activity and a similar increase in phosphate-supplemented cultures was observed in the phosphate-deregulated and arsenate-resistant strains PD-110 and AS-28.

In wild-type cells grown in the presence of 5 mM phosphate, the specific activity of glucose-6-phosphate dehydrogenase was about threefold higher than in cells grown in the absence of phosphate. Stimulation of glucose-6-phosphate dehydrogenase was also observed in short-term experiments when phosphate was added to resting cell cultures of *S. griseus*. The increase in pentose–phosphate cycle activity would support the increased formations of ribose–phosphate precursors for nucleic acid synthesis that follows phosphate addition.

The same results were observed in extracts of mutants PD-110 and AS-28 that had been grown in the absence and presence of phosphate. Evidently, these mutants are not altered in phosphate regulation of the pentose–phosphate cycle. Since they were also not altered in phosphate control of glycolysis, they are mostly likely deregulated in specific phosphate control mechanisms affecting secondary metabolism.

5. Control by Phosphate of the Primary Aromatic Pathway. Phosphate also regulates the first enzyme of the aromatic amino acid biosynthetic pathway. Extracts of *S. griseus* cells grown in 10 mM phosphate were clearly reduced in DAHP synthetase activity. The same effect was observed in short-term experiments with resting cell systems. The enzyme activity itself is not inhibited by phosphate concentrations up to 100 mM added directly to the cell-free extract. However, enzyme activity in vitro was inhibited by 0.5–1 mM of fructose-1,6-diphosphate and glucose-6-phosphate. The phosphate effect on DAHP synthetase may be explained on the basis of inhibition of enzyme activity by the increased levels of sugar–phosphate derivatives formed following phosphate addition to the culture.

The inhibitory effect of sugar–phosphates on DAHP synthetase explains in part the decreased flow of aromatic precursors leading to the biosynthesis of aromatic polyene macrolide antibiotics. Similar results were reported by Lowe and Westlake (1970) for the DAHP synthetase of the chloramphenicol producer *S. venezuelae*.

6. Control by Phosphate of Specific Enzymes of Secondary Metabolism. The activity of PABA synthetase, which behaves as a typical enzyme of secondary metabolism, is strongly repressed in cells grown in the presence of phosphate (Naharro et al., 1981). However, neither inorganic phosphate nor ATP inhibited in vitro activity. Also, neither glucose-6-phosphate nor fructose-1,6-diphosphate had any detectable effect on PABA synthetase activity in vitro. Repression of PABA synthetase activity paralleled the reduction in candicidin biosynthesis caused by phosphate (Fig. 8.7). Differential plots of PABA synthetase activity versus protein in the extracts supported a repression mechanism.

Phosphate-deregulated and arsenate-resistant mutants were less sensitive to phosphate repression of PABA synthetase than the wild type. PABA synthetase activity in mutant PD-110 was still appreciably repressed by high concentrations of phosphate, although less than in the wild type. PABA synthetase activity in arsenate-resistant mutant AS-28 was much less sensitive to phosphate repression than in the wild type. In resting cell systems, PABA

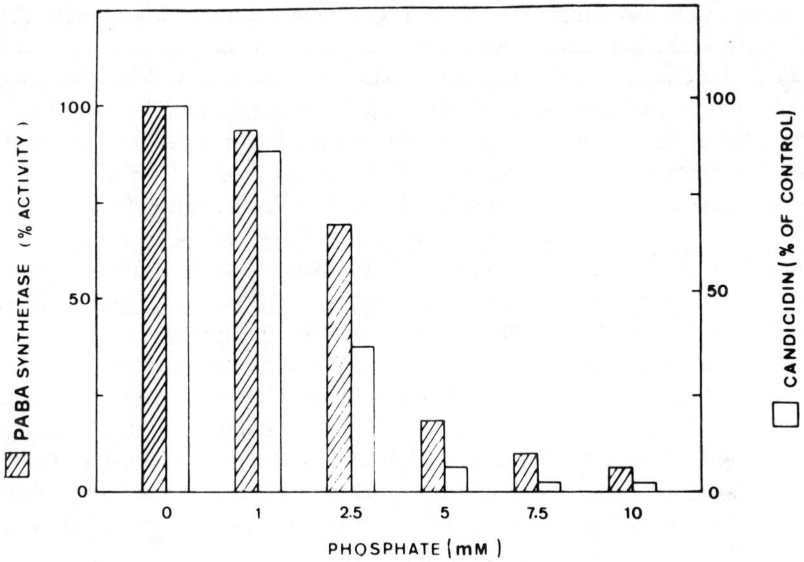

FIGURE 8.7 Effect of growth in increasing concentrations of phosphate on PABA synthetase activity and candicidin biosynthesis. Cultures were grown for 24 hours in liquid medium supplemented with phosphate at the concentrations indicated.

synthetase activity in mutant AS-28 was completely unaffected by up to 10 mM of phosphate.

Phosphate repression of PABA synthetase explains the lack of enzyme activity in inoculum cultures grown in a phosphate-rich medium. Only when phosphate is depleted does formation of PABA synthetase, and therefore candicidin biosynthesis, occur (Fig. 8.5). If phosphate-deregulated mutants have a "repressor" molecule with a weaker affinity for phosphate (or the functional corepressor), this would explain their constitutive expression of PABA, even in the absence of high phosphate concentrations.

III. BIOCHEMICAL GENETICS OF POLYENE BIOSYNTHESIS

Little is known about the genetics of polyene-producing strains. However, in recent years, some preliminary studies have been carried out on the genetics of candicidin biosynthesis by *S. griseus*.

A. PABA Synthetase–negative Mutants

Treatment of *S. griseus* with "curing" agents, which are known also to produce chromosomal rearrangements, led to the formation of up to 25%

of clones that had either lost PABA synthetase activity completely (*pab*) and were therefore unable to make candicidin, or contained only a basal level of the enzyme. The same result was obtained with UV light treatment or by protoplasting and regeneration of the protoplasts. The *pab* mutants were also unable to form extracellular proteases (*prt*) when incubated at 38°C (nonpermissive temperature) (M. T. Alegre and J.-F. Martín, unpublished). Those mutants completely deficient in PABA synthetase showed a reversion rate of less than 10^{-8}. Candicidin biosynthesis by these mutants could not be restored by addition of *p*-aminobenzoic acid. In all *pab* mutants having residual PABA synthetase activity, this activity was insensitive to phosphate (*pho* R) (J. A. Gil and J.-F. Martín, unpublished).

These results implicate an unstable genetic element that is involved in the formation of PABA synthetase and in the temperature-sensitive formation of extracellular proteases in *S. griseus*. It controls expression of the structural gene for protease (*prt*) but does not code for the protease itself because, after curing (or chromosomal rearrangement), normal expression of proteases at low (permissive) temperature still occurs. Apparently, the structural gene is not altered by the curing (rearrangement) treatment.

One intriguing question is how the pleitropic mutation (*pab, prt*) affects phosphate regulation of PABA synthetase residual activity, which becomes completely insensitive to phosphate. Our results suggest that phosphate regulation is exerted by an intracellular pleiotropic effector that controls expression of both *pab* and *prt* genes (Martín et al., 1983).

B. Extrachromosomal Genetic Element pSG1

An extrachromosomal genetic element, pSG1, has been found in cleared lysates of the candicidin producer *Streptomyces griseus* IMRU 3570. This element presents some unusual features. It is better observed when lysates are electrophoresed in 0.8% agarose gel at low voltage (0.6 volt/cm). It could not be isolated by cesium chloride–ethidium bromide centrifugation. The abnormal behavior might be related to association with chromosomal DNA, protein, or membrane fragments (Marahiel et al., 1981; Ikeda et al., 1982). Protein association with the ends of linear plasmid DNA has been found in a *Streptomyces* species producing lankacidin (Hirochika and Sakaguchi, 1982).

An extrachromosomal band similar to pSG1 has been found in *Streptomyces levoris*-producing levorin, an antibiotic almost identical to candicidin, but not in strains producing other polyene macrolide antibiotics. Those *pab* mutants lacking the enzyme or those which have a low level of phosphate-derepressed enzyme still show a faint plasmid band, indicating that the plasmid has not been completely removed by the curing treatment. Deletions or rearrangements in the extrachromosomal element or in the chromosome may therefore be responsible for the loss of PABA synthetase, as is the case in *Streptomyces reticuli* (Schrempf, 1982).

ACKNOWLEDGMENTS

This work was supported partially by grants from Antibioticos, S. A. (León, Spain) and the Comision Asesora de Investigación Científica y Técnica, Madrid.

REFERENCES

Altendorf, K. H., Bacher, A., and Lingens, F. (1969) *FEBS Lett. 3,* 319–321.
Altendorf, K. H., Gilch, B., and Lingens, F. (1971) *FEBS Lett. 16,* 95–98.
Belousova, I. I., Lishenevskaya, E. B., and Elgart, R. E. (1971) *Antibiotiki 16,* 684–687.
Birch, A. J., Holzapfel, C. W., Rickards, R. W., Djerassi, C., Suzuki, M., Westley, J., Dutcher, J. D., and Thomas, R. (1964) *Tetrahedron Lett.* (1964) 1485–1490.
Bohlmann, F., Behamlow, E. V., Neuhahn, H. J., Brandt, R., and Bethke, H. (1970) *Tetrahedron 26,* 2199–2207.
Buchanan, J. M. (1973) *Adv. Enzymol. 39,* 91–183.
Corcoran, J. W., and Chick, M. (1966) in *Biosynthesis of Antibiotics* (Snell, J. F., ed.) Vol. I, pp. 159–201, Academic Press, New York.
Dimroth, P., Walter, H., and Lynen, F. (1970) *Eur. J. Biochem. 13,* 98–110.
Dutcher, J. D. (1968) in *Encyclopedia of Chemical Technology* (Kirk, R. E., and Othmer, D. F., eds.) Vol. XVI, pp. 133–143. Interscience, New York.
Dutcher, J. D., Walters, D. R., and Wintersteiner, O. (1963) *J. Org. Chem. 28,* 995–999.
Francis, M. M., Vining, L. C., and Westlake, D. W. S. (1978) *J. Bacteriol. 134,* 10–16.
Ganju, P. L., Rengaraju, S., and Iyengar, M. R. S. (1976) Abstracts of the Fifth International Fermentation Symposium, Berlin, p. 220.
Gibson, F., and Pittard, J. (1968) *Bacteriol. Rev. 32,* 465–492.
Gibson, F., Gibson, M., and Cox, G. B. (1964) *Biochim. Biophys. Acta 82,* 637–638.
Gil, J. A., Liras, P., Naharro, G., Villanueva, J. R., and Martín, J.-F. (1980) *J. Gen. Microbiol. 118,* 189–195.
Gil, J. A., Naharro, G., Villanueva, J. R., and Martín, J.-F. (1981a) in *Actinomycetes* (Schaal, K. P., and Pulverer, G., eds.) pp. 459–463, Gustav Fischer, Stuttgart, West Germany.
Gil, J. A., Naharro, G., Villanueva, J. R., and Martín, J.-F. (1981b) in *Advances in Biotechnology* Vol. III, *Fermentation Products* pp. 141–146, (Vezina, C., and Singh, K., eds.) Pergamon Press, Toronto.
Goldberg, I., Walker, J. R., and Bloch, K. (1973) *Antimicrob. Agents Chemother. 3,* 549–554.
Gräfe, U., Bocker, H., and Thrum, H. (1975) *Z. Allg. Mikrobiol. 13,* 115–129.
Hagino, H., and Nakayama, K. (1975) *Agr. Biol. Chem. 39,* 351–361.
Hendler, S., and Srinivasan, P. R. (1967) *Biochim. Biophys. Acta 141,* 656–658.
Hirochika, H., and Sakagüchi, K. (1982) *Plasmid 7,* 59–65.
Huang, M., and Gibson, F. (1970) *J. Bacteriol. 102,* 767–773.
Huang, M., and Pittard, J. (1967) *J. Bacteriol. 93,* 1938–1942.

Hung, P. P., Marks, C. L., and Tardrew, P. L. (1965) *J. Biol. Chem. 240,* 1322–1326.
Ikeda, H., Tanaka, H., and Ōmura, S. (1982) *J. Antibiot. 32,* 497–506.
Jensen, R. A., and Rebello, J. L. (1970) *Dev. Ind. Microbiol. 11,* 105–121.
Kane, J. F., and O'Brien, H. D. (1975) *J. Bacteriol. 123,* 1131–1138.
Lee, C. H., and Schaffner, C. P. (1966) *Tetrahedron Lett. 1966,* 5837–5840.
Light, R. J. (1970) *J. Agr. Food Chem. 18,* 260–267.
Lingens, F. (1976) *Acta Microbiol. Acad. Sci. Hung. 23,* 161–166.
Linke, H. A. B., Mechlinski, W., and Schaffner, C. P. (1974) *J. Antibiot. 27,* 155–160.
Liras, P., Villanueva, J. R., and Martín J.-F. (1977) *J. Gen. Microbiol. 102,* 269–277.
Liu, C. M., McDaniel, L. E., and Schaffner, C. P. (1972a) *J. Antibiot. 25,* 116–121.
Liu, C. M., McDaniel, L. E., and Schaffner, C. P. (1972b) *J. Antibiot. 25,* 187–188.
Lowe, D. A., and Westlake, D. W. S. (1970) *Can. J. Biochem. 49,* 448–455.
Manwaring, D. G., Rickards, R. W., Gaudiano, G., and Nicolella, V. (1969) *J. Antibiot. 22,* 545–550.
Marahiel, M. A., Lurz, R., and Kleinkauf, H. (1981) *J. Antibiot. 34,* 323–330.
Martín, J.-F. (1976a) *Dev. Ind. Microbiol. 17,* 223–231.
Martín, J.-F. (1976b) in *Microbiology 1976* (Schlessinger, D., ed.) pp. 548–552, American Society for Microbiology, Washington, D.C.
Martín, J.-F. (1977) *Ann. Rev. Microbiol. 31,* 13–38.
Martín, J.-F. (1979a) in *Economic Microbiology* Vol. III, *Secondary Products of Metabolism* (Rose, A. H., ed.) pp. 355–388, Academic Press, London.
Martín, J.-F. (1979b) in *Economic Microbiology* Vol III, *Secondary Products of Metabolism* (Rose, A. H., ed.) pp. 239–291, Academic Press, London.
Martín, J.-F., and Demain, A. L. (1976) *Biochem. Biophys. Res. Comm. 71,* 1103–1109.
Martín, J.-F., and Demain, A. L. (1977a) *Can. J. Microbiol. 23,* 1334–1339.
Martín, J.-F., and Demain, A. L. (1977b) *J. Bacteriol. 132,* 590–595.
Martín, J.-F., and Demain, A. L. (1980) *Microbiol. Rev. 44,* 230–251.
Martín, J.-F., and Gil, J. A. (1979) *J. Antibiot. 32* S122–S128.
Martín, J.-F., and Liras, P. (1976) *J. Antibiot. 29,* 1306–1309.
Martín, J.-F., and Liras, P. (1982) in *Biotechnology* Vol. I, *Microbial Fundamentals* (Rehm, H. J., and Reed, G., eds.) pp. 211–233, Verlag-Chemie, Weinheim, West Germany.
Martín, J.-F., and McDaniel, L. E. (1974) *Dev. Ind. Microbiol. 15,* 324–337.
Martín, J.-F., and McDaniel, L. E. (1975a) *J. Chromatogr. 104,* 151–160.
Martín, J.-F., and McDaniel, L. E. (1975b) *Antimicrob. Agents Chemother. 8,* 200–208.
Martín, J.-F., and McDaniel, L. E. (1975c) *Antimicrob. Agent Chemother. 7,* 208–214.
Martín, J.-F., and McDaniel, L. E. (1975d) *Biochim. Biophys. Acta 411,* 186–194.
Martín, J.-F., and McDaniel, L. E. (1976) *Eur. J. Appl. Microbiol. 3,* 135–144.
Martín, J.-F., and McDaniel, L. E. (1977) in *Advances in Applied Microbiology* (Perlman, D., ed.) Vol. XXI, pp. 2–52, Academic Press, New York.

Martín, J.-F., Naharro, G., Liras, P., and Villanueva, J. R. (1979) *J. Antibiot. 32,* 600–606.

Martín, J.-F., Alegre, M. T., Castro, J. M., and Liras, P. (1983) in *Biology of Actinomycetes* (Bojalil, L. F., and Ortiz, L., eds.) Academic Press, New York. (in press)

Matern, H., Brillinger, G. U., and Pape, H. (1973) *Arch. Mikrobiol. 88,* 37–48.

Mechlinski, W. (1973) in *CRC Handbook of Microbiology* (Lechevalier, H. A., and Laskin, A., eds.) Vol. III, pp. 93–107, CRC Press, Cleveland.

Murphy, M. F., and Katz, E. (1980) *Can. J. Microbiol. 26,* 874–880.

Naharro, G., Gil, J. A., Villanueva, J. R., and Martín, J.-F. (1981) in *Actinomycetes* (Schaal, K. P., and Pulverer, G., eds.) pp. 453–457, Gustav Fischer, Stuttgart, West Germany.

Ōmura, S. (1976) *Bacteriol. Rev. 40,* 681–697.

Pape, H., and Brillinger, G. U. (1973) *Arch. Mikrobiol. 88,* 25–35.

Pazaout, J., Pazoutova, S., and Vancura, V. (1982) *Curr. Microbiol. 7,* 133–136.

Perlman, D., and Semar, J. B. (1965) *Biotechnol. Bioeng. 7,* 133–137.

Roszkowski, J., Kotiuszko, D., Rafalski, A., Morawaska, H., and Raczyńska-Bojanowska, K. (1972) *Acta Microbiol. Pol. Ser. B. 4,* 9–22.

Saltza, M. von, Dutcher, J. D., Reid, J., and Wintersteiner, O. (1963) *J. Org. Chem. 28,* 999–1004.

Sawula, R. V., and Crawford, I. P. (1973) *J. Biol. Chem. 248,* 3573–3581.

Schrempf, H. (1982) *J. Chem. Technol. Biotechnol. 32,* 292–295.

Stevens, C. L., Gupta, S. K., Glinski, R. P., Taylor, K. G., Blumbergs, P., Schaffner, C. P., and Lee, C. H. (1968) *Carbohydr. Res. 7,* 502–504.

Turner, W. B. (1971) *Fungal Metabolites* 446 pp., Academic Press, London.

Vance, D., Goldberg, I., Mitsuhashi, O., Bloch, K., Ōmura, S., and Nomura, S. (1972) *Biochem. Biophys. Res. Comm. 48,* 649–656.

Vaněk, Z., and Majer, J. (1967) in *Antibiotics* Vol. II, *Biosynthesis* (Gottlieb, D., Shaw, P. D., eds.) pp. 154–188, Springer, Berlin.

Volpe, J. J., and Vagelos, P. R. (1973) *Ann. Rev. Biochem. 42,* 21–60.

Yagil, E., and Beeri, H. (1977) *Mol. Gen. Genet. 154,* 185–189.

CHAPTER

9

Ansamycins

Giancarlo Lancini

I. INTRODUCTION

The term ansamycin (from the Latin "ansa," handle) was proposed by Prelog to designate a class of natural compounds in which an aliphatic chain forms a bridge between two nonadjacent positions of an aromatic moiety (Prelog and Oppolzer, 1973). When Prelog first defined the structure of an ansamycin (rifamycin B; Oppolzer et al., 1964), only two microbial strains producing these metabolites had been described: Siminoff et al. (1957) had reported the isolation of a complex of antibiotics, called streptovaricins, from *Streptomyces spectabilis*; Sensi et al. (1959) had isolated the rifamycins from an actinomycete classified as *Streptomyces mediterranei* by Margalith and Beretta (1960). Later, this strain was reclassified as *Nocardia mediterranea* by Thiemann because of the presence of *meso*-diaminopimelic acid in its cell wall (Thiemann et al., 1969). To date, about 20 ansamycin producers have been independently isolated during antibiotic screening programs (Table 9.1). According to the aromatic group present, we can distinguish naphthalene ansamycins and benzene ansamycins. This is not merely a chemical subdivision, since the structural differences correspond with different biological properties. Naphthalene ansamycins are antibacterial agents and, with a few exceptions, act by inhibiting bacterial RNA polymerase. Benzene an-

TABLE 9.1 Ansamycins Produced by Wild-Type Microbial Strains

Producing organism	*Products*	*Ansa Type**	*C-3 Group***	*Biological Activity*	*Reference*
Naphthalenic chromophore					
Nocardia mediterranea	Rifamycins A, B, C, D, and E	C_{17} int	—	Antibacterial	Sensi et al. (1960)
	Rifamycins Y, G	″	—	Inactive	Lancini et al. (1967) Lancini and Sartori (1976)
Streptomyces 4107A_2	Rifamycins O, B	″	—	Antibacterial	Sugawara et al. (1964)
Nocardia mediterranea NT19	Rifamycin SV	″	—	Antibacterial	Birner et al. (1972)
Micromonospora lacustris	3-Methylthio rifamycin SV	″	$-S-CH_3$	Antibacterial	Celmer et al. (1977)
Micromonospora halophitica	Halomycin B	″	—	Antibacterial	Ganguly et al. (1974)
	Halomycin A, C	″	—	Antibacterial	Ganguly et al. (1977)
Streptomyces tolypophorus	Tolypomycin Y	″	—	Antibacterial	Shibata et al (1971)
	Rifamycins B and O	″	—	Antibacterial	Hasegawa et al. (1971)
Streptomyces spectabilis	Streptovaricins	C_{17} cont	$-CH_3$	Antibacterial	Siminoff et al. (1957)
	Damavaricins	″	$-CH_3$	Antibacterial	Rinehart et al. (1976)
	Protostrepto-varicins	″	$-CH_3$	Antibacterial	Deshmukh et al. (1976)
Streptomyces B44P	Streptovaricins	″	$-CH_3$	Antibacterial	Yamazaki (1968)

TABLE 9.1 (Continued)

Producing organism	*Products*	*Ansa Type**	*C-3 Group***	*Biological Activity*	*Reference*
Streptomyces collinus	Naphthomycin	C_{23} cont	Cl	Antibacterial, antifungal	Balerna et al. (1969)
Streptomyces X12384	Naphthomycin	″	Cl	Antibacterial, antifungal	Williams (1975)
Streptomyces E 784	Actamycin	″	OH	Antibacterial, antifungal	McDonald and Rickards (1981)
Streptomyces achromogenes	Rubradirins	C_9 cont	Oxygen	Antibacterial	Hoeksema et al. (1978)
Benzenic chromophore					
Streptomyces hygroscopicus	Geldanamycin	C_{15} cont	$-OCH_3$	Antiprotozoal, antitumor	De Boer et al. (1970)
Streptomyces hygroscopicus	Herbimycins A, B	″	—	Herbicidal	Iwai et al. (1980)
Nocardia C 15003	Ansamitocins	″	Cl	Antitumor	Higashide et al. (1977)
Nocardia C 14919	Macbecins I & II	″	—	Antitumor	Muroi et al. (1980)

* C_{17} int = 17-carbon chain interrupted by oxygen; C_{17} cont = 17-carbon continuous chain
** Substituent at C-3 of chromophore (rifamycin numbering, see Fig. 9.5)

samycins are active on eukaryotic cells and interfere with the cell replication mechanism. The producing organisms are listed in Table 9.1, where the main structural characteristics of the natural ansamycins are also summarized.

Although all known ansamycins have been extensively examined for their antibacterial, antiviral, and antitumor activity, so far the only compounds that have been developed to industrial production belong to the rifamycin family. For this reason, most biosynthesis and fermentation studies deal with this family of antibiotics. However, extensive studies have been made on the biosynthesis of streptovaricins and geldanamycin and, to a lesser extent, herbimycins and actamycin. These antibiotics are described in greater detail in Section II. Reviews on ansamycins have been published by Prelog and Oppolzer (1973), Rinehart and Shield (1976), Werhli (1977), and Brufani (1977). Rifamycin biosynthesis has been reviewed by Lancini and White (1976) and Lancini and Parenti (1978); ansamycin biosynthesis has been reviewed by Lancini and Grandi (1981).

II. ANSAMYCIN FAMILIES RELEVANT TO BIOGENETIC STUDIES

A. Rifamycins

From cultures of the original strain of *N. mediterranea,* eight metabolites have been described: rifamycins A, B, C, D, E, G and Y. Among these, only three have been obtained pure and had their structure determined: these are rifamycin B (Fig. 9.1) rifamycin Y (Fig. 9.10) (Leitich et al., 1967), and rifamycin G (Fig. 9.11) (Lancini and Sartori, 1976). The *N. mediterranea* fermentation and production of rifamycins have been extensively reviewed by Sensi and Thiemann (1967). Some aspects of this will be discussed in the section dealing with control of rifamycin production. Several other rifamycins have been isolated from mutant strains of *N. mediterranea*. They will be described and discussed later in connection with the biosynthetic studies during which they were detected.

Members of the rifamycin family reported from other microorganisms include rifamycin O, isolated from *Streptomyces* 4107A_2 and rifamycin SV from *N. mediterranea* strain MT19. Kishi et al. (1972) reported that rifamycin B and rifamycin O were produced as minor components in *Streptomyces tolypophorus* fermentations, and Celmer et al. (1977) isolated rifamycin S from *Micromonospora lacustris* in which the main product was 3-methyl-rifamycin SV. Rifamycins S, SV, and O had previously been obtained from rifamycin B by Sensi et al. (1961) through a series of chemical reactions. Their structure (Fig. 9.1) was determined with that of rifamycin B by Oppolzer and Prelog (1973). Rifamycin SV was introduced into clinical practice in 1961. Rifamycin B and rifamycin S were used to prepare numerous semisynthetic derivatives, among which rifampicin is at present widely used

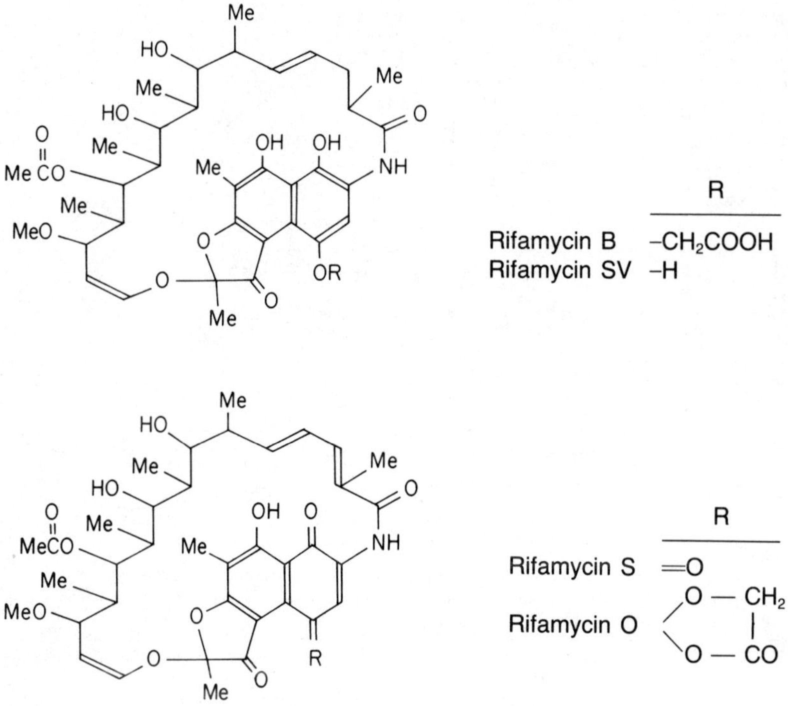

FIGURE 9.1 Rifamycins B, O, S, and SV.

for the treatment of tuberculosis and other infectious diseases. The structure–activity relationship among rifamycins has been reviewed by Lancini and Zanichelli (1977).

B. Streptovaricins

Originally studied for its antimicrobial activity (Siminoff et al., 1957), the streptovaricin complex attracted renewed interest from biologists when it was found to inhibit the reverse transcriptase activity of oncogenic viruses (Rinehart, 1972). The characteristics of the producing strain, *S. spectabilis*, have been described by Waksman (1961). The structure of streptovaricin C (Fig. 9.2) was defined by Rinehart and Antosz (1972). Other members of the complex, streptovaricins A, B, D, E, G, and I were described by Rinehart et al. (1971). Further studies of the fermentation revealed the presence of two other groups of metabolites, protostreptovaricins (Deshmukh et al., 1976) and damavaricins (Rinehart et al., 1976). Their biogenetic relationship with streptovaricins will be discussed in Section IV. A *Streptomyces* strain independently isolated by Yamazaki (1968) produced a complex of antibiotics, at least three of which were identical with streptovaricins.

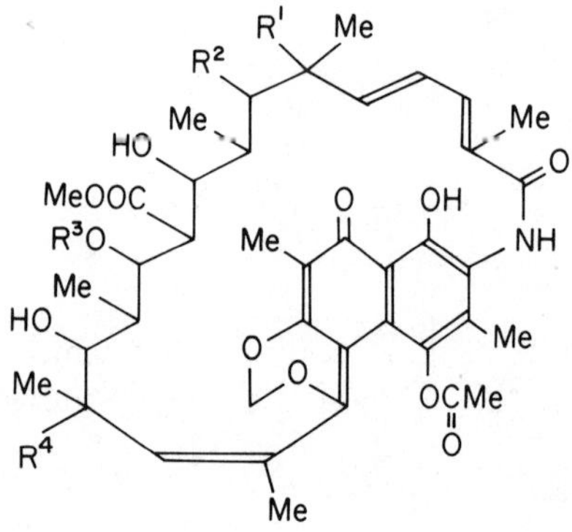

	R^1	R^2	R^3	R^4
Streptovaricin A	OH	OH	$COCH_3$	OH
B	H	OH	$COCH_3$	OH
C	H	OH	H	OH
D	H	OH	H	H
E	H	=O	H	OH
G	OH	OH	H	OH
J	H	$OCOCH_3$	H	OH

FIGURE 9.2 Streptovaricins.

C. Naphthomycin and Actamycin

Naphthomycin, for which a structure was proposed by Williams (1975) and revised by Deshmukh et al. (1976) and Brufani et al. (1979), is the first example of a class of naphthalene ansamycins differing from streptovaricins in their longer ansa chain of 23 carbons (Fig. 9.3). Recently, another member of this family, actamycin, was described and its structure (Fig 9.3) established (McDonald and Rickards, 1981; Allen et al., 1981). It is noteworthy that evidence relevant to the structure determination was obtained from the pattern of incorporation of ^{13}C-labeled propionate.

D. Geldanamycin and Herbimycins

Geldanamycin, an antibiotic active on a variety of eukaryotic cells, is produced by a strain of *Streptomyces hygroscopicus*. The structure of geldanamycin (Fig. 9.4) differs from that of naphthalene ansamycins in having not only a benzoquinone aromatic moiety but also a shorter ansa chain comprising 15 carbons instead of the 17 of streptovaricins (Sasaki et al., 1970). Recently, herbimycin, a compound closely related to geldanamycin, was isolated from *S. hygroscopicus*. As with actamycin, the incorporation of labeled precursors and biosynthetic considerations were used to define its structure (Ōmura et al., 1979).

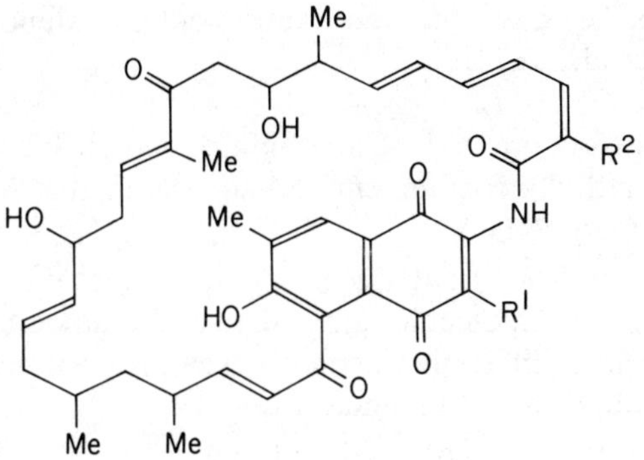

	R^1	R^2
Naphtomycin	–Cl	–Me
Actamycin	–OH	–H

FIGURE 9.3 Naphthomycin and actamycin.

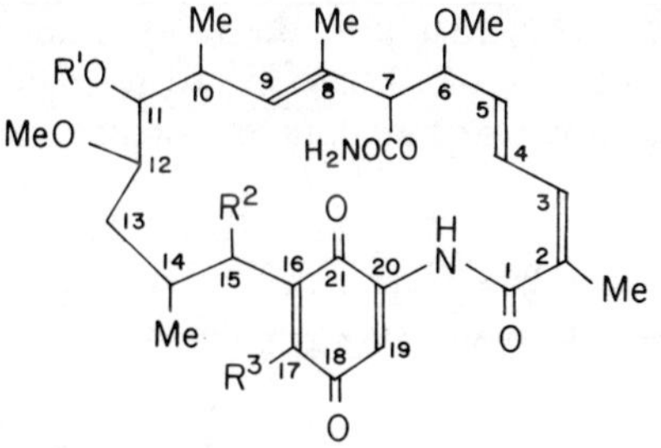

	R^1	R^2	R^3
Geldanamycin	–Me	–H	–OMe
Herbimycin A	–Me	–OMe	–H
Herbimycin B	–H	–H	–H

FIGURE 9.4 Geldanamycin and herbimycins.

III. ORIGIN OF THE ANSAMYCIN CARBON SKELETON

A. Early Experiments with Radiolabeled Precursors

The structure of the rifamycin and streptovaricin ansa chain, in which methyl groups alternate with hydroxyl groups, suggests that it may originate from the well-known mechanism of methylmalonate–malonate condensation occurring, for instance, in macrolide biosynthesis. Moreover, the stereochemistry of the aliphatic chains of rifamycins and streptovaricins is surprisingly similar to that of the erythromycins, suggesting synthesis by a similar enzyme system. Cooperative studies carried out in Prelog's and in

Lepetit's laboratories gave the first experimental evidence (Brufani et al., 1973; Karlsson et al., 1974).

1. Radioactivity from [methyl-^{14}C]methionine was exclusively incorporated into the O-methyl of rifamycin B, suggesting that all other methyls derive from methylmalonate.
2. [3-^{14}C]Propionate was efficiently incorporated into rifamycin B. Chemical degradation of the molecule showed that the amount of radioactivity incorporated into different fragments was proportional to the number of methyl groups they contained.
3. [1-^{14}C]Acetate was incorporated with low efficiency. The proportion of label in the ansa chain to that in the aromatic chromophore was about 2:1. This result is incompatible with formation of the naphthalene group from acetate-methylmalonate.
4. The incorporation of [1-^{14}C]propionate gave results difficult to interpret and apparently not in agreement with those obtained with [3-^{14}C]-propionate.

Taken together, these findings support the assumption that the ansa chain and part of the chromophore of rifamycin are formed by condensation of methylmalonate units. However, the data do not allow a detailed description of chain formation.

B. Origin of the Rifamycin Ansa Chain

Use of ^{13}C-enriched precursors made it possible to determine unambiguously the origin of the ansa carbon atoms of rifamycin (White et al., 1973). The precursors were pulse-fed to *N. mediterranea* cultures, and the rifamycin B produced was chemically converted to rifamycin S. The ^{13}C nuclear magnetic resonance spectrum of this substance was compared with the natural abundance spectrum of rifamycin S in which each peak had been assigned to the appropriate carbon atom. Incorporation of [methyl-^{13}C]-propionate confirmed that the five methyl groups on the ansa chain and the two on the chromophore came from propionate. No other carbon was enriched by this precursor. In rifamycin S biosynthesized in the presence of [2-^{13}C]propionate, all seven carbons adjacent to a methyl group were labeled as expected. However, an eighth carbon, C-28 (Fig. 9.5), was also enriched, indicating that eight propionates participate in building the molecule. This was confirmed by the results obtained with carboxyl-labeled propionate. Here again, eight atoms of rifamycin S were enriched, including, unexpectedly, C-29.

Interpretation of acetate incorporation was complicated by the metabolic conversion of this precursor, through the Krebs cycle, to succinate and then to methylmalonate. For this reason, carbons derived from [2-^{13}C]- and [3-^{13}C]propionate may be enriched also by [2-^{13}C]acetate. However, the combined results of [1-^{13}C]- and [2-^{13}C]-acetate incorporation indicate clearly that two acetate units, in addition to the acetate unit used for the acetoxy group at C-25, take part in the formation of rifamycin. In conclusion, out of 34 atoms in the rifamycin carbon skeleton, 23 are derived from eight units of propionate (one of which has lost its methyl), and four are derived from two acetate units. The remaining seven carbons, all part of the aromatic moiety, cannot originate from acetate–propionate condensations.

C. Formation of the Rifamycin Carbon Skeleton and Biosynthesis of Rifamycin W

The incorporation of acetate and propionate units into a secondary metabolite can be taken as evidence that it is biosynthesized through a methylmalonate–malonate condensation mechanism. This is confirmed in the case of rifamycin by the similar pattern of isotopic incorporation given by [2-^{13}C]malonate and [2-^{13}C]acetate (Karlsson et al., 1974). The results can be accommodated if the rifamycin molecule is built from a single chain formed by polymerization of eight methylmalonate and two malonate units (Fig. 9.5). The chain gives

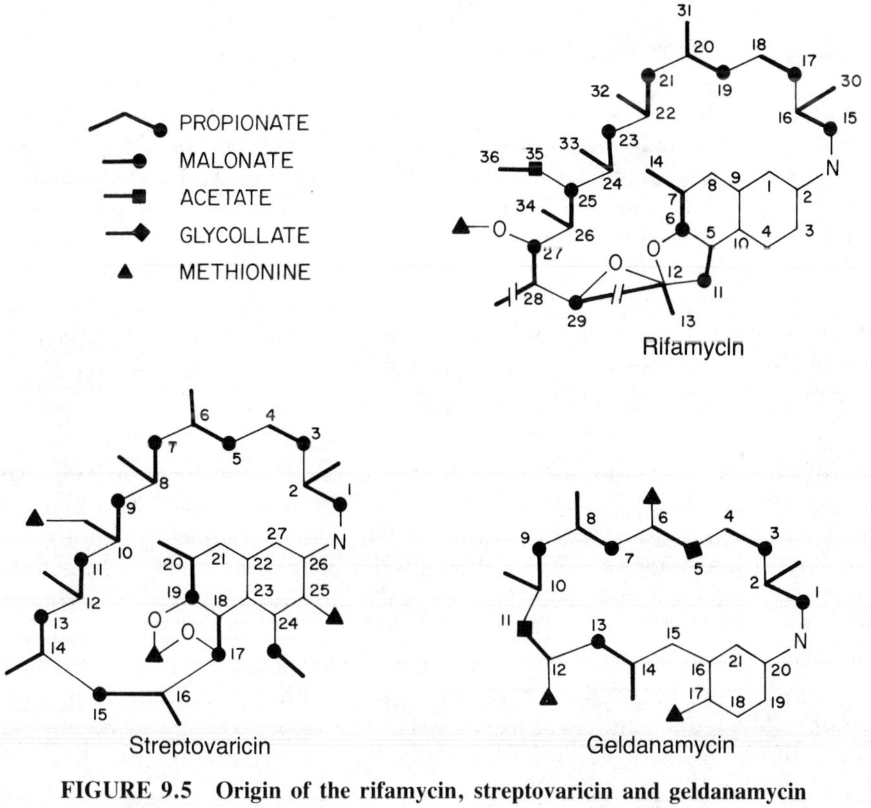

FIGURE 9.5 Origin of the rifamycin, streptovaricin and geldanamycin carbon skeletons.

rise to part of the aromatic nucleus and to all of the ansa components. As initiator of the chain, a seven-carbon aromatic moiety, possibly a benzoic acid derivative, was postulated (White et al., 1973).

After the chain is completed, the molecule is assumed to undergo two major modifications: (1) splitting of the carbon–carbon bond and insertion of an oxygen between the carboxyl and the methylene of one propionate unit, and (2) loss of a methyl group. The correctness of the assumption was demonstrated by the isolation of rifamycin W and by biosynthetic studies with this molecule. Rifamycin W is a metabolite accumulated by several mutant strains of *N. mediterranea*. Its main structural difference from rifamycin S (Martinelli et al., 1974) is a continuous carbon ansa chain; it still possesses, although hydroxylated, the methyl carbon postulated to be lost during rifamycin B formation (Fig. 9.8). Incorporation studies with ^{13}C-labeled propionate (White et al., 1974) demonstrated that rifamycin W is made from eight propionate and two acetate units connected head-to-tail to form a continuous chain, as predicted. Since rifamycin W is converted by *N. mediterranea* into rifamycin B (see Section IV), the assumptions made about the biogenesis of the latter are supported.

D. Origin of Other Ansa Chains

1. Streptovaricins. The model proposed for ansa chain formation in rifamycin appears to be valid for other ansamycin families. Incorporation studies have demonstrated that eight carbon atoms of streptovaricin D are enriched when [1-^{13}C]propionate is used as a precursor and two carbons are equally enriched by [2-^{13}C]malonate (Milavetz et al., 1973; Kakinuma et al., 1976). The labeling pattern is virtually identical with that of rifamycin W, suggesting an identical biogenesis (Fig. 9.5). The extra methyl group on the streptovaricin nucleus comes from methionine. This precursor also supplies, as expected, the methoxy and the methylene carbons. No information is available on the origin of the aromatic ring initiator.

2. Actamycin. Naphthomycin and actamycin differ from the ansamycins so far discussed mainly because of their longer ansa chain. Their general structure is fully compatible with the biosynthetic model proposed for rifamycins and streptovaricins, although the presence of only six methyls indicates that more malonate and less methylmalonate units are involved in the synthesis. In fact, incorporation experiments with [1-^{13}C]propionate revealed that six carbons of the actamycin molecule derive from the propionate carboxyl group. All of these are correctly positioned in relation to methyl groups (McDonald and Rickards, 1981). The experiments related to the origin of the aromatic chain initiator of actamycin are discussed in Section III.E.

3. Geldanamycin and Herbimycins. The applicability of the rifamycin biosynthetic scheme to benzenic ansamycins was demonstrated by incorporation studies with geldanamycin (Johnson et al., 1974). The carbons of the three methoxy groups derive from methionine. There are four other methyl groups in the geldanamycin molecule and [1-^{13}C]propionate labeled carbons at the correct positions in relation to each of them. Thus four methylmalonates appear to take part in geldanamycin synthesis. Three two-carbon units are required to complete the ansa moiety, but only one of these is labeled by [^{13}C]malonate. Good incorporation of [^{13}C]glycolate in the other units and the pattern of oxygen substituents indicate that glycolic acid, an unusual precursor, participates in the formation of geldanamycin's aliphatic chain (Fig. 9.5). Again, the assumption that the chain is initiated by a seven-carbon aromatic unit is in agreement with the experimental data discussed in the next section.

For herbimycin, only data on incorporation of [1-^{13}C]propionate are available (Ōmura et al., 1979). These conform to the pattern described for geldanamycin.

E. Origin of the Aromatic Nucleus

The incorporation data discussed above suggest that the biosynthesis of all ansamycins is initiated by a seven-carbon unit not derived from acetate–propionate condensations. The structural characteristics of this unit are intriguing, since its biosynthetic role requires an amino group in meta position to the carboxyl of benzoic acid. This feature is unknown in primary metabolism where only ortho- or para-aminobenzoates are found. Such a structure could derive from the shikimate pathway, either directly from one of the early intermediates or indirectly from the degradation of an aromatic amino acid. Several techniques have been used to investigate this point.

With classical precursors, White and Martinelli (1974) demonstrated that, after [1-^{13}C]glucose incorporation, C-10 and C-1 in the rifamycin nucleus were enriched, whereas after [1-^{13}C]glycerate incorporation, the carbons enriched were C-8 and C-3. These results clearly indicate that the seven-carbon unit comes from the shikimate pathway by way of the sequence of reactions shown in Fig. 9.6. A similar approach was used by Haber et al. (1977), who studied incorporation of [6-^{13}C]glucose into the geldanamycin aromatic ring. Strong enrichments were observed in C-17 and C-21, corresponding to C-1 and C-10 of the rifamycins. Since both C-6 and C-1 of glucose give rise to C-3 and C-7 of deoxysedoheptulosonic acid, the same biosynthetic pathway may be assumed for rifamycins and geldanamycin. In both rifamycin and geldanamycin, shikimate does not specifically label the aromatic chromophore. White and Martinelli (1974) suggested that the precursor is an intermediate before shikimate, possibly dehydroquinate or dehydroshikimate.

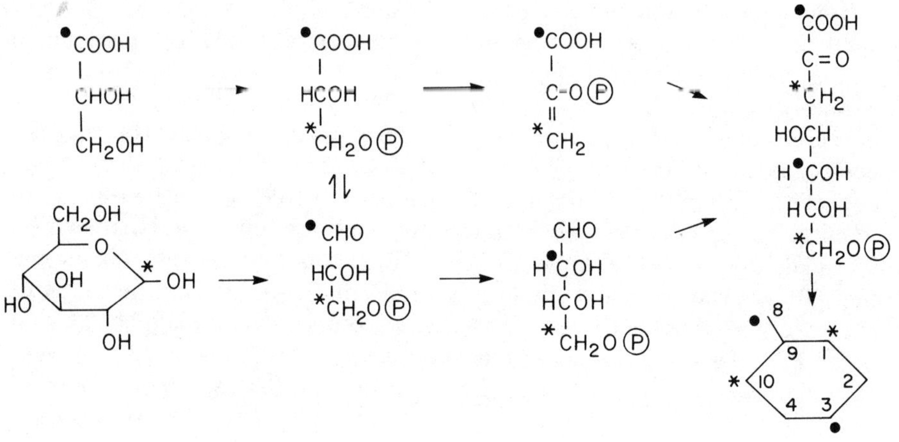

FIGURE 9.6 Incorporation of [1-^{13}C]glucose and [1-^{13}C]glycerate into the rifamycin chromophore.

Although the evidence from the shikimate experiment was inconclusive because shikimate might have entered *N. mediterranea* only after degradation (Bruggisser, 1975), the substantial correctness of the hypothesis was confirmed by genetic experiments (Ghisalba and Nüesch, 1978a, b). Two *N. mediterranea* mutants auxotrophic for aromatic amino acids were isolated: strain A-8, lacking transketolase and thus unable to synthesize 3-deoxyheptulosonic acid, and strain A-10 accumulating shikimic acid but apparently unable to convert it to shikimate-3-phosphate. Since mutant A-8 produced very little rifamycin B while strain A-10 produced about the same amount of rifamycin B as the parent strain, the authors concluded that the precursor of the aromatic ring is a molecule intermediate between 3-deoxysedoheptulosonic acid and shikimate. A new variation in the established pathway for aromatic ring biosynthesis must be assumed for conversion of the early precursor to the actual initiator of the ansamycin chain. The identity of this initiator is now established as 3-amino-5-hydroxybenzoic acid. Two independent lines of evidence are available: Kibby et al. (1980) demonstrated that 3-amino-5-hydroxy-[7-^{14}C]benzoic acid is incorporated with high efficiency and low dilution into actamycin. Ghisalba and Nüesch (1981) showed that 3-amino-5-hydroxybenzoic acid, in contrast to 13 other substituted benzoic acids, could restore the capacity to produce rifamycins in strain A-8. Stimulation of rifamycin B production was proportional to the amount of acid supplied. Further evidence was provided by isolation of the product P8/1-OG (Fig. 9.7) from several nonproducing *N. mediterranea* mutants (Ghisalba et al., 1981). Although this product is not a precursor of rifamycin, its structure clearly shows that *N. mediterranea* can use 3-amino-5-hydroxybenzoic acid to initiate a polyketide chain.

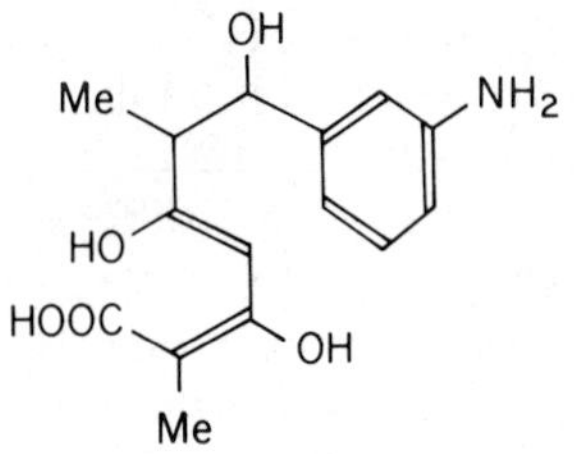

FIGURE 9.7 Metabolite P8/1-OG, a product of several mutant *N. mediterranea* strains.

IV. PATHWAY INTERMEDIATES

A. Early Intermediates in Rifamycin and Streptovaricin Biosynthesis

The key compound in rifamycin biosynthesis appears to be rifamycin S (or its hydroquinone form rifamycin SV), the first intermediate in the biogenetic line active as an antibiotic. All the final products isolated from wild-type producing strains are, in fact, derivatives of rifamycin S; we can define as early intermediates in rifamycin synthesis those compounds showing stepwise modification of the original polyketide chain en route to this molecule.

It is well established that one of these compounds is rifamycin W, a product of several *N. mediterranea* mutants. Its structure (Fig. 9.8) contains an uninterrupted carbon chain and a hydroxymethyl group on carbon 14 and is clearly intermediate between that of the original polyketide chain suggested by incorporation experiments and that of rifamycin S. Moreover, ^{14}C-labeled rifamycin W is converted into labeled rifamycin B by mycelium

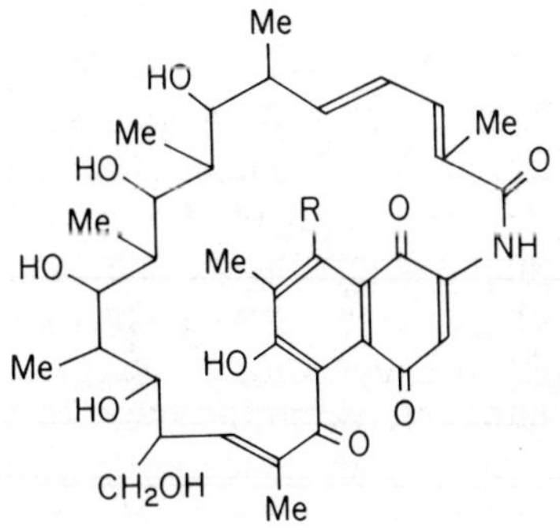

	R
Protorifamycin I	–H
Rifamycin W	–OH

FIGURE 9.8 Early intermediates in rifamycin biosynthesis.

of *N. mediterranea,* good evidence that it is a true intermediate (White et al., 1974). Protorifamycin I, the main product of another *N. mediterranea* mutant, strain F1/24, was postulated to be a precursor of rifamycin W (Ghisalba et al., 1978). Protorifamycin I differs from rifamycin W only in the absence of the hydroxyl group at C-8 (Fig. 9.8). The isolation of protorifamycin I and its conversion into rifamycin W by other *N. mediterranea* strains are intriguing, since they suggest that an oxygen is first taken away and then reintroduced at carbon-8 of the rifamycin nucleus. Although a parallel biosynthetic scheme maintaining the original oxygen could be envisaged for rifamycin W, the one proposed is supported by the observation that several natural ansamycins do not carry an oxygen in that position.

A similar scheme has been proposed for streptovaricins. Cultures of *S. spectabilis* produce streptovaricins and a series of other compounds lacking the oxygen at position 21, which corresponds to position 8 in the rifamycins. These are considered to be precursors of streptovaricins, although direct transformation evidence is not available. The simplest derivative, protostreptovaricin I, differs from protorifamycin I only in the presence of an extra methyl on the chromophore and the absence of a hydroxyl on the methyl group at C-34 (rifamycin numbering).

The conversion of rifamycin W to rifamycin S requires:

1. oxidation of the ansa chain methyl group to a carboxyl group and subsequent decarboxylation;
2. insertion of an oxygen between C-12 and C-29 in the ansa chain;
3. closure of the furanone ring; and
4. methylation of the hydroxyl at position 27 and acetylation of that at position 25.

The order in which these reactions occur is tentatively established from the following: Mutant strain S 725 of *N. mediterranea* produces rifamycin Z (Fig. 9.9), identified as the 34a-25-lactone of rifamycin W (Cricchio et al., 1981). From a recombinant strain of *N. mediterranea,* a number of new rifamycins, including rifamycin W hemiacetal and rifamycin W lactone, have been isolated (Schupp et al., 1981). Thus elimination of the methyl group seems to precede furanone ring closure and oxygen insertion. The methylation and acetylation steps, on the other hand, probably occur later because demethylrifamycin S and desacetylrifamycin S have been isolated from another mutant strain (Lancini and White, 1973). Protorifamycin I seems to be accepted as a substrate by the enzymes performing this series of reactions. In fact, it is converted into 8-deoxyrifamycin S and 8-deoxyrifamycin B, although with low efficiency.

In the streptovaricin family three reactions are needed to convert protostreptovaricin I to streptovaricin C:

1. Hydroxylation of the ansa C-14. This converts protostreptovaricin I to protostreptovaricin III.
2. Oxidation of the C-10 methyl group on the ansa chain to a carboxyl group, followed by esterification.
3. Hydroxylation at C-21.

No information is available on the order in which these reactions take place.

B. Rifamycins Not in the Main Pathway

From the mutant strain F1/24 producing protorifamycin I, Ghisalba et al. (1979) isolated two minor metabolites, protorifamycin B-M1 and proansamycin B-M1. These are closely related in structure to protorifamycin I but have a cleaved ansa chain (Fig. 9.9). Streptovaricin Y, a compound similarly characterized by an open ansa chain, was found at about the same time by Knöll et al. (1980) among the minor metabolites of *S. spectabilis*. The pattern of ansamycin carbon skeleton formation excludes the possibility that these compounds are intermediates in the biosynthetic process, and this was confirmed by experiments showing that proansamycin B-M1 is not transformed into other rifamycins by *N. mediterranea* strains (Ghisalba et al., 1979). However, the isolation of proansamycin B-M1 is of interest because this compound appears to be a degradation product of proansamycin B, the hypothetical precursor of protorifamycin I postulated by Ciba researchers to be the common precursor of all naphthalenic ansamycins (Ghisalba et al., 1978).

Several other products related to protorifamycin I were isolated from strain F1/24, namely, 30-hydroxyprotorifamycin I, 20-hydroxyprotorifamycin I, 13-hydroxyprotorifamycin I, 23-acetylprotorifamycin I, protorifamycin I lactone, and 23-ketoprotorifamycin I (Ghisalba et al., 1980). It is evident that, when a blocked mutant accumulates an intermediate, this may become the substrate for several hydroxylating enzymes. Novel rifamycins have also been isolated from a recombinant strain, R21, obtained from a cross between a rifamycin B producer and a blocked mutant that accumulates rifamycin W (Schupp et al., 1981; Traxler et al., 1981). Besides the already mentioned rifamycin W lactone and rifamycin W hemiacetal, other interesting products isolated were 3-hydroxyrifamycin S (Fig. 9.9) and 30-hydroxyrifamycin W, structurally related to rifamycin R (see Section IV.C).

C. Last Steps in Rifamycin and Streptovaricin Biosynthesis

1. Rifamycins B, L, O, and Y. The key intermediate in the biogenesis of all the final products of the original strain of *N. mediterranea* is rifamycin

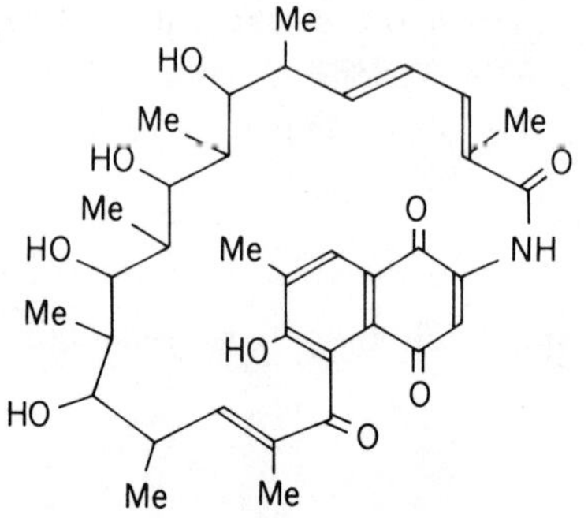

Proansamycin B
(proposed structure)

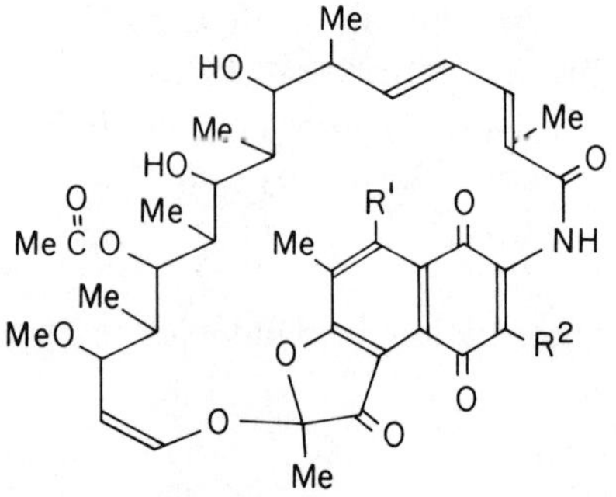

	R_1	R_2
8-Deoxyrifamycin S	–H	–H
8-Deoxy-3-hydroxyrifamycin S	–H	–OH
3-Hydroxyrifamycin S	–OH	–OH

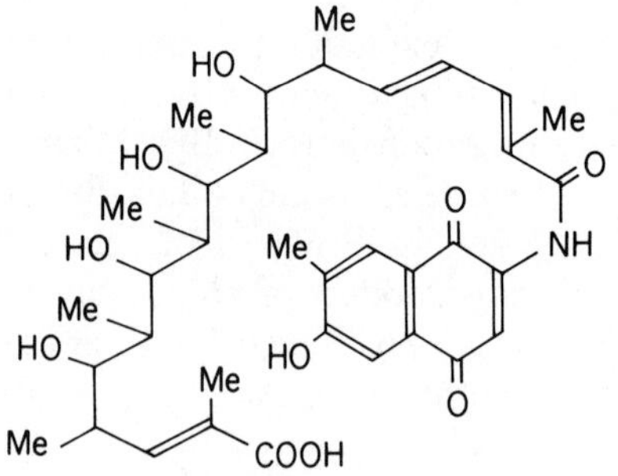

Proansamycin B-M1

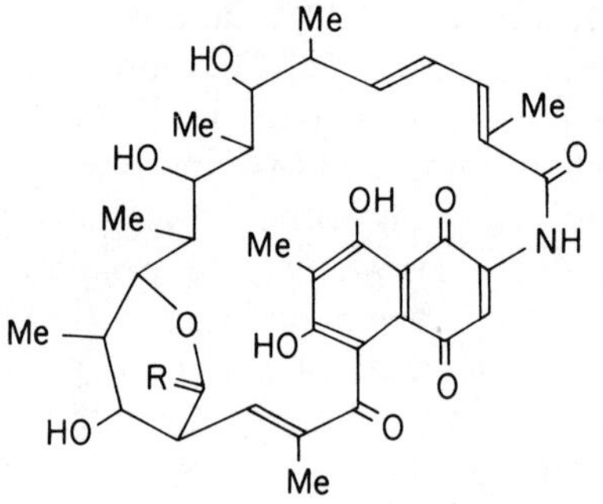

	R
Rifamycin Z	=O
Rifamycin W hemiacetal	=H,OH

FIGURE 9.9 Rifamycins from mutant strains F 1/24 and S 725 and recombinant strain R 21.

S (or its reduced form, rifamycin SV). It is in fact a precursor of rifamycin B since:

1. ^{14}C-labeled rifamycin S is converted in high yields to labeled rifamycin B by *N. mediterranea* cultures (Lancini and Sensi, 1967).
2. Mycelium suspensions efficiently transform rifamycin S into rifamycin B (Lancini et al., 1969).
3. Mutagenesis and colony selection starting from a high rifamycin B producer yielded a strain producing almost exclusively rifamycin SV (Lancini and Hengeller, 1969).

Together, these studies establish that rifamycin SV, besides being the final product of the fermentation in some *Nocardia* and *Micromonospora* wild-type strains, is the precursor of rifamycin B in *N. mediterranea*.

The origin of the glycollic acid moiety at position-4 rifamycin B is still a matter for speculation. Lancini et al. (1969) showed that acetate and glycollate are not precursors. Their results are compatible with the hypothesis that a C_3 unit, such as 3-phosphoglyceric acid or 3-phosphohydroxypyruvate is first attached to position 4 of rifamycin SV, forming the ether bond, and then decarboxylated and oxidized (Lancini and Parenti, 1978). Two other molecules with a glycollate residue at position 4 have been described: rifamycin L, in which the glycollic acid forms an ester bond (Lancini et al., 1969), and rifamycin O in which it is cyclized to form a *spiro*-lactone. Postulated biosynthetic relationships (Fig. 9.10) cannot be established with certainty. Rifamycin L labeled in the glycollic acid moiety is not converted to rifamycin B (Lancini et al., 1969). Interconversion of rifamycin B and rifamycin O occurs very easily in the presence of oxidizing or reducing agents without need of enzymatic catalysis (Sensi et al., 1961). The attractive hypothesis that rifamycin O is the immediate precursor of both rifamycin B and rifamycin L has been contradicted recently by results showing that the glycollic moieties

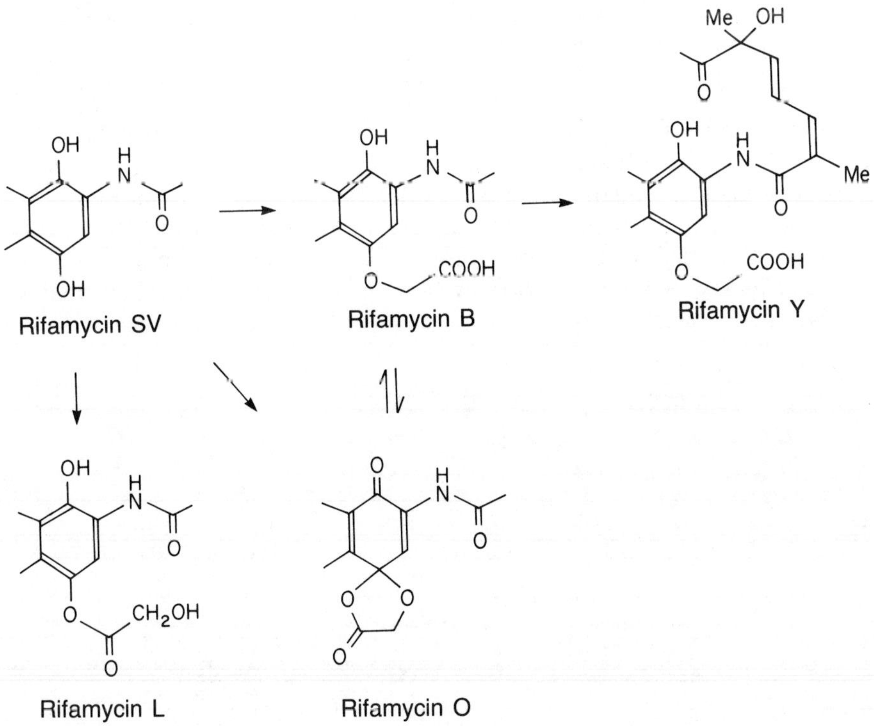

FIGURE 9.10 Last steps in rifamycin biosynthesis.

of the last two metabolites are differently labeled when ^{14}C-glycerol or ^{14}C-pyruvate is used as a precursor (Ghisalba et al., 1982).

Rifamycin B is converted to a small extent to rifamycin Y, an inactive derivative modified in the ansa chain. This was shown both by adding labeled rifamycin B to *N. mediterranea* cultures and by transformation experiments with mycelium suspensions. The transformation is not affected by barbital but is enhanced by the presence of inorganic phosphate (Lancini et al., 1967).

2. Rifamycin Complex and Rifamycin G. All the other final products of *N. mediterranea* fermentations derive also from transformations of rifamycin SV. This was shown for rifamycins A, C, D, and E (the so-called rifamycin complex) by R. J. White and G. C. Lancini through cosynthesis experiments (unpublished results). When a nonproducing *N. mediterranea* mutant and a rifamycin SV producer were grown together in the presence of barbital, rifamycin B was obtained; when barbital was omitted, rifamycins C and D were produced. The formation of rifamycin G (Fig. 9.11) from rifamycin S was demonstrated directly by adding ^{14}C-labeled rifamycin S to fermentations of *N. mediterranea*. Moreover, rifamycin S was converted in high yields to rifamycin G by washed mycelium, and recombinant strain R 21 accumulated 16,17-dehydrorifamycin G, a possible intermediate between rifamycin S and rifamycin G (Traxler et al., 1981).

3. Thiazorifamycins. Three biologically active natural rifamycins, P, Q, and Verde, were isolated from mutant strain M-36 of *N. mediterranea* (Cricchio et al., 1980a). Using as a precursor [U-^{14}C]cysteine or [3-^{14}C]-cysteine, Cricchio et al. (1980b) demonstrated that these new molecules are formed by condensation of rifamycin S and cysteine through the route outlined in Fig. 9.12. Formation of rifamycin Verde and the intermediate

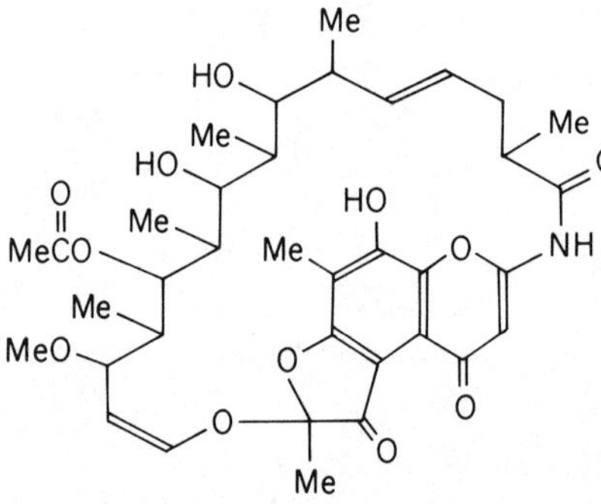

FIGURE 9.11 Rifamycin G.

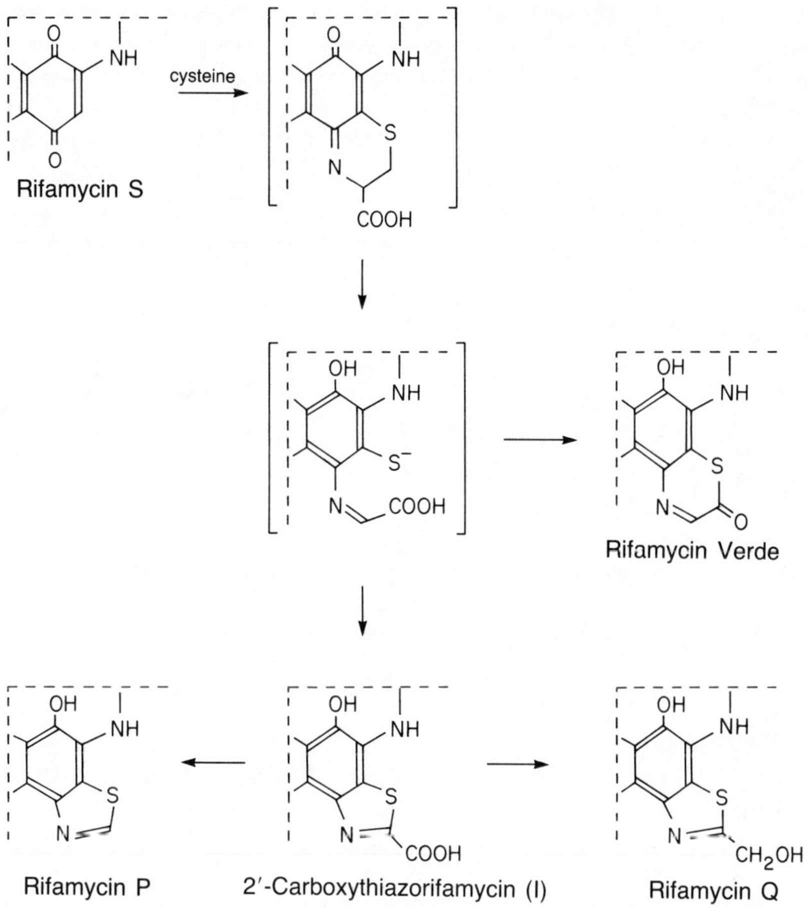

FIGURE 9.12 Conversion of rifamycin S to thiazorifamycins.

2′-carboxythiazorifamycin (I) appears to be nonenzymatic. Also, decarboxylation of I to rifamycin P is simply due to acid catalysis, whereas reduction of I to rifamycin Q requires the living microorganism. During these studies, an additional product, rifamycin R, was obtained. This is 30-hydroxyrifamycin S. No data are available on its biosynthesis, but its likely origin is by oxidation of rifamycin S (Martinelli et al., 1978).

4. Final Steps in Streptovaricin Biosynthesis. Often in the biosynthesis of secondary metabolites, the final products may be the result of more than one reaction sequence. On the basis of structural considerations a complex scheme of biogenetic interrelations has been proposed by Rinehart and Shield (1976) (see also Deshmukh et al., 1976) for the protostreptovaricins, damavaricins, and streptovaricins. Transformation experiments demonstrate that streptovaricin D is converted to streptovaricin C by cultures of *S. spectabilis*. Streptovaricin C, labeled with ^3H, is enzymatically transformed by a cell extract from *S. spectabilis* into streptovaricins A and B. In lesser amounts, streptovaricins E and G are also formed (Rinehart and Shield, 1976). Thus the last steps in streptovaricin biosynthesis appear to be:

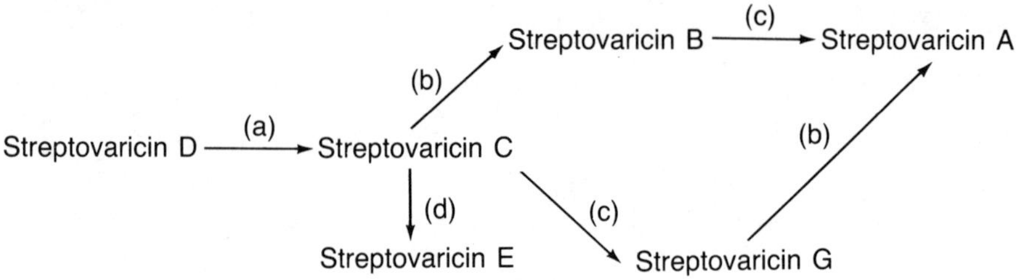

where (a) = hydroxylation at C-14,
(b) = acetylation at C-11,
(c) = hydroxylation at C-6,
(d) = oxidation at C-7.

V. CONTROL OF RIFAMYCIN BIOSYNTHESIS

A. General Aspects

Data on factors controlling the biosynthesis of ansamycins are scanty. Some information can be found in papers dealing with rifamycin production, but in general, no attempt has been made to define the biochemical mechanism regulating antibiotic synthesis. In contrast with other secondary metabolites, rifamycin production is not associated with only the resting phase of the fermentation. In fact, Margalith and Pagani (1961) noticed that, in optimal fermentation conditions, production of rifamycin B parallels mycelial growth, and the same observation was made by Küenzi et al. (1977) about rifamycin S production by a mutant of *N. mediterranea*. In complex media (see Sensi and Thiemann, 1967), antibiotic accumulation is not noticeable until 24–48 hours after inoculation. However, glucose consumption and mycelial growth appear to be also somewhat delayed. Thus some factor not specifically related to antibiotic production may be responsible for the sluggish metabolism in the initial phase of fermentation. Osmotic pressure due to the high concentration of glucose and other soluble nutrients is a possibility. Since a fast rate of production is established before glucose and inorganic nitrogen are substantially consumed, we may assume that antibiotic biosynthesis is not under stringent control from glucose or ammonia catabolite repression. Also, inorganic phosphate does not play an important role in controlling synthesis. Its only specific effect is to increase the conversion of rifamycin B to rifamycin Y (Lancini et al., 1967).

B. Inhibition by Rifamycin S

Rifamycin S inhibits its own biosynthesis when it is added to *N. mediterranea* cultures (Küenzi et al., 1977). However, it is not clear whether this is a specific repression or the result of a toxic effect on the cells. Some toxicity of rifamycin S toward *N. mediterranea* was noticed during early studies (Lancini and Sensi, 1967), and rifamycin SV added to resuspended mycelium in relatively low concentrations inhibits its own transformation into rifamycins C and D (R. J. White and G. C. Lancini, unpublished results). Production of rifamycin SV appears to be self-limiting, in contrast to that of rifamycin B, which does not inhibit growth or rifamycin synthesis (Küenzi et al., 1977). Rifamycin B tolerance might be due to its lack of antimicrobial activity, but it is more probably related to some specific mechanism of antibiotic excretion through the cell membrane. In fact, rifamycin B is unable to penetrate bacterial membranes, although it is easily excreted into the fermentation liquor, where it accumulates in large amounts. When high-producing strains of rifamycin B are prevented by mutation from converting rifamycin S to rifamycin B, we observe a tendency to form other conversion products, either enzymatically (rifamycin R and other hydroxylated compounds) or chemically (thiazorifamycins). In both cases, accumulation of the antibiotic in the cell, and thus a slow excretion process, must be postulated.

C. Role of Barbiturates

Rifamycin B production increases at the expense of rifamycin complex when barbital (diethyl barbituric acid) is added to fermentation media. Kluepfel et al. (1965) demonstrated that barbital and not its metabolites is responsible for this activity. At the same time, they could exclude barbital acting as a specific precursor for rifamycin B. When Margalith and Pagani (1961) studied the effect of other barbiturates, they found that a few of these also enhance the rifamycin B yield; others seem to have no effect or provoke a general stimulation of rifamycin production.

The action of barbital seems to be specifically related to the transformation of rifamycin SV into rifamycin B. In fact:

1. production of rifamycin SV by blocked mutants accumulating this metabolite is not influenced by the presence of barbital (Lancini and Sensi, 1967);
2. transformations of rifamycin S into rifamycin G and of rifamycin B into rifamycin Y are also indifferent to the presence of barbital (Lancini and Sartori, 1976; Lancini et al., 1967).

A reasonable hypothesis is that barbital acts as an allosteric effector of one of the enzymes in the rifamycin SV → rifamycin B transformation, increasing its efficiency. The lower production of rifamycin complex would then be due to decreased availability of rifamycin SV.

The practical importance of these studies is now somewhat diminished because of the isolation of a mutant *N. mediterranea* strain producing the same amount of rifamycin B in the presence and in the absence of barbital (White and Lancini, 1975).

REFERENCES

Allen, M. S., McDonald, I. A., and Rickards, R. W. (1981) *Tetrahedron Lett.*, 1145–1148.

Balerna, M., Keller-Schierlein, W., Martius, C., Wolf, H., and Zähner, H. (1969) *Arch. Mikrobiol. 65,* 303–359.

Birner, J., Hodgson, P. R., Lane, W. R., and Baxter, E. H. (1972) *J. Antibiot. 25,* 356–359.

Brufani, M. (1977) in *Top. Antibiot. Chem. 1,* 91–212.

Brufani, M., Kluepfel, D., Lancini, G. C., Leitich, J., Mesentsev, A. S., Prelog, V., Schmook, F. P., and Sensi, P. (1973) *Helv. Chim. Acta 56,* 2315–2323.

Brufani, M., Cellai, L., and Keller-Schierlein, W. (1979) *J. Antibiot. 32,* 167–168.

Bruggisser, S. (1975) Ph.D. dissertation 5435, Federal School of Technology, Switzerland.

Celmer, W. D., Sciavolino, F. C., Routien, J. B., and Cullen, W. P. (1977) U.S. Patent 4,013,789.

Cricchio, R., Antonini, P., Lancini, G. C., Tamborini, G., White, R. J., and Martinelli, E. (1980a) *Tetrahedron 36,* 1415–1421.

Cricchio, R., Antonini, P., and Sartori, G. (1980b) *J. Antibiot. 33,* 842–846.

Cricchio, R., Antonini, P., Ferrari, P., Ripamonti, A., Tuan, G., and Martinelli, E. (1981) *J. Antibiot. 34,* 1257–1260.

DeBoer, C., Meulman, P. A., Wnuk, R. J., and Peterson, D. H. (1970) *J. Antibiot. 23,* 442–447.

Deshmukh, P. V., Kakinuma, K., Ameel, J. J., Rinehart, K. L., Jr., Wiley, P. F., and Li, L. H. (1976) *J. Amer. Chem. Soc. 98,* 870–872.

Ganguly, A. K., Szmulewicz, S., Sarre, O. Z., Greeves, D., Morton, J., and McGlotten, J. (1974) *J. Chem. Soc. Chem. Comm.* 395–396.

Ganguly, A. K., Liu, Y. T., Sarre, O. Z., and Szmulewicz, S. (1977) *J. Antibiot. 30,* 625–627.

Ghisalba, O., Fuhrer, H., Richter, W. J., and Moss, S. (1981) *J. Antibiot. 34,* 58–63.

Ghisalba, O., and Nüesch, J. (1978a) *J. Antibiot. 31,* 202–214.

Ghisalba, O., and Nüesch, J. (1978b) *J. Antibiot. 31,* 215–225.

Ghisalba, O., and Nüesch, J. (1981) *J. Antibiot. 34,* 64–71.

Ghisalba, O., Traxler, P., and Nüesch, J. (1978) *J. Antibiot. 31,* 1124–1131.

Ghisalba, O., Traxler, P., Fuhrer, H., and Richter, W. J. (1979) *J. Antibiot. 32,* 1267–1272.

Ghisalba, O., Traxler, P., Fuhrer, H., and Richter, W. J. (1980) *J. Antibiot. 33,* 847–856.

Ghisalba, O., Roos, R., Schupp, T., and Nüesch, J. (1982) *J. Antibiot. 35,* 74–82.

Haber, A., Johnson, R. D., and Rinehart, K. L., Jr. (1977) *J. Amer. Chem. Soc. 99,* 3541–3544.

Hasegawa, T., Higashide, E., and Shibata, M. (1971) *J. Antibiot. 24,* 817–822.
Higashide, E., Asai, M., Ootsu, K., Tanida, S., Kozai, Y., Hasegawa, T., Kishi, T., Sugino, Y., and Yoneda, M. (1977) *Nature 270,* 721–722.
Hoeksema, H., Chidester, C., Mizsak, S. A., and Baczynski, L. (1978) *J. Antibiot. 31,* 1067–1069.
Iwai, Y., Nakagawa, A., Sadakane, N., Ōmura, S., Oiwa, H., Matsumoto, S., Takahashi, M., Ikai, T., and Ochiai, Y. (1980) *J. Antibiot. 33,* 1114–1119.
Johnson, R. D., Haber, A., and Rinehart, K. L., Jr. (1974) *J. Amer. Chem. Soc. 96,* 3316–3317.
Kakinuma, K., Milavetz, B. I., and Rinehart, K. L. (1976) *J. Org. Chem. 41,* 1358–1364.
Karlsson, A., Sartori, G., and White, R. J. (1974) *Eur. J. Biochem. 47,* 251–256.
Kibby, J. J., McDonald, I. A., and Rickards, R. W. (1980) *J. Chem. Soc. Chem. Comm.* 768–769.
Kishi, T., Yamana, H., Muroi, M., Harada, S., Asai, M., Hasegawa, T., and Mizuno, K. (1972) *J. Antibiot. 25,* 11–15.
Kluepfel, D., Lancini, G. C., and Sartori, G. (1965) *Appl. Microbiol. 13,* 600–604.
Knöll, W. M. J., Rinehart, K. L., Wiley, P. F., and Li, L. H. (1980) *J. Antibiot. 33,* 249–251.
Küenzi, M. T., Gruner, J., Fiechter, A., and Nüesch, J. (1977) *Experientia 33,* 141.
Lancini, G. C., and Grandi, M. (1981) in *Antibiotics* Vol. IV, *Biosynthesis,* (Corcoran, J. W., ed.) pp. 12–40, Springer, Berlin.
Lancini, G. C., and Hengeller, C. (1969) *J. Antibiot. 22,* 637–638.
Lancini, G. C., and Parenti, F. (1978) in *Fifth FEMS Symposium: Antibiotics and Other Secondary Metabolites. Biosynthesis and Production* (Hütter, R., Leisinger, T., Nüesch, J., and Wehrli, W., eds.) pp. 129–139, Academic Press, New York.
Lancini, G. C., and Sartori, G. (1976) *J. Antibiot. 29,* 466–468.
Lancini, G. C., and Sensi, P. (1967) in *Proceedings of the Fifth International Congress on Chemotherapy* Vol. 1 (Spitay, K. H., ed.) pp. 41–47, Wiener Medizinische Akademie, Vienna.
Lancini, G. C., and White, R. J. (1973) *Proc. Biochem. 8,* 14–16.
Lancini, G. C., and White, R. J. (1976) in *Second International Symposium on Genetics of Industrial Microorganisms* (McDonald, K. D. ed.) pp. 139–153, Academic Press, New York.
Lancini, G. C., and Zanichelli, W. (1977) in *Structure-Activity Relationships among the Semi-Synthetic Antibiotics* (Perlman, D., ed.) pp. 531–600, Academic Press, New York.
Lancini, G. C., Thiemann, J. E., Sartori, G., and Sensi, P. (1967) *Experientia 23,* 899–900.
Lancini, G. C., Gallo, G. G., Sartori, G., and Sensi, P. (1969) *J. Antibiot. 22,* 369–377.
Leitich, J., Prelog, V., and Sensi, P. (1967) *Experientia 23,* 505–507.
Margalith, P., and Beretta, G. (1960) *Mycopathol. Mycol. Appl. 13,* 4.
Margalith, P., and Pagani, H. (1961) *Appl. Microbiol. 9,* 325–333.
Martinelli, E., Gallo, G. C., Antonini, P., and White, R. J. (1974) *Tetrahedron 30,* 3087–3091.
Martinelli, E., Antonini, P., Cricchio, R., Lancini, G. C., and White, R. J. (1978) *J. Antibiot. 31,* 949–951.
McDonald, I. A., and Rickards, R. W. (1981) *Tetrahedron Lett.,* 1149–1152.

Milavetz, B., Kakinuma, K., Rinehart, K. L., Jr., Rolls, J. P., and Haak, W. J. (1973) *J. Amer. Chem. Soc. 95,* 5793–5795.
Muroi, M., Haibara, K., Asai, M., and Kishi, T. (1980) *Tetrahedron Lett.* 309–312.
Ōmura, S., Nakagawa, A., and Sadakane, N. (1979) *Tetrahedron Lett.* 4323–4326.
Oppolzer, W., and Prelog, W. (1973) *Helv. Chim. Acta 56,* 2287–2314.
Oppolzer, W., Prelog, V., and Sensi, P. (1964) *Experientia 20,* 336–339.
Prelog, V., and Oppolzer, W. (1973) *Helv. Chim. Acta 56,* 2279–2286.
Rinehart, K. L., Jr. (1972) *Accounts Chem. Res. 5,* 57–64.
Rinehart, K. L., Jr., and Antosz, F. J. (1972) *J. Antibiot. 25,* 71–73.
Rinehart, K. L., and Shield, L. S. (1976) *Fortschr. Chem. Org. Naturst. 33,* 231–307.
Rinehart, K. L., Jr., Maheshwari, M. L., Antosz, F. J., Mathur, H. H., Sasaki, K., and Schacht, R. J. (1971) *J. Amer. Chem. Soc. 93,* 6273–6274.
Rinehart, K. L., Jr., Antosz, F. J., Deshmukh, P. V., Kakinuma, K., Martin, P. K., Milavetz, B. I., Sasaki, K., Witty, T. R., Li, L. H., and Reusser, F. (1976) *J. Antibiot. 29,* 201–203.
Sasaki, K., Rinehart, K. L., Jr., Slomp, G., Grostic, M. F., and Olson, E. C. (1970) *J. Amer. Chem. Soc. 92,* 7591–7593.
Schupp, T., Traxler, P., and Auden, J. A. L. (1981) *J. Antibiot. 34,* 965–970.
Sensi, P., and Thiemann, J. E. (1967) *Prog. Ind. Microbiol. 6,* 21–59.
Sensi, P., Margalith, P., and Timbal, M. T. (1959) *Il Farmaco (Ed. Sci.) 14,* 146–147.
Sensi, P., Greco, A. M., and Ballotta, R. (1960) *Antibiot. Ann. 1959–1960,* 262–270.
Sensi, P., Ballotta, R., Greco, A. M., and Gallo, G. G. (1961) *Il Farmaco (Ed. Sci.) 16,* 165–180.
Shibata, M., Hasegawa, T., and Higashide, E. (1971) *J. Antibiot. 24,* 810–816.
Siminoff, P., Smith, R. M., Sokolski, W. T., and Savage, G. M. (1957) *Am. Rev. Tuberc. Pulm. Dis. 75,* 576–583.
Sugawara, S., Karasawa, K., Watanabe, M., and Hidaka, T. (1964) *J. Antibiot. Ser. A 17,* 29–32.
Thiemann, J. E., Zucco, G., and Pelizza, G. (1969) *Arch. Mikrobiol. 67,* 147–155.
Traxler, P., Schupp, T., Fuhrer, H., and Richter, W. J. (1981) *J. Antibiot. 34,* 971–979.
Waksman, S. A. (1961) *The Actinomycetes* Vol. II, p. 276, William and Wilkins, Baltimore.
Wehrli, W. (1977) *Topics in Current Chem. 72,* 21–49.
White, R. J., and Lancini, G. C. (1975) U.S. Patent 3,871,965.
White, R. J., and Martinelli, E. (1974) *FEBS Lett. 49,* 233–236.
White, R. J., Martinelli, E., Gallo, G. G., Lancini, G. C., and Beynon, P. (1973) *Nature 243,* 273–277.
White, R. J., Martinelli, E., and Lancini, G. C. (1974) *Proc. Nat. Acad. Sci. U.S.A. 71,* 3260–3264.
Williams, T. H. (1975) *J. Antibiot. 28,* 85–86.
Yamazaki, H. (1968) *J. Antibiot. 21,* 204–208.

CHAPTER
10

Tetracyclines

Vladislav Běhal
Michal Bučko
Zdeněk Hošťálek

I. INTRODUCTION

Tetracyclines are broad-spectrum antibiotics widely used in medical practice. The first representative of the group to be discovered was chlortetracycline (Duggar, 1948). Streptomycetes isolated from natural sources and laboratory-constructed strains subsequently yielded a number of other tetracyclines; the commonest ones are listed in Table 10.1. The main producers of tetracycline antibiotics are *Streptomyces aureofaciens,* which produces chlortetracycline and tetracycline (Minieri et al., 1954), and *Streptomyces rimosus,* which produces oxytetracycline (Finlay et al., 1950); their mutants produce 6-demethyl- (McCormick et al., 1957) and decarboxamido-derivatives of tetracyclines (Hochstein et al., 1960; M. W. Miller and Hochstein, 1962). Another antibiotic, minocycline (Barringer et al., 1974), was prepared semisynthetically. Among synthetic tetracycline analogs, 6-thiotetracycline, which contains a sulfur atom in the molecule, has high antibacterial activity (Kirchlechner and Rogalski, 1980).

Heating of chlortetracycline and tetracycline in hydrochloric acid yields corresponding anhydrotetracyclines, intensively colored substances with an absorbance maximum at 440 nm. This property is used in spectrophotometric

TABLE 10.1 Structural Formulae for the Major Tetracyclines

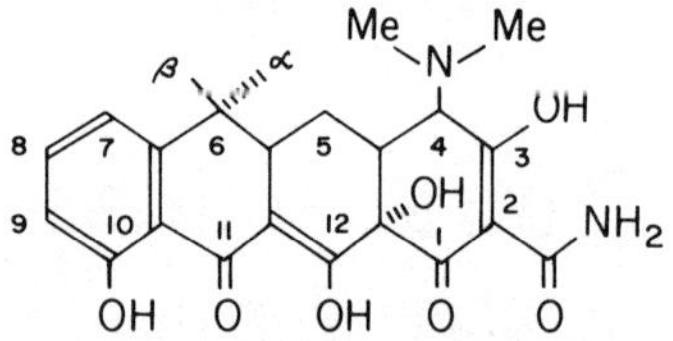

	5	6α	6β	7
Tetracycline	H	CH_3	OH	H
Chlortetracycline	H	CH_3	OH	Cl
6-Demethyltetracycline	H	H	OH	H
6-Demethylchlortetracycline	H	H	OH	Cl
Oxytetracycline	OH	CH_3	OH	H
6-Deoxyoxytetracycline	OH	CH_3	H	H
Minocycline	H	H	H	$N(CH_3)_2$

determination of the two antibiotics (Levine et al., 1949). With iron, oxytetracycline forms a yellow-brown complex with an absorbance maximum at 490 nm, which may be used for its chemical determination (Monastero et al., 1951).

Tetracyclines exert their antibacterial action by inhibiting the synthesis of bacterial proteins. The antibiotic binds to ribosomes, mediated probably by divalent cations. At higher concentrations, tetracyclines bind metal ions, which react with the tricarbonylmethane function in the A ring of the molecule (Williamson and Everett, 1975).

II. BIOSYNTHESIS

Tetracyclines are derived from an oligoketide (a more suitable term for this group of compounds than polyketide) synthesized by a condensation of acetate units (P. A. Miller et al., 1956; Snell et al., 1960; Gatenbeck, 1961; Birch et al., 1962). Eight molecules of malonyl-CoA and the CoA ester of a starter group (malonate semiamide in the case of tetracyclines) participate in the assembly of a 19-membered tetracyclic carbon skeleton (Fig. 10.1).

A. Formation of the Starter Group

The detailed synthesis of malonate semiamide-CoA has not yet been unambiguously elucidated. The carbon of the carboxamide group can come from carbon dioxide (Gatenbeck, 1961), but the carboxyl group of pyruvate was incorporated to the same extent as bicarbonate (Běhal et al., 1974). Asparagine was also used for the biosynthesis of the starter group. Podojil et al. (1973) observed that the [U-^{14}C]-labeled amino acid was preferentially

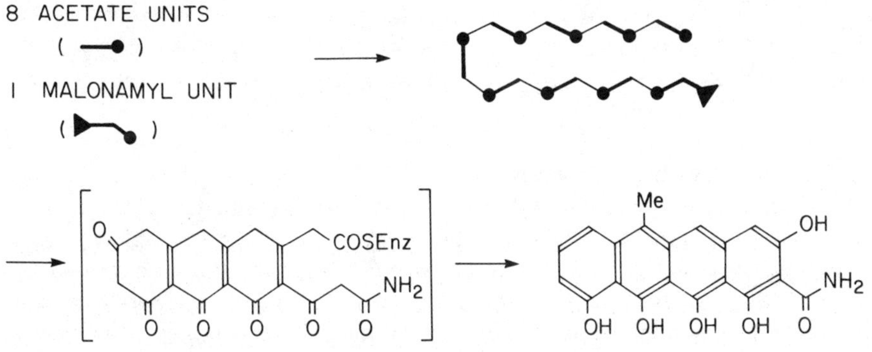

FIGURE 10.1 Biosynthesis of the tetracycline skeleton.

incorporated. Two possibilities for its conversion to malonyl semiamide-CoA are shown in Fig. 10.2. Běhal et al. (1974) demonstrated that [1,3-^{14}C]-malonate semiamide methylester was uniformly incorporated into the tetracycline molecule, while [1,3-^{14}C]malonate semiamide did not penetrate into the cell. Labeled glutamate, alanine, and propionate were efficiently incorporated into tetracycline; all carbon atoms of the skeleton were enriched, indicating that incorporation took place only after randomization.

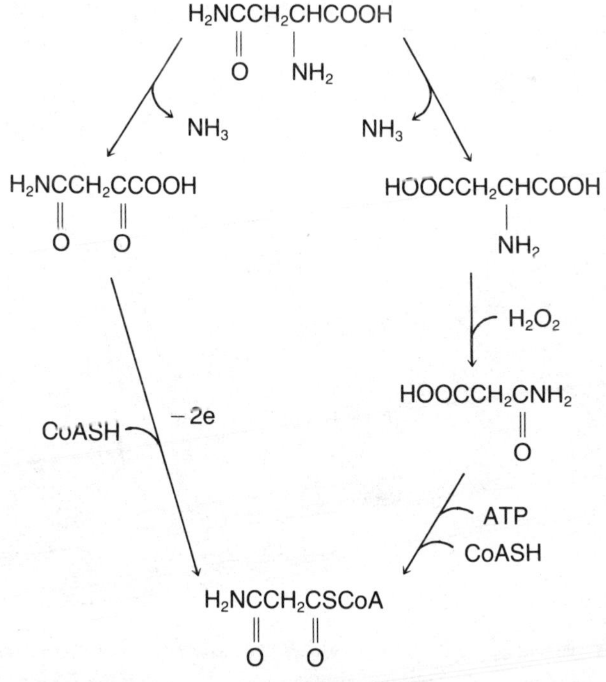

FIGURE 10.2 Possible routes to malonate semiamide-CoA from asparagine.

Hochstein et al. (1960), using a mutant of *S. rimosus,* prepared 2-acetyl-2-decarboxamidooxytetracycline. This attests to the low specificity of the tetracycline-synthesizing enzyme system in regard to the starter group.

B. Formation of Malonyl-CoA

In the generally accepted pathway, malonyl-CoA is formed by carboxylation of acetyl-CoA. However, the specific activity of acetyl-CoA carboxylase in vegetatively growing cultures of *S. aureofaciens* was considerably higher than it was when the cultures entered the period of intensive chlortetracycline production (Běhal and Vaněk, 1970). Similar behavior was found with pyruvate kinase and the pyruvate dehydrogenase complex. On the other hand, the activity of phosphoenolpyruvate carboxylase decreased slightly toward the end of the growth phase and increased again in the production phase. The differences were more marked in a good producing strain than in one whose production was low (Fig. 10.3) (Běhal et al., 1977). The existence of an alternative pathway for malonyl-CoA formation via oxaloacetate was established with a cell-free preparation from *S. aureofaciens.* From phosphoenolpyruvate and ^{14}C-bicarbonate, this synthesized oxaloacetate, which was decarboxylated to malonate (Fig. 10.4). The results of in vivo experiments

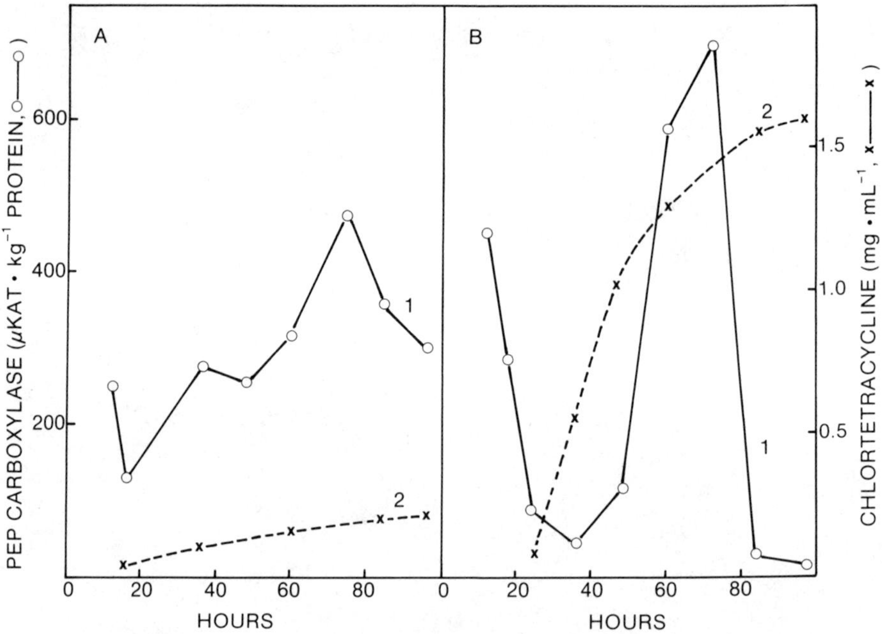

FIGURE 10.3 Specific activity of phosphoenolpyruvate carboxylase (1) and production of chlortetracycline (2) in (A) a low-producing strain and (B) a high-producing strain.

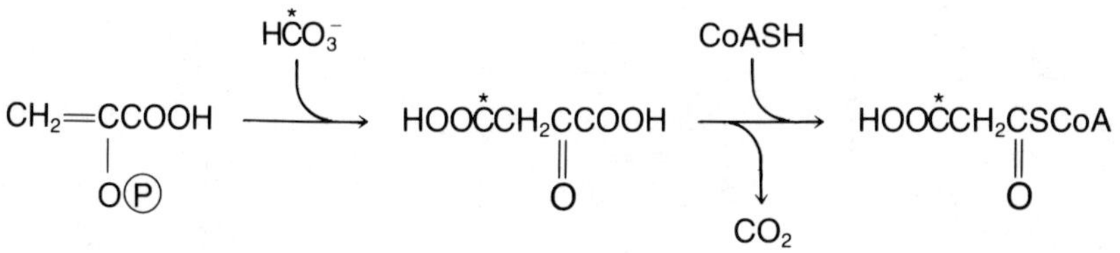

FIGURE 10.4 Postulated formation of malonyl-CoA from phosphoenolpyruvate via oxaloacetate.

also excluded formation of malonyl-CoA solely by the direct pathway from pyruvate via acetylCoA. With acetate labeled in the methyl carbon (and also with [3-^{14}C]pyruvate and [2-^{14}C]malonate), 28% of the radioactivity in tetracycline was present in carbons derived from the acetate carboxyl. This indicates that 53–78% of the acetate units used for tetracycline biosynthesis passed through other metabolic pathways prior to incorporation. Similar conclusions can be drawn from the results of Catlin et al. (1969), who found 15.8% of the radioactivity in oxytetracycline labeled from [2-^{14}C]acetate in carbons originating from the acetate carboxyl group.

C. Transformations of the Tetracycline Skeleton

Condensation of malonyl-CoA with malonate semiamide-CoA gives rise to a 19-carbon chain, probably bound to an enzyme template (McCormick, 1965). This chain is aromatized to yield a hypothetical tricyclic nonaketide. At this stage, the hydroxyl group on C-8 is probably reduced, and methylation at C-6 takes place as shown in Fig. 10.5 (Vaněk et al., 1971). Closure of ring A gives 6-methylpretetramide. When a cyclized oligoketide chain is released from the enzyme template prior to closure of the A ring, substances like protetrone (McCormick, 1966; McCormick and Jensen, 1968), ekatetrone (Podojil et al., 1978; Přikrylová et al., 1978), and an oxanthranol derivative (McCormick et al., 1968b) are formed.

According to McCormick (1965), the initial intermediate in the biosynthesis of tetracyclines is 6-methylpretetramide. This is converted in vivo to tetracycline and chlortetracycline. By using blocked mutants, intermediates in the biosynthetic pathway have been obtained. When pretetramide was used for the conversion, 6-demethyltetracycline derivatives were formed (McCormick et al., 1963; McCormick, 1969). 6-Demethyltetracycline is produced by a mutant, *S. aureofaciens* S 604, and can also be obtained by inhibiting methylation with ethionine (Hendlin et al., 1962). This implies that the C-6 methyl comes from the methyl group of methionine and is introduced by transmethylation. However, in *S. rimosus*, methylation at C-6 is not inhibited by ethionine (P. A. Miller et al., 1964).

In chlortetracycline biosynthesis, 6-methylpretetramide is first hydroxylated at C-4 and the hydroxyl group is then oxidized to a ketone. The

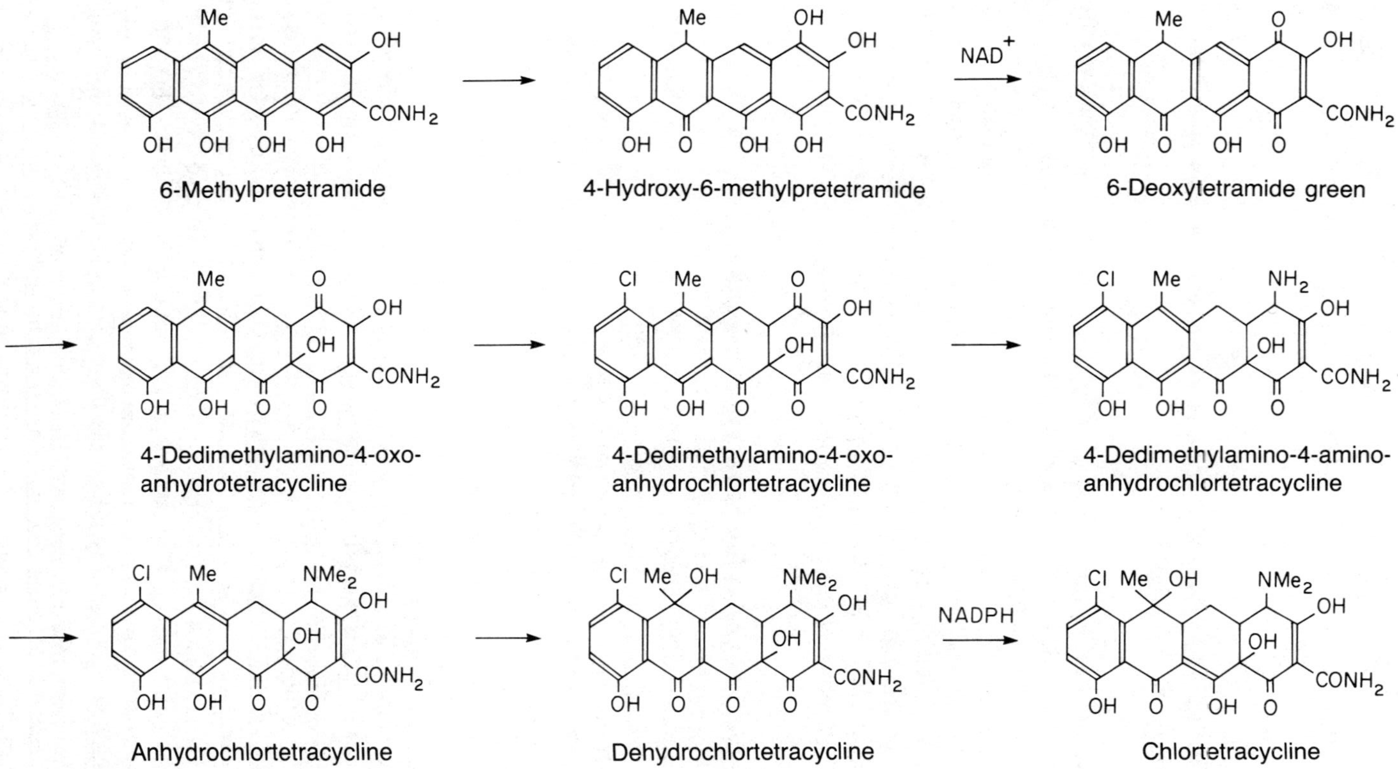

FIGURE 10.5 Transformations of the tetracycline skeleton.

oxidized compound, 6-deoxytetramide green, has not yet been isolated, but 4-hydroxy-6-methylpretetramide was identified by McCormick and Jensen (1965) in the mutant *S. aureofaciens* V 655 and later was transformed to chlortetracycline (McCormick et al., 1965).

Further hydroxylation occurs at position 12a; it gives rise to 4-oxo-4-dedimethylaminoanhydrotetracycline, which was isolated from *S. aureofaciens* mutants T 219 and S 2308 (McCormick, 1967, 1969). Chlorination of this substance yields 4-oxo-4-dedimethylaminochloranhydrotetracycline, also isolated by McCormick (1969). Chlorination does not occur if the medium lacks sufficient chloride ions (Minieri et al., 1954). It can be competitively inhibited by bromide ions; under these conditions, 7-bromotetracycline is formed along with tetracycline (Gourevitch et al., 1955; Sensi et al., 1955; Doerschuk et al., 1959).

The amino group on C-4 is introduced by transamination; 4-dedimethylamino-4-aminoanhydrotetracycline was isolated by P. A. Miller et al. (1964) in the soluble fraction of disintegrated cells of *S. aureofaciens* BC 41 cultivated in the presence of ethionine. Cultivation of *S. rimosus* on a synthetic medium containing methionine analogs yielded a mixture of demethylanhydrotetracyclines: 4-dedimethylamino-4-amino-5a,6-anhydrotetracycline, 4-dedimethylamino-4-methylamino-5a,6-anhydrotetracycline, and their 6-demethyl analogs. 4-Dedimethylamino-4-amino-6-demethyl-5a,6-anhydrochlortetracycline was used by McCormick et al. (1968a) as an intermediate in the preparation of 6-demethylchlortetracycline.

Methylation of the amino group proceeds in two steps by transmethylation from methionine. 4-Dedimethylamino-4-amino-5a,6-anhydrotetracycline and its demethyl derivative were converted in a *S. aureofaciens* cell-free system containing S-adenosylmethionine to 5a,6-anhydrotetracycline or 6-demethyl-5a,6-anhydrotetracycline (P. A. Miller et al., 1964). Ethionine in this case did not inhibit methylation, but it interfered with methionine synthesis.

Oxidation of anhydrotetracyclines at C-6 requires oxygen and NADPH. The final reaction, reduction of the 5a,11a double bond, proceeds at the expense of NADPH (P. A. Miller et al., 1964).

III. ENZYMES OF TETRACYCLINE BIOSYNTHESIS

Vaněk et al. (1971) assume that tetracycline-synthesizing enzymes form a multienzyme complex similar to that participating in fatty acid synthesis or 6-methylsalicylic acid synthesis (Lynen and Tada, 1961; Dimroth et al., 1970). The terminal reactions of tetracycline synthesis were detected in cell-free systems from *S. aureofaciens* and *S. rimosus* (P. A. Miller et al., 1964, 1965). 4-Dedimethylamino-4-amino-5a,6-anhydrotetracycline is methylated in two steps by a methyl group transferred from methionine (P. A. Miller et al., 1964) to yield anhydrotetracycline. This is a common precursor of both tetracycline and oxytetracycline (Fig. 10.6). P. A. Miller et al.

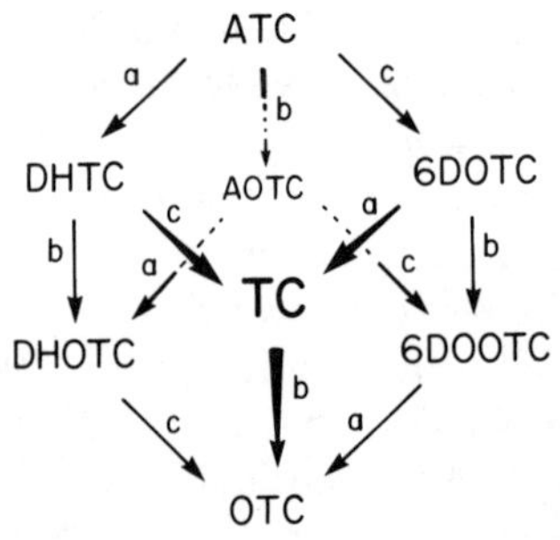

FIGURE 10.6 Possible pathways for the conversion of anhydrotetracycline to oxytetracycline: the metabolites are: ATC = anhydrotetracycline, DHTC = 5a,(11a)-dehydrotetracycline, AOTC = 5-hydroxyanhydrotetracycline, 6DOOTC = 6-deoxy-5-hydroxytetracycline, 6DOTC = 6-deoxytetracycline, TC = tetracycline, DHOTC = 5a,(11a)-dehydro-5-hydroxytetracycline, OTC = 5-hydroxytetracycline.
Reactions: a = 6-hydroxylation; b = 5-hydroxylation; c = 5a,(11a)-reduction.

(1965) proposed the sequence anhydrotetracycline → dehydrotetracycline → dehydrooxytetracycline → oxytetracycline as the only possible metabolic pathway for oxytetracycline formation. However, V. Běhal (unpublished data) found that anhydrooxytetracycline was hydroxylated at C-6 by cell-free preparations of both *S. rimosus* and *S. aureofaciens*.

The activities of the final enzymes of tetracycline synthesis can be measured quantitatively. Methylation of 4-dedimethylamino-4-aminoanhydrotetracycline to anhydrotetracycline by the enzyme S-adenosylmethionine:dedimethylamino-4-aminoanhydrotetracycline N-methyltransferase is assayed by measuring incorporation of the labeled methyl group of S-adenosylmethionine catalyzed by a crude enzyme preparation of *S. aureofaciens* or *S. rimosus* (P. A. Miller and Hash, 1975a). The enzyme activity of the lyophilized cell-free preparation was stable for several months at −10°C. The enzyme methylated 7-chloro-, 6-demethyl-, 2-decarboxamido-2-nitrile-, and 4-methyl-derivatives of dedimethylamino-4-aminotetracycline.

The penultimate reaction of tetracycline synthesis is hydration of anhydrotetracycline, catalyzed by anhydrotetracycline oxygenase. The reaction proceeds at the expense of NADPH and atmospheric oxygen. The activity of the enzyme can be measured spectrophotometrically by the decrease in absorbance at 440 nm as anhydrotetracycline is converted to dehydrotetracycline (Běhal et al., 1979a). The enzyme is localized in the cytoplasm, and the optimum pH of the reaction is in the range of 7.4–7.5. The enzyme is relatively stable. It remains active during storage for several weeks at −25°C (Běhal et al., 1979a).

Reduction of the 5a,11a-double bond by NADP:tetracycline 5a(11a)-dehydrogenase can be assayed by measuring the increase in antibacterial activity due to formation of chlortetracycline. The substrate, 7-chloro-5a(11a)dehydrotetracycline, shows no activity against *Staphylococcus aureus* (P. A. Miller and Hash, 1975b).

IV. PHYSIOLOGY OF TETRACYCLINE PRODUCERS

Industrial fermentations to produce tetracycline antibiotics employ media that ensure high yields. Complex media containing natural materials such as saccharides, fats, soybean or peanut flour, cornsteep liquor, and molasses are used. The ingredients provide any trace substances necessary for satisfactory growth and high production. To obtain amino acids from proteins, *S. rimosus* produces exocellular proteases (Pokorný et al., 1979). These include serine proteases and metalloproteases, active over a wide pH range and possessing different substrate specificities.

Laluce and Molinari (1977) achieved increased production of extracellular proteases in *S. aureofaciens* by replacing sucrose with maltose in the cultivation medium. A sharp maximum of caseinolytic activity occurred at the end of the growth phase. The authors separated the proteases into four protein fractions (Laluce and Molinari, 1979).

A. Metabolism of Sugars

Horvath and Szentirmai (1962) confirmed the presence in *S. rimosus* of a number of enzymes of sugar metabolism. Enzymes of the hexose monophosphate pathway were found in *S. aureofaciens* by Boretti et al. (1956). Carbon sources used in tetracycline biosynthesis include starch, sucrose, or maltose; *S. aureofaciens* can grow and produce chlortetracycline on a defined medium with glycerol as the carbon source and ammonium ions as the nitrogen source (P. A. Miller et al., 1956). Chlortetracycline yields depend on the rate of sugar utilization, higher yields being attained in media with slowly assimilated saccharides. The negative effect of glucose on chlortetracycline production observed by Rokos and Procházka (1957) can be explained as catabolite repression of the synthesis of enzymes of the tetracycline biosynthetic pathway (Hošťálek, 1980).

Sugar phosphorylation in tetracycline producers proceeds at the expense of ATP as well as polyphosphates. Comparison of the activities of ATP glucokinase and polyphosphate glucokinase in *S. aureofaciens* (Kulayev et al., 1976; Hošťálek et al., 1976) showed that ATP glucokinase was active predominantly in the growth phase, while polyphosphate glucokinase participated in sugar metabolism only in the period of chlortetracycline synthesis. Glycolytic dissimilation of sugars in *S. aureofaciens* provides pyruvate, which is accumulated in the medium. Its concentration in complete medium after 12 h of growth was 100 μg/mL (Doskočil et al., 1959). On further cultivation, the concentration decreased considerably. Běhal (1968) found a pyruvate concentration of up to 4500 μg/mL in a synthetic medium in which *S. aureofaciens* was cultivated for 48 h. Growth of the culture was suppressed, and chlortetracycline production was about ten times lower than in complex medium.

When the activity of tricarboxylic acid cycle enzymes in *S. aureofaciens* was examined (Hošťálek et al., 1969a), a high-producing strain was found to have twofold to fivefold lower activity than a low-producing strain. Benzylthiocyanate reduced the activity of tricarboxylic acid cycle enzymes, especially in the low-producing strain. The results imply that the tricarboxylic acid cycle competes with chlortetracycline biosynthesis by draining off metabolic intermediates that provide building units for the antibiotic molecule.

B. Fatty Acids

Another enzymic process requiring acetyl-CoA and malonyl-CoA is fatty acid synthesis. The amount of lipid, which is presumed to constitute part of the cell membranes in *S. aureofaciens,* is about 4% of the dry weight (Běhal et al., 1969a). Fatty acids were found to undergo a considerable turnover, 60% of those synthesized being decomposed during the cultivation. For this reason, it is unlikely that the synthesis of fatty acids sequesters enough malonyl-CoA to affect significantly the biosynthesis of chlortetracycline (Běhal and Jílek, 1969). Furthermore, a considerable proportion of the fatty acids contains carbon residues of amino acids formed by protein degradation; almost half of the fatty acids are branched (Běhal et al., 1969a).

Batrakov et al. (1977) showed that the proportion of branched fatty acids increased in all types of lipids of *Actinomyces streptomicini* throughout the cultivation. This indicates that streptomycetes use carbon residues of amino acids from degraded proteins extensively for fatty acid biosynthesis. Aspartic acid added to the medium in which *S. aureofaciens* was cultivated for 6 h was incorporated into fatty acids to a level only threefold lower than was acetic acid (Běhal, 1968). Fatty acid synthesis in *S. aureofaciens* is most intensive in the first part of the growth phase, particularly in high-producing strains. During the production phase, synthesis of fatty acids is low, and the content of lipids in the cells is also low; this indicates that lipids are not reserve substances in *S. aureofaciens* (Běhal et al., 1969b).

In tetracycline producers, the NADPH necessary for fatty acid synthesis is obtained by oxidation of glucose-6-phosphate. The presence of glucose-6-phosphate dehydrogenase has been demonstrated in a cell-free preparation from *S. rimosus* (P. A. Miller et al., 1965). In addition, NADPH is regenerated by the action of malate dehydrogenase; in *S. aureofaciens,* the enzyme is especially active in the growth phase (Jechová et al., 1969).

C. Bioenergetics

Energy levels in the cell probably play a significant role in the expression of secondary metabolism. An insufficient supply of energy may be a crucial factor triggering the biosynthesis of oligoketides. The oligoketide is not reduced during its formation from acetate units, and the structure formed by cyclization of the 19-membered carbon chain is only partially reduced. Biosynthesis of the tetracycline skeleton is thus energetically less exacting than biosynthesis of a fatty acid of the same chain length.

Janglová et al. (1969) showed that in *S. aureofaciens* the ATP level was depressed during the production phase and was especially low in a higher-producing strain (Fig. 10.7). Since the decrease in ATP level correlated with increased ATP-diphosphohydrolase activity, this enzyme was probably responsible (Vaněk et al., 1973). ATP inhibits allosterically the activity of some enzymes of primary metabolism in *S. aureofaciens*. The activity of the tricarboxylic acid cycle is controlled via citrate synthase (Hošťálek et al., 1969b) and phosphoenolpyruvate carboxylase (Vořísek et al., 1969, 1970). However, the concentration of ATP required to inhibit these enzymes is tenfold higher than the maximum found in even the low-producing strain. Thus this regulatory mechanism probably plays little part in the expression of chlortetracycline biosynthesis.

The adenylate energy charge (Atkinson and Walton, 1967) did not change in value throughout cultivation of either the high- or the low-producing strain. However, appreciable differences were found in the total adenylate

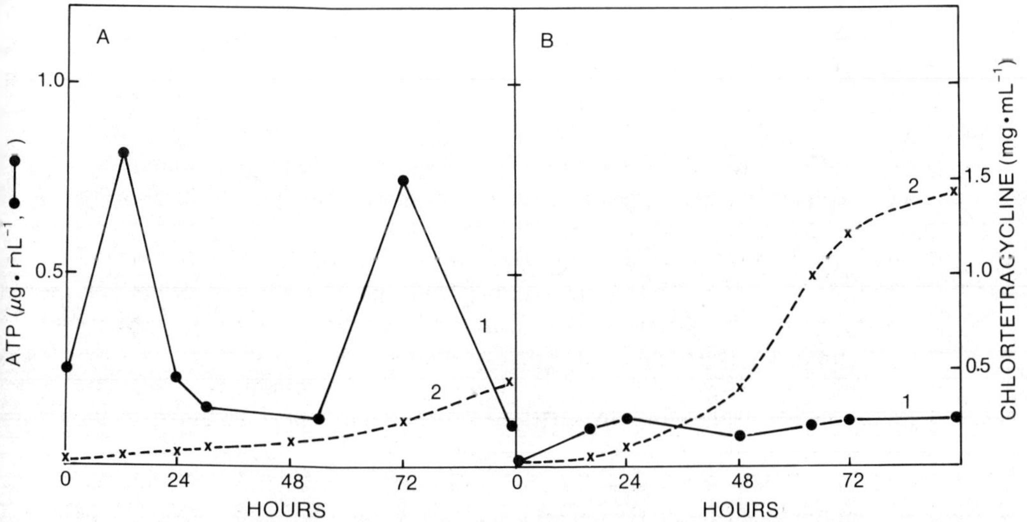

FIGURE 10.7 Changes in the concentration of intracellular ATP (1) and chlortetracycline (2) during growth of (A) low-producing and (B) high-producing strains of *S. aureofaciens*.

level; this was about tenfold greater in the low producer than in the high producer (Čurdová et al., 1976). Anhydrotetracycline oxygenase activity and the rate of chlortetracycline synthesis increase rapidly after 12 h of cultivation, coinciding with a drop in adenylate concentration in the cells. The subsequent rise in adenylate level at the end of cultivation is in keeping with a decrease in the level of the enzyme and with the cessation of chlortetracycline production. The regulatory factor controlling expression of antibiotic synthesis is apparently the adenylate pool level, not the energy charge of the system (Čurdová et al., 1976; see also Chapter 3).

V. GENETIC REGULATION

Physical and chemical mutagens can be used to prepare biochemical variants of tetracycline producers either possessing increased productivity or blocked in the biosynthetic pathway (for a review, see Hošťálek et al., 1974). Blocked mutants either accumulate the intermediate that serves as substrate for the missing enzyme or use it in an alternative metabolic pathway. Those which accumulate unchanged intermediates have been used in the cosynthesis procedure to determine the sequence of reactions in tetracycline biosynthesis (McCormick, 1965, 1966). The range of theoretically possible alternative pathways for the biosynthesis of tetracyclines, giving rise to both established antibiotic end products and shunt products, has been elaborated by Vaněk et al. (1971).

Genetic maps of *S. rimosus* and *Streptomyces coelicolor* are very similar (Friend and Hopwood, 1971; Alačevič et al., 1973). The map of the *S. rimosus* genome includes some markers for oxytetracycline biosynthesis as well as those from primary metabolic pathways. The genes for enzymes converting anhydrotetracycline to oxytetracycline were localized in a small segment of the circular chromosome of *S. rimosus* between *proA* and *adeA*. Genes associated with the initial steps in oxytetracycline biosynthesis were localized on the opposite side of the chromosome between *ribB* and *sysD* (Rhodes et al., 1981). The fertility of *S. rimosus* is controlled by a plasmid (Friend et al., 1978). Extrachromosomal DNA may also carry genes participating indirectly in tetracycline synthesis, but the evidence is insecure (see Chapter 1). Boronin et al. (1972) found lowered resistance to oxytetracycline in a strain of *S. rimosus* isolated after treatment with acridine under plasmid-curing conditions. It should be noted that resistance of the organism to its own product is of prime importance in antibiotic overproduction. Genetic work with the chlortetracycline producer *S. aureofaciens* is exceptionally difficult because of marked variability and instability of strains. This is enhanced in nutritional or resistance mutants (Blumauerová et al., 1972). Knowledge of the genetic regulation of tetracycline synthesis is still meager, even though it should have practical applications in industrial production of antibiotics (Hopwood, 1978).

The mechanism through which the information for tetracycline biosynthesis is expressed is not yet clear. This is mainly due to current ignorance of the role of tetracyclines in the biology of the producer. The manner in which biosynthesis of an excessive amount of the antibiotic is controlled deserves particular attention. Like a number of other antibiotics, tetracyclines are synthesized in detectable amounts only in the second half of the growth phase, which is characterized by an equal increment and loss of biomass. The significance of this turnover is uncertain, since both *S. aureofaciens* and *S. rimosus* are filamentous microorganisms. The cell population is heterogeneous, and even during rapid growth, the culture contains older cells that already produce tetracyclines.

It is not yet clear what signal the expression of tetracycline biosynthesis responds to. Solution of the problem is hindered by the fact that tetracyclines interact with the synthesis of nucleic acids and proteins in the producer (Mikulík et al., 1971). They form inactive aggregates with *S. aureofaciens* ribosomes in vitro (Ludvík et al., 1971). Despite this interference, some protein is synthesized during tetracycline production (Běhal et al., 1979b; Šimúth et al., 1979). RNA synthesis decreases sharply in the second half of the growth phase and is inhibited by addition of exogenous chlortetracycline; in a low-producing strain, the concentration needed is much lower than in a good producer (Mikulík and Vaněk, 1979; Šimúth et al., 1979). The possibility that tetracyclines play a regulatory role in transcription in *S. aureofaciens* is supported by evidence that polynucleotide phosphorylase from *S. aureofaciens* is competitively inhibited by chlortetracycline (Šimúth et al., 1975). Also, the synthesis of DNA, which in the production phase occurs at less than one tenth the rate that prevails in the growth phase, is strongly inhibited by exogenously supplied chlortetracycline (Běhal et al., 1979b). Nevertheless, changes during cultivation are not necessarily caused by the chlortetracycline formed and may be the result of culture aging. Coincident with or closely following the drop in RNA synthesis in high-producing strains, the activity of most enzyme sequences, such as those of the tricarboxylic acid cycle and fatty acid synthesis, also decreases (Hošťálek et al., 1969a; Běhal and Jílek 1969).

Initiation factor IF-3 from *Escherichia coli* stimulates formation of the initiation complex in *S. aureofaciens* (Mikulík and Vaněk, 1979). Elongation factors isolated from the two microorganisms are also mutually replaceable in the formation of a ternary complex with Phe-tRNA. Despite this similarity (also observed in immunological comparisons), they show differences in heat stability, in binary EF-Tu-GTP complex formation, and in the GTPase activity of this complex (Weiser, 1980; Weiser et al., 1981).

Highly phosphorylated nucleotides function as regulatory substances in the metabolism of tetracycline producers. In *S. aureofaciens,* large differences were found in the rate of synthesis of these substances during cultivation, the rate being lower in a low-producing strain than in a good producer (Šimúth et al., 1979). Exogenously added chlortetracycline reduced the rate of guanosine tetraphosphate synthesis. *S. aureofaciens* also contained guanosine pentaphosphate. The concentration of highly phosphorylated nucleotides was highest during transition from the growth to the stationary phase; it was also higher during cultivation in a medium in which no tetracycline production took place (Mikulík and Vaněk, 1979; Šťastná and Mikulík, 1981).

VI. OVERPRODUCTION OF TETRACYCLINES AND ITS REGULATION

A. Genetic Aspects

Under optimum cultivation conditions, wild tetracycline-producing strains synthesize several tens of micrograms of the antibiotic per milliliter of medium. Values achieved currently in industrial productions are 1000-fold higher and are the result of multistage selection. Cultivation conditions are adjusted to suit the changed properties of higher producers so that they do not become a limiting factor in production. The metabolism of high-producing variants differs substantially from that of parent strains, although the range of metabolic pathways is essentially the same; the main difference is in the regulation of pathways associated with antibiotic production. Under natural conditions the wild-type strain synthesizes only as much of the antibiotic as it needs, whether for interspecies competition, as a signal molecule, or for other physiological roles. Wild strains are inhibited or killed by the antibiotic concentrations synthesized by high-producing mutants. To enable strains to produce excessive amounts of the antibiotic, regulatory mechanisms restricting antibiotic production have to be removed or modified.

High tetracycline production is determined by three main conditions: high activity of the enzyme system synthesizing the antibiotic, a sufficient supply of building units, and resistance to the amount of antibiotic produced. The third condition is important because tetracycline acts as an inhibitor of protein synthesis and thus of the formation of antibiotic synthetases. A certain proportion of RNA synthesis and protein synthesis in *S. aureofaciens* is resistant to chlortetracycline (Mikulík and Vaněk, 1979; Šimúth et al., 1979), but this resistance appears later in the growth phase. It probably requires induction of a system protecting protein synthesis in the producer. Borisoglebskaya et al. (1979) assume that the formation of such a protective substance (probably of proteinaceous nature) in *S. rimosus* is mediated by plasmid DNA.

B. Physiological Aspects

A basic requirement for high yields is the creation of cultivation conditions that permit sufficient growth and, at the same time, promote maximum expression of tetracycline synthesis. These conditions differ for individual genotypes but are always based on specific nutrient limitations. Most often, limitation of phosphate is used; this makes possible, with excess nitrogen and carbon sources, the achievement of maximum productivity (Doskočil et al., 1958, 1959). Zaitseva and Orlova (1959) give 50–60 mg P/mL as an optimum concentration of inorganic phosphate for oxytetracycline biosynthesis by *S. rimosus*. The synthesis of chlortetracycline in *S. aureofaciens* is more sensitive to phosphate, and the maximum yield is attained at concentrations below 30 μg P/mL (Makarevich et al., 1975).

As was shown by Běhal et al. (1979a), lower amounts of ATC oxygenase are synthesized in *S. aureofaciens* cultivated in media with high phosphate concentrations. Tetracycline production is directly proportional to the activity of the enzyme (Fig. 10.8), which apparently reflects the activity of the whole tetracycline synthetase complex. During cultivation under phosphate limitation, phosphate is exhausted during the first half of the growth phase (Běhal et al., 1982; Běhal and Hošťálek, 1981); synthesis of ATC oxygenase occurs only after phosphate disappearance. If more phosphate is added to the depleted medium while ATC oxygenase is being made, synthesis of the enzyme stops, and its activity decreases until the phosphate is again exhausted. After phosphate exhaustion, the amount of the enzyme in the cells rises again. Adding phosphate in the production phase does not significantly reduce formation of the antibiotic. The rate of protein synthesis decreases after phosphate exhaustion. If the level of phosphate in the medium is

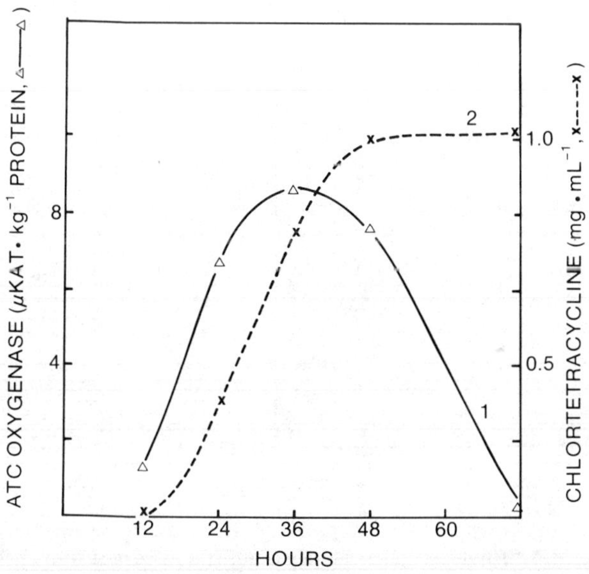

FIGURE 10.8 Specific activity of ATC oxygenase (1) and chlortetracycline yield (2).

maintained above the limiting value by constant phosphate addition, no decrease in protein synthesis is observed.

Physiological conditions suitable for biosynthesis of chlortetracycline and tetracycline can be modified by adding benzylthiocyanate to the medium (Pecák et al., 1958). However, this compound has no effect on the biosynthesis of oxytetracycline in *S. rimosus*. The mechanism of action of benzylthiocyanate is not yet known; in *S. aureofaciens,* it slows down sugar consumption and reduces the growth rate (Erban et al., 1981). Its site of action is probably the cell surface, where it reduces the rate of sugar uptake (Hošťálek, 1964a, b, c). Altering the physiological state of *S. aureofaciens* with benzylthiocyanate brings about increased synthesis of ATC-oxygenase. Synthesis of the enzyme (and hence antibiotic production) is stimulated most effectively when benzylthiocyanate is added in the first half of the growth phase (Běhal et al., 1982). The stimulatory effect is also much more marked in low- than in high-producing strains (Fig. 10.9). An observable rise in ATC-oxygenase synthesis takes place 2 h after addition of benzylthiocyanate (Běhal, 1982).

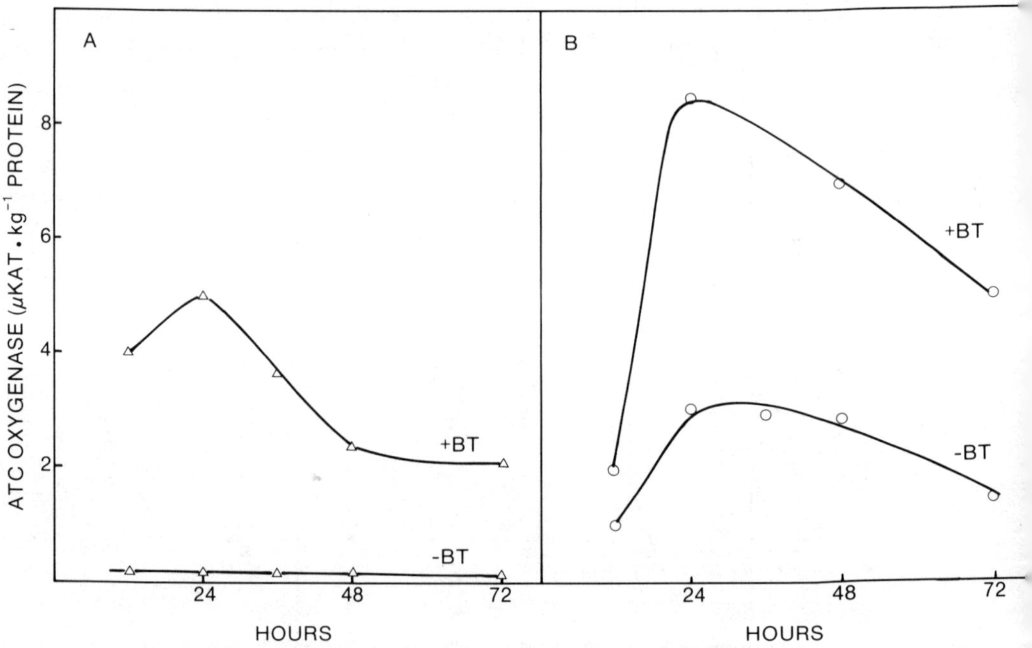

FIGURE 10.9 Effect of benzylthiocyanate (BT, 2 μmol/mL medium) on the activity of ATC oxygenase in (A) low-producing and (B) high-producing strains of *S. aureofaciens*.

C. Bioengineering Aspects

High yields of tetracycline antibiotics on an industrial scale require harmonious adjustment of all important parameters of the fermentation process such as media composition, pH values, culivation temperature, aeration, and mixing. The parameters are monitored and adjusted by devices that automatically maintain the metabolism of the culture within the limits of an optimum model.

Long-term productivity and a high production rate are maintained by semicontinuous or continuous addition of nutrients—normally the carbon and nitrogen sources. It is especially important to maintain the carbon source above a certain limiting concentration; if it falls below this value, irreversible changes occur in the producer's metabolism, and production

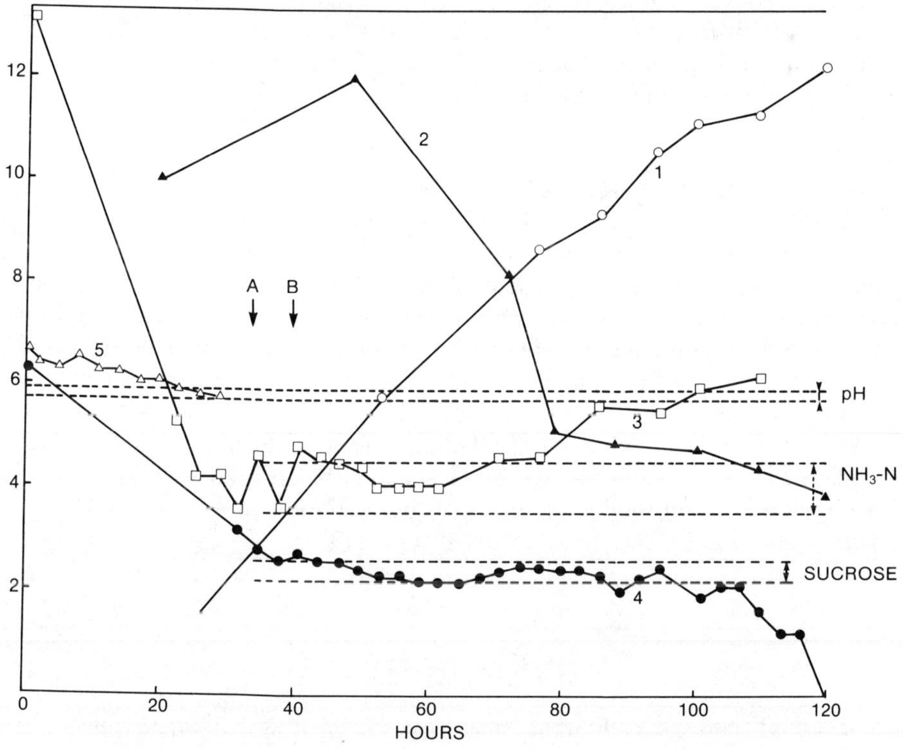

FIGURE 10.10 Parameters of an industrial fermentation of *S. aureofaciens:* 1=CTC production ($mg \cdot mL^{-1}$); 2=ATC oxygenase ($pkat \cdot mg^{-1}$ protein x 2); 3=NH_3-nitrogen ($mg \cdot mL^{-1}$ x 0.1); 4=sucrose ($mg \cdot mL^{-1}$ x 10); 5=pH. Aqueous ammonia was added after 27 h to maintain the pH at 5.7–5.9; sucrose was added from 33 h to 115 h to hold the concentration at 22–25 $mg \cdot mL^{-1}$. Ammonium sulfate supplements were added at points A and B.

ceases. The level of nitrogen source is sustained by adding inorganic nutrients such as ammonium sulfate or ammonia liquor. Ammonia liquor serves, together with calcium carbonate, as a regulator of culture pH; an optimum range of pH in *S. aureofaciens* is 5.6–6.2 (Upiter et al., 1974). Calcium carbonate contributes to high yields also by forming an insoluble complex with tetracyclines, thus reducing the effect of the antibiotic on the producer. A typical course for a chlortetracycline fermentation is shown in Fig. 10.10.

The requirement for aeration of the culture changes during cultivation and is different in seed and production fermentors. *Streptomyces aureofaciens* requires more intensive aeration than *S. rimosus* (Orlova et al., 1959). In the initial phase of cultivation, *S. aureofaciens* exhibits high sensitivity to the oxygen supply rate; a short interruption inhibits chlortetracycline production irreversibly (Matelová et al., 1955). Aeration intensity also has to match the level of phosphate in the medium (Doskočil et al., 1958, 1959). An important factor in the overproduction of tetracycline antibiotics is the design of fermentors and impellers; systems that shear or otherwise damage the mycelium are unsuitable.

VII. CONCLUDING REMARKS

In the space available, the full scope of problems associated with tetracycline biosynthesis could not be presented; in particular, it has been necessary to limit references to the available literature. In the field of fermentation technology, the reader is referred to an older review by DiMarco and Pennella (1959) and a recent one by Podojil et al. (1982). Studies on biosynthetic pathways and intermediates are summarized in reviews by McCormick (1967, 1969) and Hutchinson (1981). Genetic aspects of tetracycline biosynthesis have been summarized by Hošťálek et al. (1974); extensive surveys of the literature on the biosynthesis of these antibiotics were published by Turley and Snell (1966) and Hošťálek et al. (1979).

REFERENCES

Alačevič, M., Strašek-Vešligaj, M., and Sermonti, G. (1973) *J. Gen. Microbiol. 77,* 173–185.

Atkinson, D. E., and Walton, G. M. (1967) *J. Biol. Chem. 242,* 3239–3246.

Barringer, W. C., Schultz, W., Sieger, G. M., and Nash, R. A. (1974) *Am. J. Pharmacy 146,* 179–191.

Batrakov, S. G., Kornitskaya, E. Y., Sarkisyan, Š. T., and Bergel'son, L. D. (1977) *Izv. Akad. Nauk SSSR Ser. Biol.* (2), 226–234.

Běhal V. (1968) Ph.D. thesis, Institute of Microbiology Czechoslovate Academy of Sciences, Prague.

Běhal, V. (1982) in *Overproduction of Microbial Products* (Krumphanzl, V., Sikyta, B., and Vaněk, Z., eds.) pp. 301–309, Academic Press, London.

Běhal, V., and Jílek, M. (1969) *Folia Microbiol.* *14,* 211–214.

Běhal, V., and Vaněk, Z. (1970) *Folia Microbiol.* *15,* 354–357.

Běhal, V., Procházková, V., and Vaněk, Z. (1969a) *Folia Microbiol.* *14,* 112–116.

Běhal, V., Cudlín, J., and Vaněk, Z. (1969b) *Folia Microbiol.* *14,* 117–120.

Běhal, V., Podojil, M., Hošťálek, Z., Vaněk, Z., and Lynen, F. (1974) *Folia Microbiol.* *19,* 146–150.

Běhal, V., Jechová, V., Vaněk, Z., and Hošťálek, Z. (1977) *Phytochemistry* *16,* 347–350.

Běhal, V., Hošťálek, Z., and Vaněk, Z. (1979a) *Biotechnol. Lett.* *1,* 177–182.

Běhal, V., Vaněk, Z., Hošťálek, Z., and Ramadan, A. (1979b) *Folia Microbiol.* *24,* 211–216.

Běhal, V., and Hošťálek, Z. (1981) Abstracts of the FEMS Symposium on Overproduction of Microbial Products, Hradec Králové, Czechoslovakia, p. 133.

Běhal, V., Prušáková-Grégrová, J., and Hošťálek, Z. (1982) *Folia Microbiol.* *27,* 102–106.

Birch, A. J., Snell, J. F., and Thomson, P. J. (1962) *J. Chem. Soc.* 425–429.

Blumauerová, M., Hošťálek, Z., and Vaněk, Z. (1972) in *Fermentation Technology Today* (Terui, G., ed.) pp. 223–232, Society of Fermentation Technology, Osaka.

Boretti, G., DiMarco, A., Julita, P., Raggi, F., and Bardi, U. (1956) *Giorn. Microbiol.* *1,* 406–416.

Borisoglebskaya, A. N., Perebityuk, A. N., and Boronin, A. M. (1979) *Antibiotiki* (Moscow) *24,* 883–889.

Boronin, A. M., and Sodovnikova, L. G. (1972) *Genetika* *8,* 174–176.

Catlin, E. R., Hassall, C. A., and Parry, D. R. (1969) *J. Chem. Soc. C,* 1363–1366.

Čurdová, E., Křemen, A., Vaněk, Z., and Hošťálek, Z. (1976) *Folia Microbiol.* *21,* 481–487.

DiMarco, A., and Pennella, P. (1959) in *Progress in Industrial Microbiology* Vol. I (Hockenhull, O. J. D., ed.) pp. 45–92, Heywood, London.

Dimroth, P., Walter, H., and Lynen, F. (1970) *Eur. J. Biochem.* *13,* 98–110.

Doerschuk, A. P., McCormick, J. R. D., Goodman, J. J., Szumski, S. A., Growitch, J. A., Miller, P. A., Bitler, B. A., Jensen, E. R., Matrishin, M., Petty, M. A., and Phelps, A. S. (1959) *J. Amer. Chem. Soc.* *81,* 3069–3075.

Doskočil, J., Sikyta, B., Kašparová, J., Doskočilová, D., and Zajíček, J. (1958) *J. Gen. Microbiol.* *18,* 302–314

Doskočil, J., Hošťálek, Z., Kašparová, J., Zajíček, J., and Herold, M. (1959) *J. Biochem. Microbiol. Technol. Eng.* *1,* 261–271.

Duggar, B. M. (1948) *Ann. N.Y. Acad. Sci.,* *51,* 177–181.

Erban, V., Novotná, J., Běhal, V., and Hošťálek, Z. (1981) Abstracts of the FEMS Symposium on Overproduction of Microbial Products, Hradec Králové, Czechoslovakia 1981, p. 134.

Finlay, A. C., Hobby, G. L., P'an, S. Y., Regna, P. P., Routien, J. B., Seeley, D. B., Shull, G. M., Sobin, B. A., Solamons, I. A., Vinson, J. W., and Kane, J. H. (1950) *Science* *111,* 85.

Friend, E. J., and Hopwood, D. A. (1971) *J. Gen. Microbiol.* *68,* 187–197.

Friend, E. J., Warren, M., and Hopwood, D. A. (1978) *J. Gen. Microbiol.* *106,* 201–206.

Gatenbeck, S. (1961) *Biochem. Biophys. Res. Comm. 6,* 422–426.
Gourevitch, A., Misick, M., and Lein, J. (1955) *Antibiot. Chemother. 5,* 448–452.
Hendlin, D., Dulaney, E. L., Drescher, D., Cook, T., and Chaiet, L. (1962) *Biochim. Biophys. Acta 58,* 635–636.
Hochstein, F. A., von Wittenau, M. S., Tanner, F. W., Jr., and Murai, K. (1960) *J. Amer. Chem. Soc. 82,* 5934–5937.
Hopwood, D. A. (1978) in *Biotechnology* (Proceedings of the 1st European Congress on Biotechnology, 1978, Dechema Monografia) pp. 199–206, Verlag Chemie, Weinheim, West Germany.
Horváth, I., and Szentirmai, A. (1962) *Acta Microbiol. Acad. Sci. Hung. 9,* 105–116.
Hošťálek, Z. (1964a) *Folia Microbiol. 9,* 79–88.
Hošťálek, Z. (1964b) *Folia Microbiol. 9,* 89–95.
Hošťálek, Z. (1964c) *Folia Microbiol. 9,* 96–102.
Hošťálek, Z. (1980) *Folia Microbiol. 25,* 445–450.
Hošťálek, Z. Tinterová, M., Jechová, V., Blumauerová, M., Suchý, J., and Vaněk, Z. (1969a) *Biotechnol. Bioeng. 11,* 539–548.
Hošťálek, Z., Ryabushko, T. A., Cudlín, J., and Vaněk, Z. (1969b) *Folia Microbiol. 14,* 121–127.
Hošťálek, Z., Blumauerová, M., and Vaněk, Z. (1974) in *Advances in Biochemical Engineering* Vol. III (Ghose, T. K., Fiechter, A., and Blakebrough, N., eds.) pp. 13–67, Springer, Berlin.
Hošťálek, Z., Tobek, I., Bobyk, M. A., and Kulayev, I. S. (1976) *Folia Microbiol. 21,* 131–138.
Hošťálek, Z., Blumauerová, M., and Vaněk, Z. (1979) in *Economic Microbiology* Vol. III, *Secondary Products of Metabolism* (Rose, A. H., ed.) pp. 293–353, Academic Press, New York.
Hutchinson, C. R. (1981) in *Antibiotics* Vol. IV, *Biosynthesis* (Corcoran, J. W., ed.) pp. 1–11, Springer, Berlin.
Janglová, Z., Suchý, J., and Vaněk, Z. (1969) *Folia Microbiol. 14,* 208–210.
Jechová, V., Hošťálek, Z., and Vaněk, Z. (1969) *Folia Microbiol. 14,* 128–134.
Kirchlechner, R., and Rogalski, W. (1980) *Tetrahedron Lett.,* 247–250.
Kulayev, I. S., Bobyk, M. A., Tobek, I., and Hošťálek, Z. (1976) *Biokhimiya 41,* 343–348.
Laluce, C., and Molinari, R. (1977) *Biotechnol. Bioeng. 19,* 1863–1884.
Laluce, C., and Molinari, R. (1979) *Biotechnol. Bioeng. 21,* 915–938.
Levine, J., Garlock, E. A., and Fischbach, H. (1949) *J. Amer. Pharm. Assoc. 38,* 473–475.
Ludvík, J., Mikulík, K., and Vaněk, Z. (1971) *Folia Microbiol. 16,* 479–480.
Lynen, F., and Tada, M. (1961) *Angew. Chem. 73,* 513–519.
Makarevich, V. G., Upiter, G. D., Slugina, M. D., Tarasova, S. S., and Gravit, N. F. (1975) *Antibiotiki* (Moscow) *20,* 295–299.
Matelov́, V., Musílková, M., Nečásek, J., and Šmejkal, F. (1955) *Preslia* (Prague) *27,* 27–34.
McCormick, J. R. D. (1965) in *Biogenesis of Antibiotic Substances* (Vaněk, Z., and Hošťálek, Z., eds.) pp. 73–91, Publishing House of the Czechoslovak Academy of Sciences, Prague.
McCormick, J. R. D. (1966) in *Antibiotics: Advances in Research, Production and Clinical Use* (Herold, M., and Gabriel, Z., eds.) pp. 556–574, Butterworth, London.

McCormick, J. R. D. (1967) in *Antibiotics* Vol. II, *Biosynthesis* (Gottlieb, D., and Shaw, P. D., eds.) pp. 113–122, Springer, Berlin.

McCormick, J. R. D. (1969) in *Genetics and Breeding of Streptomyces* (Sermonti, G., and Alačevič, M., eds.) pp. 163–179, Yugoslav Academy of Sciences and Arts, Zagreb.

McCormick, J. R. D., and Jensen, E. R. (1965) *J. Amer. Chem. Soc. 87,* 1794–1795.

McCormick, J. R. D., and Jensen, E. R. (1968) *J. Amer. Chem. Soc. 90,* 7126–7127.

McCormick, J. R. D., Sjolander, N. O., Hirch, U., Jensen, E. R., and Doerschuk, A. P. (1957) *J. Amer. Chem. Soc. 79,* 4561–4562.

McCormick, J. R. D., Johnson, S., and Sjolander, N. D. (1963) *J. Amer. Chem. Soc. 85,* 1692–1693.

McCormick, J. R. D., Joachim, U. H., Jensen, E. R., Johnson, S., and Sjolander, N. O. (1965) *J. Amer. Chem. Soc. 87,* 1793–1794.

McCormick, J. R. D., Jensen, E. R., Johnson, S., and Sjolander, N. O. (1968a) *J. Amer. Chem. Soc. 90,* 2201–2202.

McCormick, J. R. D., Jensen, E. R., Arnold, N. H., Corey, H. S., Joachim, U. H., Johnson, S., Miller, P. A., and Sjolander, N. O. (1968b) *J. Amer. Chem. Soc. 90,* 7127–7129.

Mikulík, K., Karnetová, J., Quyen, N., Blumauerová, M., Komersová, I., and Vaněk, Z. (1971) *J. Antibiot. 24,* 801–809.

Mikulík, K., and Vaněk, Z. (1979) in *Regulation of Secondary Product and Plant Hormone Metabolism* (Luckner, M., and Schreiber, K., eds.) pp. 199–208, Pergamon Press, New York.

Miller, M. W., and Hochstein, F. A. (1962) *J. Org. Chem. 27,* 2525–2528.

Miller, P. A., and Hash, J. H. (1975a) *Meth. Enzymol. 43,* 603–606.

Miller, P. A., and Hash, J. H. (1975b) *Meth. Enzymol. 43,* 606–607.

Miller, P. A., McCormick, J. R. D., and Doerschuk, A. P. (1956) *Science 123,* 1030–1031.

Miller, P. A., Saturnelli, A., Martin, J. H., Mitscher, L. A., and Bohonos, N. (1964) *Biochem. Biophys. Res. Comm. 16,* 285–291.

Miller, P. A., Hash, J. H., Lincks, M., and Bohonos, N. (1965) *Biochem. Biophys. Res. Comm. 18,* 325–331.

Minieri, P. P., Firman, M. C., Histretta, A. G., Abbey, A., Brickner, C. E., Rigler, N. E., and Sokol, H. (1954) *Antibiot. Ann. 1953–1954,* 81–87.

Monastero, F., Means, J. A., Grenfell, F. C., and Hedger, F. H. (1951) *J. Amer. Pharm. Assoc. Sci. Ed. 40,* 241–245.

Orlova, N. V., Popova, L. A., Makarevich, V. G., and Verchovtseva, T. P. (1959) *Trudy Inst. Mikrobiol. Akad. Nauk SSSR 61,* 251–263.

Pecák, V., Čížek, S., Musil, J., Čerkes, L., Herold, M., Bělík, E., and Hoffman, J. (1958) *J. Hyg. Epidemiol. Microbiol. Immunol. 2,* 111–115.

Podojil, M., Vaněk, Z., Běhal, V., and Blumauerová, M. (1973) *Folia Microbiol. 18,* 415–417.

Podojil, M., Vaněk, Z., Přikrylová, V., and Blumauerová, M. (1978) *J. Antibiot. 31,* 850–854.

Podojil, M., Blumauerová, M., Čulík, K., and Vaněk, Z. (1982) in *Antibiotic Production by Fermentation* (Vandamme, E. J., ed.) Marcel Dekker, New York.

Pokorný, M., Vitole, L., Turk, V., Renko, M., and Žuvanic, J. (1979) *Eur. J. Appl. Microbiol. Biotechnol. 8,* 81–90.

Přikrylová, V., Podojil, M., Sedmera, P., Vokoun, J., Vaněk, Z., and Hassall, C. H. (1978) *J. Antibiot. 31,* 855–862.

Rhodes, P. M., Winskill, N., Friend, E. J., Warren, M. W. (1981) *J. Gen. Microbiol. 124,* 329–338.

Rokos, J., and Procházka, P. (1957) *Českosl. Mikrobiol. 2,* 251–253.

Sensi, P., DeFerrari, G. A., Gallo, G. G., Roland, G. (1955) *Il Farmaco (Ed. Sci.) 10,* 337–345.

Šimúth, J., Zelinka, J., and Polek, B. (1975) *Biochim. Biophys. Acta 379,* 397–407.

Šimúth, J., Hudec, J., Hoang Thang Chan, Dányi, O., and Zelinka, J. (1979) *J. Antibiot. 32,* 53–58.

Snell, J. F., Birch, A. J., and Thompson, P. L. (1960) *J. Amer. Chem. Soc. 82,* 2402.

Šťastná, J., and Mikulík, K. (1981) in *Actinomycetes* (Schaal, K. P., and Pulverer, G., eds.) *Zbl. Bakt. Suppl. 11,* Gustav Fischer, Stuttgart, West Germany.

Turley, R. H., and Snell, J. F. (1966) in *Biosynthesis of Antibiotics* Vol. I (Snell, J. F., ed.) pp. 94–122, Academic Press, New York.

Upiter, G. D., Makarevich, V. G., Tarasova, S. S., Biryukov, V. V. (1974) *Antibiotiki* (Moscow) *19,* 17–21.

Vaněk, Z., Cudlín, J., Blumauerová, M., and Hošťálek, Z. (1971) *Folia Microbiol. 16,* 225–240.

Vaněk, Z., Hošťálek, Z., Blumauerová, M., Mikulík, K., Podojil, M., Běhal, V., and Jechová, V. (1973) *Pure Appl. Chem. 34,* 463–486.

Voříšek, J., Powell, A. J., and Vaněk, Z. (1969) *Folia Microbiol. 14,* 398–405.

Voříšek, J., Powell, A. J., and Vaněk, Z. (1970) *Folia Microbiol. 15,* 153–159.

Weiser, J., (1980) Ph.D. thesis, Institute of Microbiology, Czechoslovak Academy of Sciences, Prague.

Weiser, J., Mikulík, K., and Bosch, L. (1981) *Biochem. Biophys. Res. Comm. 99,* 16–22.

Williamson, D. E., and Everett, G. W., Jr. (1975) *J. Amer. Chem. Soc. 97,* 2397–2405.

Zaitseva, Z. M., and Orlova, N. V. (1959) *Mikrobiologiya* (Moscow) *28,* 216–223.

CHAPTER

11

Anthracyclines

Richard J. White

I. INTRODUCTION

In 1950, Hans Brockmann described the isolation of rhodomycin, a new antibiotic produced by an actinomycete (Brockmann and Bauer, 1950). Rhodomycin was the first example of an anthracycline, a class of antibiotic with an anthraquinone chromophore and one or more sugars attached. In the subsequent 30 years, a large number of anthracyclines have been discovered; these differ from one another primarily in the nature of the substituent sugar (or sugars) and in the hydroxylation pattern of the anthraquinone. The latter variations are responsible for the different colors of these compounds, which range from red through orange to yellow.

The first anthracycline characterized by Brockmann possessed potent antibacterial activity but was also highly toxic. It was not until the early 1960s that the antitumor activity of these compounds was established with the simultaneous discovery of daunorubicin by groups at Farmitalia in Italy (Grein et al., 1963) and at Rhone-Poulenc in France (Dubost et al., 1963); daunorubicin is a contraction of the original Italian (daunomycin) and French (rubidomycin) names. During attempts to develop variants of the daunorubicin structure directed toward improving the antitumor activity, a new antibiotic called adriamycin was isolated from a mutant strain of the daunorubicin-producing organism (Arcamone et al., 1969). Currently, adriamycin and daunorubicin are the only commercially established and important anthracyclines that are marketed in non-Communist countries. A closely related drug, carminomycin, was discovered in the Soviet Union (Gause et al., 1974), where it is currently available for clinical use. The structures of daunorubicin, adriamycin, and carminomycin are shown in Fig. 11.1. This chapter will be limited to a discussion of these three anthracyclines.

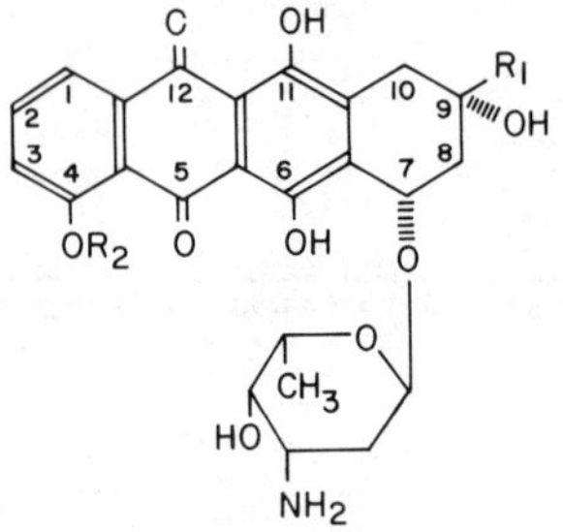

	R_1	R_2
	13 14	
Daunorubicin	—COMe	—Me
Adriamycin	—$COCH_2OH$	—Me
Carminomycin	—COMe	—H

FIGURE 11.1 Structures of daunorubicin, adriamycin, and carminomycin.

Anthracyclines exhibit a broad spectrum of biological activity, inhibiting the growth of a number of bacteria (mostly gram-positive species), fungi, viruses, and animal cells. This biological activity is considered to be the result of intercalation with double-stranded DNA. Daunorubicin, adriamycin, and carminomycin are mutagenic in the Ames test (Benedict et al., 1977). They induce lysogenic bacteria (Anderson et al., 1980), and are carcinogenic in animals (Marquardt et al., 1976).

II. CLINICAL UTILITY

The commercial importance of anthracyclines derives from their antitumor activity. Daunorubicin, the first clinically useful anthracycline, has good activity against acute leukemia (Von Hoff et al., 1978). Although still used to treat leukemia, it has, to a large extent, been eclipsed by the closely related adriamycin (Carter, 1975). Adriamycin (doxorubicin) is active against a broader spectrum of cancers, including some of the clinically refractory solid tumors, such as breast cancer. At present, adriamycin is one of the most widely prescribed agents in cancer chemotherapy.

III. MICROORGANISMS THAT PRODUCE DAUNORUBICIN, ADRIAMYCIN, AND CARMINOMYCIN

The two organisms originally reported to produce daunorubicin were *Streptomyces peucetius* and *Streptomyces coeruleorubidus* (Table 11.1). Those discovered subsequently are all streptomycetes. The closely related carminomycin, on the other hand, is produced by *Actinomadura carminata*. As is common with commercially important antibiotics, details of their discovery and, in particular, the screening methodology employed have not been published. The promising clinical activity of daunorubicin spurred an active interest in the potential of the anthracyclines as a class of antitumor agents. Programs in semisynthesis, total synthesis, and genetic manipulation of the producing organisms have all been tried in the effort to obtain improved derivatives. The major objectives were to increase the potency of the agent, to broaden its spectrum of activity, and to decrease its toxicity. The daunorubicin-producing *S. peucetius* is known to exhibit a high degree of morphological and cultural variability (Grein et al., 1980). Such variability could be enhanced with mutagens, and adriamycin was isolated from one such variant subsequently called *S. peucetius* var. *caesius* (Arcamone et al., 1969). Adriamycin, in addition to being derived directly from fermentation, can be made by semisynthesis from daunorubicin. Furthermore, both these anthracyclines can be made by total synthesis; at present, this does not appear to be an economically viable alternative to their production by fermentation.

TABLE 11.1 Organisms Producing Daunorubicin and Related Anthracyclines

Product	*Species*	*Origin of Culture*
Daunorubicin		
	Streptomyces bifurcus 23219	(Rhône-Poulenc)
	S. coeruleorubidus 8899 and 31723	(Rhône-Poulenc)
	S. coeruleorubidus	(Gause)
	S. griseoruber	(Takeda)
	S. griseus 32041	(Rhône-Poulenc)
	S. griseus IMET SA, 3933, 5570, 10086, 10431	(Strauss Jena)
	S. peucetius	(Farmitalia)
	S. viridochromogenes	(Pfizer)
Adriamycin		
	S. peucetius var. *caesius*	(Farmitalia)
Carminomycin		
	Actinomadura carminata	(Gause)

From Lunel et al. (1977).

IV. FERMENTATION

Daunorubicin, adriamycin, and carminomycin are produced by standard methods of submerged culture. A number of studies to improve the fermentation by environmental manipulation have been described. In the earliest of these, Manafova (1966) reported that a medium containing 3% glucose as a carbon source, 3% soybean meal as a nitrogen source, 0.3% sodium chloride, and 0.3% calcium carbonate was most favorable for antibiotic production. Addition of potassium phosphate suppressed synthesis of the antibiotic, while iron stimulated it. A more detailed study reported by Blumauerová et al. (1977) showed that 3.5% soluble starch was better than glucose for certain strains of *S. coeruleorubidus*. Attempts to stimulate synthesis by adding 1-propanol, sodium propionate, and bacitracin to fermentations were unsuccessful but the titers of antibiotic achieved in these studies are uncertain, since the bioassay did not discriminate between the different active components present in the broth. In a more recent study, the production of daunorubicin in 10,000-liter fermentors was described in some detail (Hamilton et al., 1981). A time course for a typical fermentation is shown in Figure 11.2. Fermentation titers cited in the patent literature for daunorubicin and adriamycin have been in the 10–30 $\mu g \cdot mL^{-1}$ range. However, it is likely that amounts in excess of 1 $mg \cdot mL^{-1}$ are currently produced in commercial processes. The bulk of the anthracycline produced is associated with the biomass, only a small percentage being released into the supernatant fluid. For this reason, recovery procedures include extraction of whole broth with an organic solvent such as butanol or release of antibiotic from the mycelium by treatment with acid. Purification is achieved by a series of liquid/liquid extractions or by ion exchange chromatography (Pandey et al., 1979). At some stage in the process, an acid hydrolysis is necessary if, as is usual, daunorubicin is produced in a substituted form (for example, a baumycin of the type described by Pandey et al. (1979)).

V. BIOSYNTHESIS OF DAUNORUBICIN

The origin of the carbon skeleton of the aglycone moiety of daunorubicin has been studied using ^{13}C-labeled precursors and ^{13}C-nuclear magnetic resonance (^{13}C-NMR) spectroscopy. Paulick et al. (1976) prepared ^{13}C-enriched antibiotic by fermentation of *S. peucetius* in the presence of $^{13}CH_3CO_2Na$, $CH_3{}^{13}CO_2Na$, and $^{13}CH_3{}^{13}CO_2Na$. The labeled daunorubicins were extracted, hydrolyzed, and acetylated before their ^{13}C-NMR spectra were recorded. Peaks in the natural abundance spectrum could all be assigned to specific carbons of daunorubicinone tetraacetate. Comparison of the spectra of natural abundance and enriched samples of daunorubicinone tetraacetate showed that eight carbons (2, 4, 5, 6, 7, 15, 18, and 19) were enriched by

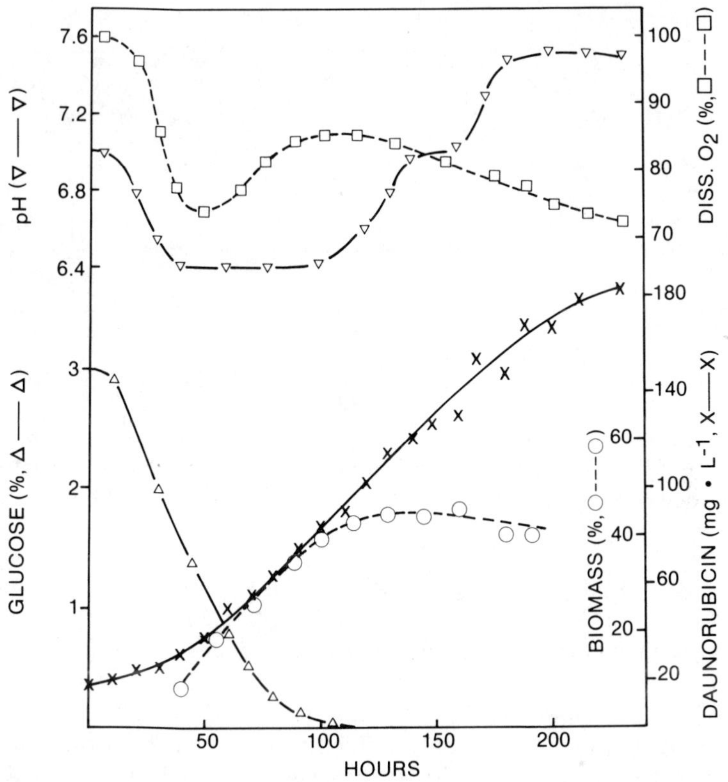

FIGURE 11.2 Time course of a typical daunorubicin fermentation in a 10,000-L fermentor. From Hamilton et al. (1981).

$CH_3{}^{13}CO_2Na$, while nine (1, 3, 8, 10, 11, 12, 16, 17, and 20) were enriched by $^{13}CH_3CO_2Na$. It is possible to establish acetate connectivity patterns in polyketides by studying the incorporation of doubly labeled acetate. The ^{13}C-NMR spectrum of $^{13}CH_3{}^{13}CO_2Na$-enriched daunorubicinone tetraacetate showed that all acetate-derived carbons except for C-10 were incorporated as intact acetate units. Since C-10 was enriched by $^{13}CH_3CO_2Na$, the carboxyl carbon of that acetate pair must have been lost during biosynthesis. Thus the aglycone of daunorubicin must have as a precursor a polyketide that has an extra carbon; the nature of this precursor will be discussed in a subsequent section.

Three carbons of the aglycone do not come directly from acetate; these are C-9, C-13, and C-14. By analogy with other established polyketide biosynthetic schemes, they could be predicted to arise from a propionate starter unit. This has been confirmed by Shaw et al. (1979) using $CH_3CH_2{}^{13}CO_2Na$, which specifically enriched C-9 of both daunorubicinone

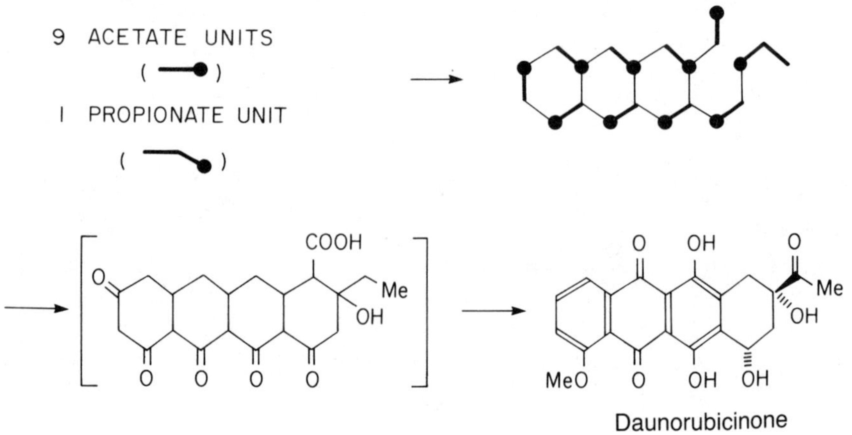

FIGURE 11.3 Biosynthetic origin of the carbon skeleton of daunorubicinone. Single arrows represent an undetermined number of steps.

and adriamycinone. The origin of the carbon skeleton of daunorubicinone is summarized in Figure 11.3 and consists of the sequential, linear condensation of a single propionate starter unit and nine acetates.

Although ^{13}C-acetate incorporation studies for adriamycin have not been reported, this compound is undoubtedly produced via the same metabolic route as daunorubicin. The only difference between the structures is the replacement of a hydrogen with a hydroxyl (Figure 11.1); modifications of this type usually occur after the initial polyketide condensation product has been formed.

Only limited data have been published on the biosynthesis of daunosamine, the amino sugar attached by glycosidic linkage to C-7 of the anthracycline chromophore. It appears to derive directly from the carbon skeleton of glucose without significant rearrangement (Paranosenkova and Karpov, 1976). The methoxyl carbon on C-4 of daunorubicin and adriamycin presumably derives from S-adenosylmethionine, a common source of methoxyl groups in secondary metabolites. Fermentation of *S. coeruleorubidus* in the presence of sulfonamides (Lunel et al., 1977) or ethionine (Blumauerová et al., 1979b) gave compounds lacking the C-4 methoxyl and thus belonging to the carminomycin series. Similarly, Cassinelli et al. (1978) have described a blocked mutant of *S. peucetius* that produces carminomycin and its 13-dihydro derivative.

VI. GENETICS OF ANTHRACYCLINE BIOSYNTHESIS

Only fragmentary information is available on the genetics of anthracycline synthesis. No systematic attempt to produce a genetic map of a daunorubicin-

or adriamycin-producing organism has been reported. The only information available is on the isolation and characterization of a variety of mutants that differ from their parent organisms in colony morphology and pigmentation. These mutants have been acquired to serve two purposes:

1. They were isolated during yield improvement programs in an attempt to achieve higher antibiotic titers.
2. They were blocked mutants isolated in attempts to establish the biosynthetic pathway leading to daunorubicin.

A. Improved Mutants

Blumauerová et al. (1978a, b) isolated a series of morphological variants of *S. coeruleorubidus* strain JA 10092. These differed in their ability to produce aerial mycelia and spores on glucose- and starch-based media. As a result of a six-step selection (including induced and spontaneous variation), combined with altered fermentation technology, an overall yield increase of 40-fold was claimed. The exact degree of improvement obtained in this study is not clear, as the assay measured total antibacterial activity and did not discriminate between the different active components present. The improved mutants produced increased amounts of daunorubicin glycosides and decreased amounts of the free aglycones. They also differed from the wild type in their higher demand for oxygen and need for increased glucose concentration. The concomitant increase in productivity and change in morphology indicate that factors influencing sporulation also exert an effect on secondary metabolism. However, sporulation does not appear to be directly associated with biosynthesis of the antibiotics (Blumauerová et al., 1978b). The ability of glucose to suppress aerial mycelium formation in certain of the mutants has been interpreted as indicating that catabolite repression is involved in control of the developmental process (Blumauerová et al., 1978a).

McGuire et al. (1979, 1980a) have recently described a directed screen for selecting mutants with increased productivity using cerulenin, an antibiotic that specifically inhibits fatty acid biosynthesis (Ōmura, 1976). The biosynthesis of polyketides is five to ten times more sensitive to inhibition by cerulenin than is growth of the producing organism (Ōmura and Takeshima, 1974). Mutagen-treated spores of a daunorubicin producer were spread on agar plates containing cerulenin at a concentration (50 μM) just sufficient to suppress pigment formation. Normally, colonies are red owing to the presence of anthracyclines. Rare colonies still pigmented in the presence of cerulenin were easily identified against a background of predominantly nonpigmented colonies. One mutant (C-5) isolated by this procedure gave threefold to fivefold higher titers of daunorubicin. Unexpectedly, cerulenin affected anthracycline and lipid synthesis similarly in the mutant and parental strains. Thus the mutation does not appear to have made polyketide synthesis more resistant. It may be that, by suppressing the overall level of anthracycline synthesis, cerulenin made it easy to detect mutations that increased pigment production and that resistance to the inhibitor was not involved.

B. Mutants Blocked in Anthracycline Biosynthesis

Blumauerová et al. (1978c) classified a series of *S. coeruleorubidus* mutants into five groups (A–E) on the basis of the anthracyclines accumulated (Table 11.2). These mutants were also studied in cosynthesis experiments (see Chapter 2 for this procedure). With secondary metabolites, an important restriction to cosynthesis can be the failure of a blocked mutant to accumulate the true substrate of the enzyme for which activity has been lost. In such situations, a shunt metabolite may be formed instead. Perhaps because of these limitations, the cosynthesis data (Table 11.2) divided the five phenotypically distinguishable groups (A–E) into only two subgroups (A and B, C, D, E). Mutants in group A produced trace amounts of the same anthracyclines as the parent strain in normal fermentations.

Umezawa and co-workers (Yoshimoto et al., 1980a) have worked with another strain of *S. coeruleorubidus* (ME-130A4) and described two different types of mutants: a series of baumycin-negative mutants that produced no detectable anthracycline-related pigments (IN-372, IU-222, IU-479) and a mutant (4N-140) that accumulated 13-deoxydaunorubicin. A group working at the Frederick Cancer Research Center (McGuire et al., 1981) has characterized four mutant types derived from an unidentified species of *Streptomyces*. These are described in Table 11.3. One of the mutants (A-21) accumulated a novel anthracycline, maggiemycin, the structure of which is shown in Figure 11.4 (Pandey et al., 1981). The parent strain does not normally accumulate detectable amounts of daunorubicin but produces baumycins A1/A2 (Pandey et al., 1979). Mutant 6-3 is apparently blocked in the conversion of daunorubicin to the baumycins and accumulates significant quantities of daunorubicin itself, suggesting that this compound is an intermediate in baumycin biosynthesis. The Farmitalia group working with *S. peucetius* has isolated a series of blocked mutants (Table 11.4). As has already been mentioned, adriamycin was originally discovered during a mutational program directed at finding improved anthracyclines and now considerably surpasses the parent molecule, daunorubicin, in extent of clinical usage.

The carminomycin series of anthracyclines can be obtained from blocked mutant B441FI of *S. peucetius* (Cassinelli et al., 1978), by using methylation inhibitors such as ethionine (Blumauerova et al., 1979b) or sulfonamides (Lunel et al., 1977), and by direct fermentation of *Actinomadura carminata* (Gause et al., 1974). An 11-deoxy series of daunorubicin-related anthracyclines is accumulated by a strain isolated during a mutational program with *S. peucetius* (Cassinelli et al., 1980; Grein et al., 1980). Surprisingly, this

TABLE 11.2 Characteristics of Mutant Groups of *S. coeruleorubidus*

		Metabolites Identified[1]							
Mutant Group	*Strain*	*Glycosides*	*13-Dihydro-daunomycinone*	*7-Deoxy-13-dihydro-daunomycinone*	*ε-Rhodo-mycinone*	*Aklavinone*	*7-Deoxy-aklavinone*	*Bisanhydro-aklavinone*	*Cosynthetic[2] Activity*
Wild type	JA10092	+	+ +	+ +	+ +	—	—	—	
A	24-1 24-203 146-154	±	±	±	—	—	—	—	B, C, D, E
B	24-2 24-64	±	+ +	+ +	±	—	—	—	A
C	96-13 96-85 98-139	—	—	—	—	—	—	—	A
D	24-41	—	—	—	+ +	—	±	—	A
E	24-74	—	—	—	—	+	+	±	A

1. + = produced, + + = increased level, ± = traces, — = undetected.
2. Daunorubicin glycosides produced during mixed fermentation with a member of a different mutant group.

TABLE 11.3 Characteristics of Mutants of *Streptomyces* sp. 40-1

Strain	*Anthracyclines Accumulated*
Wild type	ε-Rhodomycinone, baumycins A1/A2
NZ4	No detectable anthracycline related pigments
R28	ε-Rhodomycinone
6-3	ε-Rhodomycinone, daunorubicin
A-21	Maggiemycin

From McGuire et al. (1981).

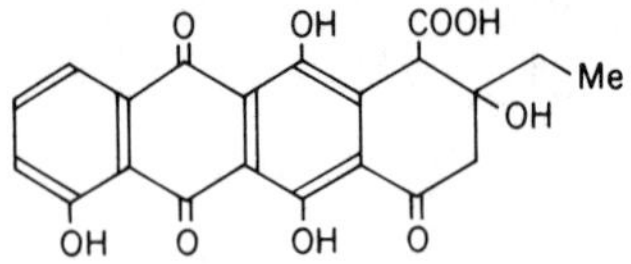

FIGURE 11.4 The structure of maggiemycin. From Pandey et al. (1981).

TABLE 11.4 Characteristics of Mutants of *Streptomyces peucetius*

Strain	*Anthracyclines Accumulated*
Wild type	Daunorubicin
var. *caesius*	Adriamycin
var. *carneus*	13-Dihydrodaunorubicin, daunosaminyldaunorubicin
B441 F1	13-Dihydrocarminomycin
B211 F1*	11-Deoxydaunorubicin, 11-deoxy-13-dihydrodaunorubicin, 11, 13-dideoxydaunorubicin, 11-deoxyadriamycin

* This strain has been classified as a *Micromonospora*.

organism was classified taxonomically as a *Micromonospora*. It seems unlikely that a chance contaminant would produce such closely related anthracyclines, especially since the "mutant" also produced antifungal polyenes characteristic of *S. peucetius*. No other accounts of anthracycline production by *Micromonospora* have been reported, according to the Italian workers.

VII. BIOTRANSFORMATION OF ANTHRACYCLINES

Several groups have studied anthracycline biotransformations to delineate the biosynthetic pathway leading to daunorubicin and its higher glycosides. Such experiments have used radiolabled substrates with strains that can still synthesize the end product or unlabeled substrates with mutants that no longer synthesize the end product in question. The successful biotransformation of a substrate to the desired anthracycline is taken as evidence for its role as an intermediate in the normal pathway. Examination of the proposed origin of daunorubicin's carbon skeleton (Figure 11.3) indicates that the aglycone is derived from a precursor with an extra carbon atom that is subsequently lost. Several anthracyclines with structures of the predicted type are known and one of these (ε-rhodomycinone) is frequently a major co-metabolite in daunorubicin fermentations (Kern et al., 1977; McGuire et al., 1979, 1980b); its structure is shown in Figure 11.5.

Two groups have observed the transformation of ε-rhodomycinone to daunorubicin by a number of mutants (McGuire et al., 1980b; Yoshimoto et al., 1980a). Yoshimoto et al. (1980b) were also able to transform ε-rhodomycinone and aklavinone to daunorubicinone and proposed a pathway (Figure 11.5) in which alklavinone is first converted to ε-rhodomycinone by introduction of a hydroxyl group at C-11. Subsequent loss of the C-10 carbomethoxy group, oxidation at C-13, and methylation of the C-4 hydroxyl yields daunorubicinone. Yoshimoto et al. (1980a) did not detect any conversion of daunorubicinone to daunorubicin. A third group (Blumauerová et al., 1979a, c) was equally unable to demonstrate the biotransformation of ε-rhodomycinone or aklavinone to daunorubicin. However, they suggested that daunorubicinone is an intermediate in the biosynthesis of daunorubicin. Yoshimoto et al. (1980a) postulated that substitution at two positions, C-4 and C-13, on the anthracyclinone influences whether or not glycosylation will occur. Of the two, C-4 is the most important, and a free hydroxyl there—for example, as in carminomycinone—is a strict requirement for glycosylation. Methylation at C-4 (as in daunorubicinone) precludes the possibility of subsequent glycosylation. The major features of the two alternative pathways proposed for the biosynthesis of daunorubicin and other anthracyclines from the hypothetical decaketide precursor are shown in Schemes 11.1 and 11.2.

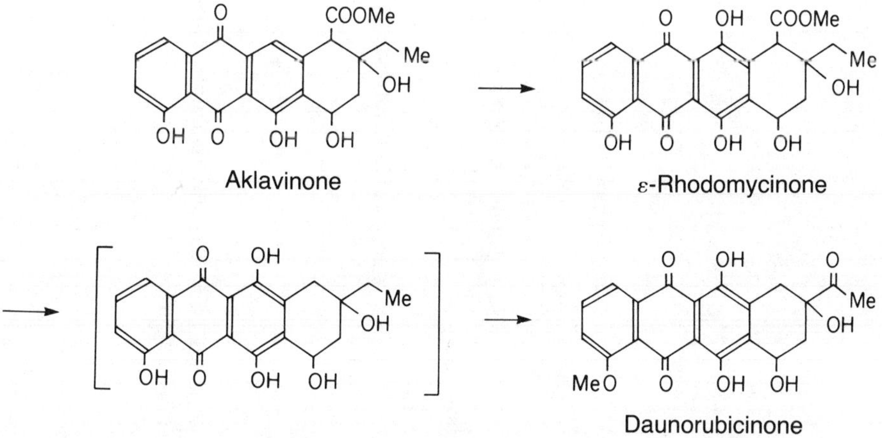

FIGURE 11.5 A possible biosynthetic pathway for the synthesis of daunorubicinone from aklavinone by *S. coeruleorubidus*. From Yoshimoto et al. (1980b).

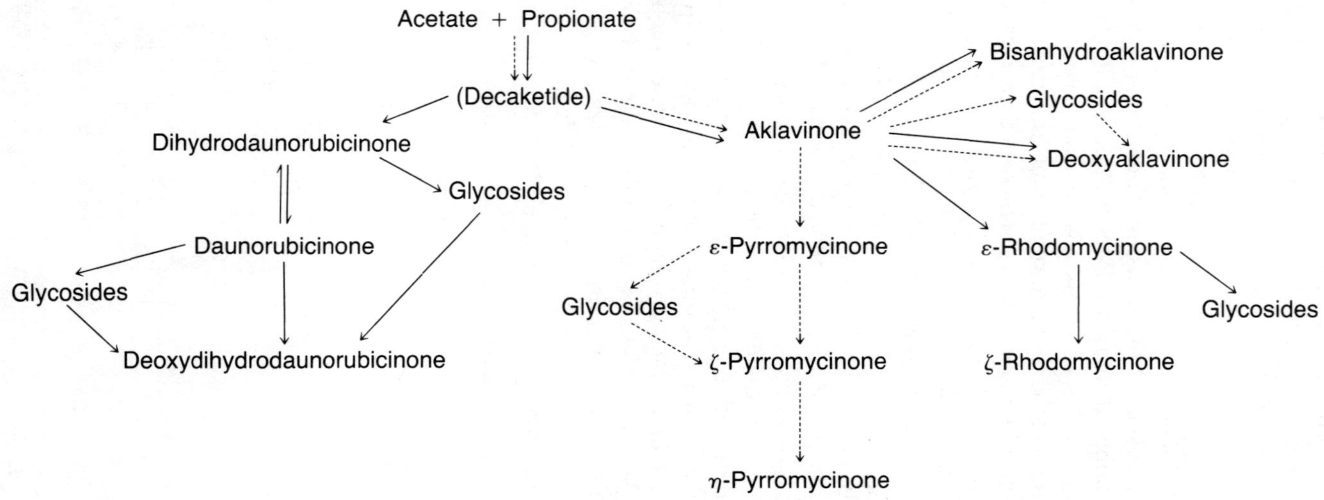

SCHEME 11.1 Tentative pathways of anthracycline biosynthesis in *S. coeruleorubidus* (⟶) and *S. galilaeus* (⇢). From Blumauerová et al. (1979d).

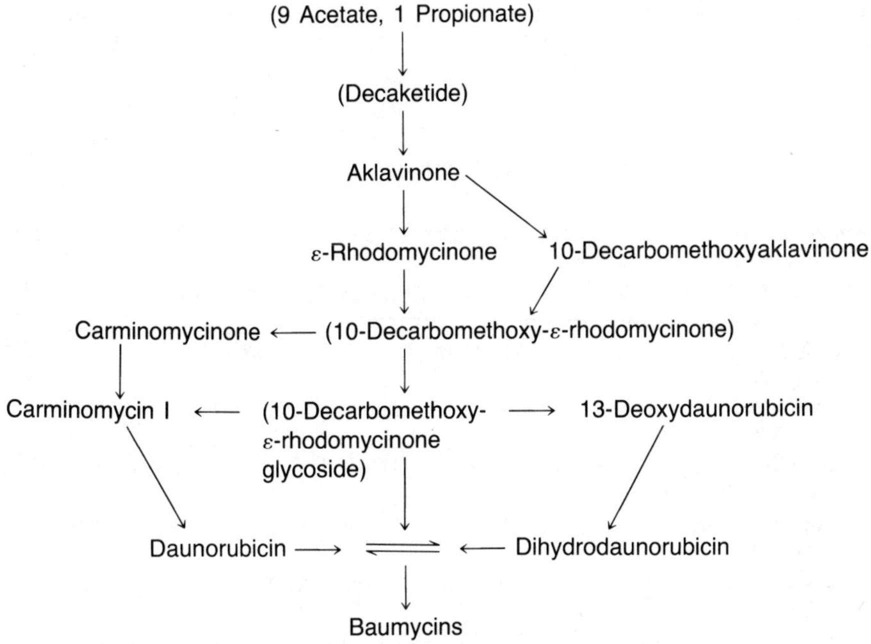

SCHEME 11.2 Proposed biosynthetic pathway leading to daunorubicin and baumycins in *S. coeruleorubidus*. From Yoshimoto et al. (1980a).

Daunorubicin-producing organisms synthesize complex mixtures of related metabolites. Daunorubicin itself may be a relatively minor component and is frequently present as an acetal of the baumycin type (McGuire et al., 1980c; Komiyama et al., 1977). These substituted glycosides, which are formed from daunorubicin (Yoshimoto et al., 1980a), can be facilely cleaved by mild acid hydrolysis in good yield to the monosaccharide. This is the basis of industrial processes for recovering total derivable daunorubicin from fermentations (Pandey et al., 1979). More vigorous acid hydrolysis will cleave the monoglycoside to daunorubicinone and daunosamine. In the case of *S. coeruleorubidus,* a total of 12 anthracycline-related products have been identified in fermentations: three aglycones (ε-rhodomycinone, 13-dihydrodaunorubicinone, 7-deoxy-13-dihydrodaunorubicinone) and nine glycosides (baumycins A1, A2, B1, and B2; N-acetyldaunorubicin, N-formyldaunorubicin, N-formyldihydrodaunorubicin, daunorubicin, 13-dihydrodaunorubicin) (Yoshimoto et al., 1980a). The amount of 7-deoxy-13-dihydrodaunorubicinone in fermentation broths is changed dramatically by the degree of aeration (Kern et al., 1977; Mateju et al., 1978). This compound increases in titer toward the end of fermentations and can become the major product if broths are kept without agitation overnight. It thus seems to be a degradative product rather than a biosynthetic intermediate. It is interesting to note that 7-deoxy-13-dihydrodaunorubicinone has been discovered as a mammalian (Takanashi and Bachur, 1975) and as a microbial (Wiley and Marshall, 1975) metabolite of daunorubicin.

REFERENCES

Anderson, W. A., Moreau, P. L., Devoret, R., and Maral, R. (1980) *Mutation Res.* *77*, 197–208.

Arcamone, F., Cassinelli, G., Fantini, G., Grein, A., Orezzi, P., Pol, C., and Spalla, C. (1969) *Biotechnol. Bioeng.* *11*, 1101–1110.

Benedict, W. F., Baker, M. S., Haroun, L., Choi, E., and Ames, B. N. (1977) *Cancer Res.* *37*, 2209–2213.

Blumauerová, M., Mateju, J., Stajner, K., and Vaněk, Z. (1977) *Folia Microbiol.* *22*, 275–285.

Blumauerová, M., Pokorny, V., Stajner, J., Hošťalék, Z., and Vaněk, Z. (1978a) *Folia Microbiol.* *23*, 177–182.

Blumauerová, M., Pokorny, V., Šťastná, J., Hošťalék, Z., and Vaněk, Z. (1978b) *Folia Microbiol.* *23*, 249–254.

Blumauerová, M., Pokorny, V., Šťastná, J., Hošťalék, Z., and Vaněk, Z. (1978c) *Folia Microbiol.* *23*, 255–260.

Blumauerová, M., Kralovkova, E., Mateju, J., Jizba, J., and Vaněk, Z. (1979a) *Folia Microbiol.* *24*, 117–127.

Blumauerová, M., Jizba, J., Stajner, K., and Vaněk, Z. (1979b) *Biotechnol. Lett.* *1*, 471–476.

Blumauerová, M., Kralovcova, E., Mateju, J., Hošťalék, Z., and Vaněk, Z. (1979c) in *Genetics of Industrial Microorganisms* (Sebek, O. K., and Laskin, A. I., eds.) pp. 90–96, American Society for Microbiology, Washington, D.C.

Brockmann, H., and Bauer, K. (1950) *Naturwiss.* *37*, 492–493.

Carter, S. K. (1975) *J. Nat. Cancer Inst.* *55*, 1265.

Cassinelli, G., Grein, A., Masi, P., Suarato, A., Bernardi, J., Arcamone, F., DiMarco, A., Casazza, A. M., Pratesi, G., and Soranzo, C. (1978) *J. Antibiot.* *31*, 178–184.

Cassinelli, G., DiMatteo, F., Forenza, S., Ripamonti, M. C., Rivola, G., Arcamone, F., DiMarco, A., Casazza, A. M., Soranzo, C., and Pratesi, G. (1980) *J. Antibiot.* *33*, 1468–1473.

Dubost, M., Ganter, P., Maral, R., Ninet, L., Pinnert, S., Preud'homme, J., and Werner, G. H. (1963) *C. R. Acad. Sci., Paris* *257*, 1813–1815.

Gause, G. F., Brazhnikova, M. G., and Shorin, V. A. (1974) *Cancer Chemother. Rep.* *58* (Part 1), 255–256.

Grein, A., Spalla, C., DiMarco, A., and Canevazzi, G. (1963) *Giorn. Microbiol.* *11*, 109–118.

Grein, A., Merli, S., and Spalla, C. (1980) *J. Antibiot.* *33*, 1462–1467.

Hamilton, B., White, R., McGuire, J., Montgomery, P., Stroshane, R., Kalita, C., and Pandey, R. (1981) in *Advances in Biotechnology* Vol. I, *Scientific and Engineering Principles* (Moo-Young, M., Robinson, C. W., and Vezina, C., eds.) pp. 63–68, Pergamon Press, Toronto.

Kern, D. L., Bunge, R. H., French, J. C., and Dion, H. W. (1977) *J. Antibiot. 30*, 432–434.

Komiyama, T., Matsuzawa, Y., Oki, T., and Inui, T. (1977) *J. Antibiot. 30*, 619–621.

Lunel, J., Florent, J., Mancy, D., and Renaut, J. (1977) Abstracts of the 174th National Meeting of the American Chemical Society, MICR 041.

Manafova, N. A. (1966) *Antibiotiki 11*, 872–877.

Marquardt, H., Philips, F. S., and Sternberg, S. S. (1976) *Cancer Res. 36*, 2065–2069.

Mateju, J., Vokoun, J., Blumauerová, M., and Vaněk, Z. (1978) *Folia Microbiol. 23*, 246–248.

McGuire, J. C., Hamilton, B. K., and White, R. J. (1979) *Proc. Biochem. 14*, 2–5.

McGuire, J. C., Glotfelty, G., and White, R. J. (1980a) *FEMS Microbiol. Lett. 9*, 141–143.

McGuire, J., Thomas, M., Pandey, R., Toussaint, M., and White, R. (1981) in *Advances in Biotechnology* Vol. III, *Fermentation Products* (Moo-Young, M., Vezina, C., and Singh, K., eds.) pp. 117–122, Pergamon Press, Toronto.

McGuire, J. C., Thomas, M. C., Stroshane, R., Hamilton, B. K., and White, R. J. (1980b) *Antimicrob. Agents Chemother. 18*, 454–464.

Ōmura, S. (1976) *Bacteriol. Rev. 40*, 681–697.

Ōmura, S., and Takeshima, H. (1974) *J. Biochem. 75*, 193–195.

Pandey, R. C., Kalita, C. C., White, R. J., and Toussaint, M. W. (1979) *Proc. Biochem. 14*, 6–13.

Pandey, R. C., Toussaint, M. W., McGuire, J. C., and Thomas, M. C. (1981) Abstract 182 of the 21st Interscience Conference on Antimicrobial Agents and Chemotherapy, Chicago, Illinois.

Paranosenkova, V. I., and Karpov, V. L. (1976) *Antibiotiki 21*, 299–301.

Paulick, R. C., Casey, M. L., and Whitlock, H. W. (1976) *J. Amer. Chem. Soc. 98*, 3370–3371.

Shaw, G. J., Milne, G. W. A., and Minghetti, A. (1979) *Phytochemistry 18*, 178–179.

Takanashi, S., and Bachur, N. (1975) *J. Pharmacol. Exp. Ther. 195*, 41–49.

Von Hoff, D. D., Rozensweig, M., and Slavik, M. (1978) in *Adv. Pharmacol. Chemother. 15*, 1–50.

Wiley, P., and Marshall, V. (1975) *J. Antibiot. 28*, 838–840.

Yoshimoto, A., Oki, T., Takeuchi, T., and Umezawa, H. (1980a) *J. Antibiot. 33*, 1158–1166.

Yoshimoto, A., Oki, T., and Umezawa, H. (1980b) *J. Antibiot. 33*, 1199–1201.

CHAPTER
12

Chloramphenicol

Vedpal Malik

I. INTRODUCTION

Chloramphenicol (Figure 12.1) was originally isolated from *Streptomyces venezuelae* cultures by workers at Parke Davis & Co. (Ehrlich et al., 1948), the University of Illinois (Gottlieb et al., 1948), and the National Institute of Health in Tokyo (Umezawa et al., 1948). Other organisms known to produce chloramphenicol are *Kitasatoa* spp., *Streptomyces phaeochromogenes, Streptomyces omiyaensis, Streptomyces viridogriseum* var. *kofuense* (Tamura et al., 1971), and the moon snail *Lunatia heros* (Price et al., 1981). A family of chloramphenicol analogs called corynecins has been isolated from the cultures of *Corynebacterium hydrocarboclastus* (Nakano et al., 1976, 1977b). These are N-acyl derivatives of D-(-)*threo-p*-nitrophenylserinol (Figure 12.1). Analogs of chloramphenicol are also produced by *Nocardia* cultures (Suzuki, et al., 1974). Interestingly, *Streptomyces fulvoviolaceus* produces an antagonist of chloramphenicol, 2-amino-3-carboxy-1,4-benzoqinone-5,6-epoxide (Imagawa et al., 1979).

Chloramphenicol has remarkable therapeutic effectiveness and inhibits a wide range of gram-positive and gram-negative bacteria, as well as rickettsiae and chlamydiae. Controversy exists regarding its clinical utility because of hematologic side effects, but its use is warranted to treat infections of *Salmonella typhi* and *Hemophilus influenzae*.

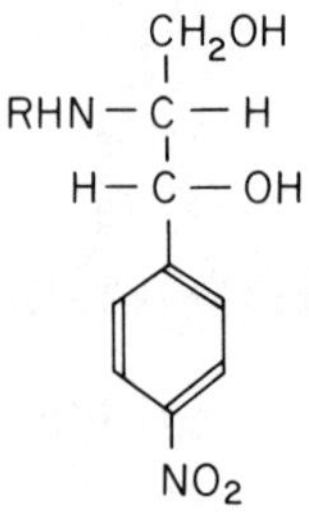

	R
p-Nitrophenylserinol	H—
Corynecin I	MeCO—
Corynecin II	$MeCH_2CO$—
Corynecin III	$(Me)_2CHCO$—
Chloramphenicol	Cl_2CHCO—

FIGURE 12.1 Structures of *p*-nitrophenylserinol, chloramphenicol, and corynecins.

The crystalline antibiotic is a neutral substance with a bitter taste. It is commercially available as the free compound and its esters. Esters, such as chloramphenicol palmitate or succinate, are almost inactive, but hydrolysis in vivo releases the antibiotic (Kono et al., 1972). The structure of chloramphenicol allows four possible stereo isomers. Only the D-*threo*-isomer shows high antimicrobial activity, and it is this form which is produced by actinomycetes. Its relatively simple structure is accessible to chemical synthesis, and it is commercially produced by this route. Current annual production of chloramphenicol is 20 metric tons. The required capacity forecast for 1983 is 50 metric tons. Because of the increasing cost of chemical synthesis, an efficient fermentation process for the commercial production of chloramphenicol would now be timely. A beginning has already been made toward understanding the genetics, biosynthesis, and regulatory systems controlling microbial production. Full understanding of the cellular mechanisms in chloramphenicol-producing organisms may make it possible to manipulate control processes and thus provide the basis for a commercial fermentation.

II. GROWTH AND CHLORAMPHENICOL PRODUCTION

Like most secondary metabolic processes, chloramphenicol production is a variable trait of the producing organism. In a rich medium *S. venezuelae* grows rapidly, produces little antibiotic during early exponential phase growth, and reaches stationary phase without having accumulated a high titer. A nutritionally poor medium containing, for example, glycerol and serine as carbon and nitrogen sources supports antibiotic synthesis in parallel with growth. During the extended growth phase, high titers build up. Lactate, which is poorly utilized by *S. venezuelae,* increases chloramphenicol production by threefold to fourfold when included in the medium. The mechanism of its action is not known, but conceivably it affects the balance between the glycolytic and pentose phosphate pathways and channels precursors into the shikimate pathway. The shikimate pathway is the route to chloramphenicol in *S. venezuelae*.

Westlake et al. (1968) showed that chloramphenicol production by *S. venezuelae* is influenced by the type and concentration of nitrogen source used in the medium. Nitrogen sources that support slow growth give the highest yields.

There is a complex relationship between chloramphenicol production and the cell ultrastructure of its producer. Bewick et al. (1976) have studied the effects of adding chloramphenicol to cultures of *S. venezuelae* during growth in media supporting high and low levels of antibiotic production. Addition of the antibiotic induced morphological and ultrastructural abnormalities similar to those reported for sensitive bacteria with an increase in the frequency of mesosomes and electron-transparent areas. The effect was less pronounced when chloramphenicol synthesis occurred, possibly because permeability barriers to external chloramphenicol were established. Electron-transparent areas were frequent and may have been associated with storage and excretion of precursors of chloramphenicol (Bewick and Williams, 1978).

As with chloramphenicol in *S. venezuelae,* production of corynecin parallels growth in *Corynebacterium hydrocarboclastus* (Nakano et al., 1977). A "semisynthetic" medium designed for the production of corynecins contained sucrose as the preferred carbon source, a low phosphate concentration, and high potassium chloride, inositol, fructose, and yeast extract concentrations. When the pH of the medium was kept in the neutral range, about 4 $g \cdot L^{-1}$ of corynecins were obtained (Nakano et al., 1974). Corynecin I is less toxic to *C. hydrocarboclastus* than other components of the corynecin complex it produces. The content of corynecin I varies with the carbon source used, but under any fermentation conditions, significant amounts of corynecin II and corynecin III accumulate. A mutant with an increased ratio of corynecin I synthesized double the amount of corynecins compared with the parent strain.

III. BIOSYNTHESIS

Isotopic labeling experiments have outlined much of the pathway of chloramphenicol biosynthesis (Figure 12.2). Primary metabolic precursors are assembled and restructured into aromatic compounds via the normal shikimate pathway (Gibson and Pittard, 1968; O'Neill et al., 1973).

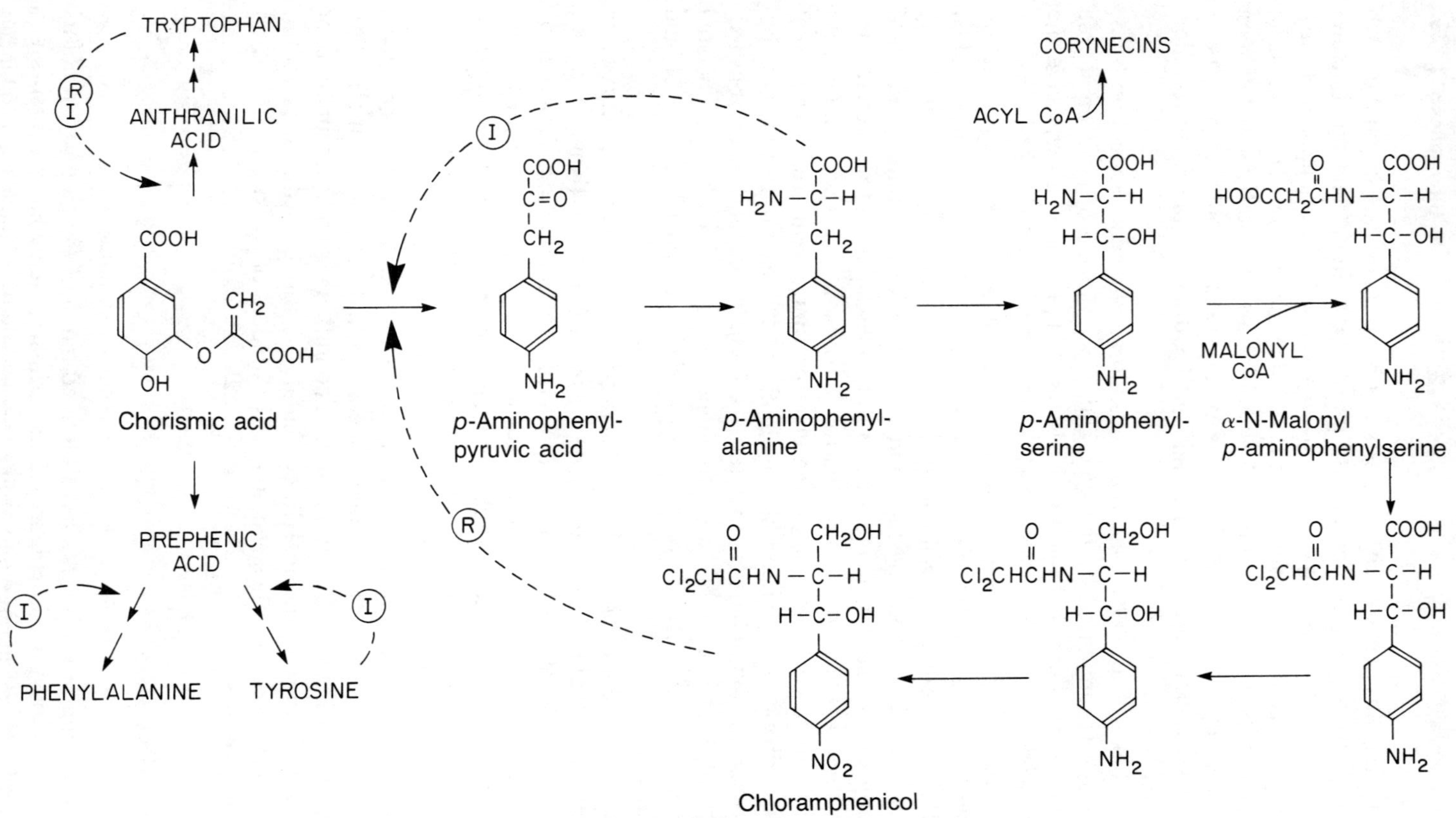

FIGURE 12.2 Biosynthesis of chloramphenicol and its regulation. Ⓡ = repression; Ⓘ = inhibition.

p-Aminophenylpyruvic acid derived from chorismic acid is the first identified intermediate on the branch leading specifically to chloramphenicol. Its formation involves an amination step that may be similar to that encountered in the synthesis of anthranilate (Gibson and Pittard, 1968; Zalkin, 1973) and *p*-aminobenzoate (Huang and Gibson, 1970; Lawrence et al., 1974).

IV. ENZYMOLOGY

A. 3-Deoxy-D-Arabinoheptulosonate-7-Phosphate Synthetase

This is the first enzyme in the shikimate pathway, and in many organisms, it is a control point regulating the flow of central pathway intermediates toward aromatic amino acid synthesis. 3-Deoxy-D-arabinoheptulosonate-7-phosphate (DAHP) synthetase is generally controlled by feedback from branch point intermediates or end products of the shikimate pathway. Jensen and Stenmark (1970) reported that, in streptomycetes, the activity of DAHP synthetase is controlled solely by tryptophan. Görisch and Lingens (1971) confirmed that DAHP synthetase from *Streptomyces aureofaciens* is a single enzyme, feedback inhibited only by tryptophan. Using a different assay, Lowe and Westlake (1971) found that DAHP synthetase of the chloramphenicol-producing *S. venezuelae,* a single protein of a 88,000 daltons, was neither repressed nor feedback-inhibited by aromatic amino acids. Previous reports of feedback inhibition by L-tryptophan were attributed to the presence of periodic acid in the enzyme assay. The absence of control over DAHP synthetase activity is not special to streptomycetes. The same enzyme from *Micrococcus luteus* or *Nocardia mediterranea* is under no apparent feedback control (Jensen and Stenmark, 1970; Christ et al., 1981). DAHP synthetase from both streptomycetes and *M. luteus* is, however, inhibited by an excess of the substrate, erythrose-4-phosphate. In addition, DAHP synthetase from the chloramphenicol producer is repressed by *p*-hydroxybenzoic acid. Not surprisingly, chloramphenicol did not repress nor feedback inhibit DAHP synthetase, since it is never found inside the cells of the producing organism. Precursors en route to chloramphenicol could have a role in regulating the activity or synthesis of DAHP synthetase in the producing organism, but this has not been tested.

B. Chorismate Mutase

Chorismate mutase is a branch point enzyme leading specifically to the biosynthesis of phenylalanine and tyrosine. It catalyzes the conversion of chorismic acid to prephenic acid. Görisch and Lingens (1972) purified this enzyme tenfold from *S. venezuelae* cells that were not producing chloramphenicol. Neither tryptophan, shikimic acid, anthranilic acid, phenylpyruvic acid, nor *p*-hydroxyphenylpyruvic acid inhibited its activity. The enzyme had a molecular weight of 55,000, a broad pH optimum between 6.6 and 7.0, and a K_M of 3.8×10^{-4} M. Lowe and Westlake (1972) studied chorismate mutase from a chloramphenicol-producing strain of *S. venezuelae* and concluded that the activity was not subject to feedback inhibition or repression by end products or intermediates of the shikimate pathway. Chorismate mutase of *S. aureofaciens* has been purified to homogeneity (Görisch and Lingens, 1974). The cell extracts were treated with phenylmethane sulfonyl fluoride to inhibit serine proteases that otherwise modify and inactivate chorismate mutase. The enzyme was similar in size (60,000 daltons) and properties to the partially purified preparation from *S. venezuelae*. It was not inhibited by end products or intermediates of the aromatic pathway and was not associated with the activity of prephenate dehydrogenase or prephenate dehydratase.

C. Prephenate Dehydratase

This enzyme, which catalyzes the conversion of prephenic acid to phenylalanine, is represented in *S. venezuelae* by a single protein of 220,000 daltons. (Lowe and Westlake, 1972). It is sensitive to feedback inhibition by phenylalanine.

D. Prephenate Dehydrogenase

Lowe and Westlake (1972) were not able to detect this enzyme in a chloramphenicol-producing strain of *S. venezuelae*. Whether the enzyme is exceptionally labile or whether tyrosine is formed by the pretyrosine pathway in this organism has not been established.

E. Anthranilate Synthetase

In *S. venezuelae,* anthranilate synthetase is feedback-inhibited by tryptophan. A tryptophan auxotroph that produced drastically lower yields of chloramphenicol accumulated anthranilic acid and had tenfold higher levels of anthranilate synthetase than its parent. A prototrophic revertant that had regained normal chloramphenicol production also had normal levels of anthranilate synthetase and did not accumulate anthranilic acid. These studies indicate that chorismate is diverted toward anthranilic acid formation if the feedback control of anthranilate synthetase exerted by tryptophan is removed. Presumably, it is channeled toward chloramphenicol when intracellular levels are elevated owing to decreased utilization by competing pathways.

In *S. venezuelae,* anthranilate synthetase is composed of two nonidentical subunits. The major subunit (MW 72,000) converts chorismic acid to anthranilic acid, using ammonia to supply the amino group. The smaller subunit (MW 28,000–29,000) confers on the enzyme the ability to use glutamine instead of ammonia as a substrate. Synthetase activity with glutamine reached its maximum at pH 7.2–7.6, whereas with ammonia it increased linearly through pH 9.0 without reaching a maximum. Activity was increased and stabilized by adding glutamine and magnesium chloride to the buffer system. Both activities of the enzyme were inhibited by anthranilic acid and by tryptophan. Synthesis was repressed by histidine, anthranilic acid, tryptophan, and *p*-aminobenzoic acid. When activity was repressed by anthranilic acid and by tryptophan, there was a concomitant increase in the activity of arylamine synthetase, an enzyme involved in chloramphenicol production. However, stimulating arylamine synthetase did not increase antibiotic synthesis (Francis et al., 1978).

F. Arylamine Synthetase

Arylamine synthetase converts chorismic acid to *p*-aminophenylpyruvic acid and is the branch point enzyme in the shikimate pathway that diverts intermediates to chloramphenicol (A. Jones et al., 1978). Like anthranilate synthetase, arylamine synthetase requires chorismic acid and glutamine.

Anthranilate and *p*-aminobenzoate synthetase share the same protein subunit in *Bacillus subtilis* (Kane et al., 1972) and *Acinetobacter calcoaceticus* (Sawula and Crawford, 1973). It is possible that, in *S. venezuelae,* arylamine synthetase, anthranilate synthetase, and *p*-aminobenzoate synthetase share a common glutamine-binding subunit. Recruitment of the same protein subunit by various enzymes would have regulatory implications and deserves further study.

Like many other enzymes of secondary metabolism, arylamine synthetase is unstable and is detected in the cell only during certain periods of growth. During growth of *S. venezuelae* in media supporting chloramphenicol synthesis, the activity of arylamine synthetase increases to a maximum and then decreases quickly to a very low level. In contrast, the activity of anthranilate synthetase remains stable. The fall in arylamine synthetase activity could be due to increased enzyme turnover associated with decreasing intracellular concentration of the substrate. The balance between degradation and synthesis of such enzymes may depend on stabilization by accumulated substrates. This has not been well studied but may have a role in regulating the activity of these proteins. The decrease in activity of arylamine synthetase has also been attributed to repression of enzyme synthesis due to the increasing concentration of chloramphenicol in the culture. It was noted that arylamine synthetase activity decreased rapidly when 100 $mg \cdot L^{-1}$ of chloramphenicol was added to a culture (A. Jones and Westlake, 1974).

In chloramphenicol-producing cultures, development of resistance to the antibiotic is calibrated to the gradually accumulating extracellular concentration of toxic metabolite. Resistance seems to be mainly due to impermeability of the cell membrane to the secreted antibiotic, which is thereby prevented from reentering the cell. The absence of intracellular chloramphenicol in producing cultures suggests that the antibiotic does not act directly as a repressor. However, chloramphenicol could inhibit the final steps in its own synthesis and cause a buildup within the cell of intermediates, which could act as repressors. *p*-Aminophenylalanine and L-*threo*-nitrophenylserinol have been shown to repress arylamine synthetase. The potent antibacterial *p*-methylthio analog of chloramphenicol has been shown to inhibit growth and chloramphenicol production in *S. venezuelae*. This analog is readily permeable and strongly represses arylamine synthetase. The *p*-methylsulfonyl analog of chloramphenicol does not enter cells and does not affect growth, chloramphenicol production, or arylamine synthetase synthesis (Malik and Vining, 1972a, b; A. Jones and Westlake, 1974). There are numerous analogs of chloramphenicol that may not be toxic to the producing organism. Some of these may be gratuitous inducers that could be used to study the molecular biochemistry of chloramphenicol production.

G. Aminotransferase

A. Jones et al. (1978) suggested that synthesis of *p*-aminophenylalanine from *p*-aminophenylpyruvic acid is catalyzed by a multispecific aminotransferase. This enzyme, purified 16-fold from *S. venezuelae,* had a molecular weight of 90,000 and a broad pH optimum at 8.0. Because of the unavailability of *p*-aminophenylpyruvic acid as a substrate, the enzyme was assayed in the direction opposite to that required for biosynthesis. Phenylalanine and tyrosine as well as *p*-aminophenylalanine were active amino donors. Since the relative activity with these substrates did not change during purification, the enzyme was probably a single protein. With phenylalanine as substrate, the preferred amino accepting cosubstrate was α-ketoglutarate, and the reaction required pyridoxal phosphate.

Streptomyces venezuelae possesses two aminotransferases that use aromatic amino acids as substrates. The one eluted at high potassium chloride concentration from DEAE cellulose was active with *p*-aminophenylalanine and was responsible for about 85% of the phenylalanine aminotransferase activity. In keeping with the general lack of repression of the shikimate pathway in *S. venezuelae,* the enzyme was not repressed by aromatic amino acids. An enzyme somewhat similar to this exists in *Escherichia coli* (Mavrides and Orr, 1974, 1975), but it is repressed by aromatic amino acids.

H. Chloroperoxidase

Chloramphenicol-producing organisms must possess a chlorinating enzyme, since the antibiotic contains a dichloroacetyl substituent. Extracts of *S. venezuelae* cells contain peroxidase activity, which could be detected with *o*-dianisidine as the substrate. However, all efforts to show chlorination of monochlorodimedone by crude extracts were unsuccessful. Since addition of small amounts of *S. venezuelae* cell extracts to the purified chloroperoxidase from *Caldariomyces fumago* totally inhibited its activity, *S. venezuelae* may possess a potent inhibitor. Alternatively, its exceptionally high level of catalase activity may compete successfully for the peroxide (V. S. Malik, unpublished).

V. CHLORAMPHENICOL RESISTANCE IN THE PRODUCING ORGANISM

A. Inducibility

Chloramphenicol is a potent inhibitor of protein synthesis in prokaryotic organisms. Producing cultures must therefore have mechanisms for escaping its toxic effect. In fact, *S. venezuelae* shifts from a state of chloramphenicol sensitivity when no antibiotic is produced to chloramphenicol resistance when antibiotic is being produced (Malik, 1972; Malik and Vining, 1972a). Growth and protein synthesis are immediately inhibited when *S. venezuelae* is first exposed to chloramphenicol. They resume after a lag that is proportional to the amount of chloramphenicol supplement in the growth medium. High concentrations of chloramphenicol also affect the growth rate. When subcultured in the presence of chloramphenicol, *S. venezuelae* grows without a lag, but this resistance is lost after subculturing once in the absence of the antibiotic. Thus resistance is inducible and is not due to selection of chloramphenicol-resistant mutants (Malik and Vining, 1971).

B. Chloramphenicol Inactivation

Most bacterial resistance to chloramphenicol is due to an R-factor coded enzyme, chloramphenicol acetyltranferase (Davies and Smith, 1978; Mlynarczyk et al., 1979). This enzyme has also been found in *Streptomyces* (Shaw and Hopwood, 1976; Nakano et al., 1977a; Wright and Hopwood, 1977; Elkersch and Plouride, 1976; Freeman et al., 1977). However, none of these species produces chloramphenicol and the enzymes could not be detected in *S. venezuelae* cultures that produce the antibiotic; the latter detoxify intracellular chloramphenicol by hydrolytic removal of the N-dichloroacetyl group (Malik and Vining, 1970). The *p*-nitrophenylserinol formed is either quickly reacylated and excreted as N-acetyl-*p*-nitrophenylserinol or further metabolized via *p*-nitrobenzoic acid and other aromatic compounds. Diacetylchloramphenicol has never been detected. Although the constitutive chloramphenicol hydrolase can protect *S. venezuelae* against the toxic effects of its antibiotic, inducible resistance is not entirely due to inactivation because inhibitory concentrations of chloramphenicol accumulate in the medium during production of the antibiotic.

C. Permeability Barrier

Resistance to chloramphenicol in the producer is most likely due to alteration of the permeability of the cell membrane during antibiotic production. Intracellular antibiotic is quickly metabolized. A permeability barrier between external chloramphenicol and the internal catabolic enzymes limits the extent of chloramphenicol metabolism by producer cells and protects the excreted antibiotic from degradation at the same time that it protects the producer. In *S. venezuelae*, the decrease in permeability to chloramphenicol is closely associated with the onset of antibiotic biosynthesis. It seems likely that the exclusion mechanism is either part of the biosynthetic process or coordinately controlled with it.

In *S. venezuelae,* antibiotic-sensitive variants occur spontaneously, but they are very unstable (V. S. Malik, unpublished). The frequency with which such variants arose from the wild type resembled that with which chloramphenicol-sensitive variants appeared in *Streptomyces coelicolor* A 3(2) and *Streptomyces lividans* 66 (Freeman et al., 1977). In the latter two organisms, chloramphenicol-sensitive strains also proved difficult to preserve. Reversion to the chloramphenicol-resistant phenotype proved that the acquisition of antibiotic sensitivity was not caused by permanent loss of a plasmid. However, segregation of chloramphenicol sensitivity versus chloramphenicol resistance in the two closely related *S. coelicolor* and *S. lividans* species did not support a simple chromosomal linkage (Freeman and Hopwood, 1978). In streptomycetes, several different mechanisms may be responsible for the genetic instability of their resistance to various antibiotics (Sermonti et al., 1977, 1980; Malik, 1982). Expression of antibiotic resistance that is altered with such high frequency is possibly due to transposition or rearrangement of the genome. The source of the unstable resistance to antibiotics in *Streptomyces* may not be enzymatic inactivation but more likely an alteration of membrane permeability that limits uptake, as is found in some other organisms (Irvin and Ingram, 1980; Süssmuth et al., 1979). Extensive genetic analysis of antibiotic resistance may help in industrial strain improvement programs

VI. GENETICS OF STREPTOMYCES VENEZUELAE

A. Chromosomal Mapping

The recombination system in *S. venezuelae* is similar to those reported in other *Streptomyces* (Malik, 1979b). Crosses between some mutants were strongly polar in the sense that one parent contributed most of the markers in the progeny, suggesting that a fertility system like that in *Streptomyces coelicolor* may be present. There is striking correspondence between the genetic maps of various *Streptomyces,* implying stability of linkage relationships during evolution.

Circular genetic maps of marker loci have been constructed by two independent groups using different strains of *S. venezuelae*. Francis et al. (1975) constructed a linkage map for 17 marker loci (Fig. 22.3) using a strain (species 3022a) isolated from a soil sample at the UpJohn Company in 1954. Akagawa et al. (1975) constructed chromosomal maps of strain

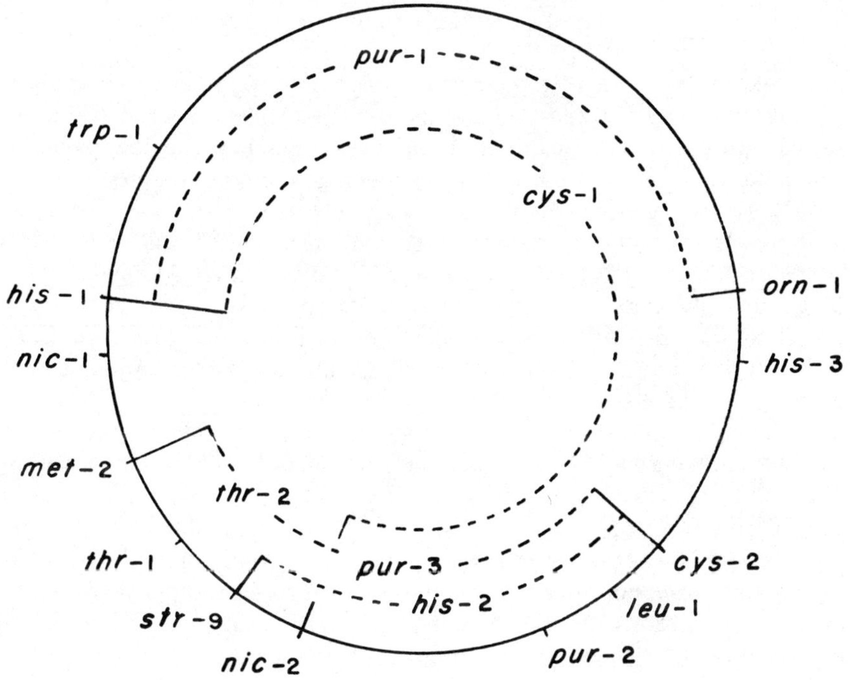

FIGURE 12.3 Linkage map of *Streptomyces venezuelae*. From Francis et al. (1975).

ISP 5230 of *S. venezuelae*, which is descended from the type culture isolated by Ehrlich et al. (1948).

B. Transduction

Phage infection of *S. venezuelae* is not uncommon, and the life cycle of one of the viruses has been described (L. A. Jones and Bradley, 1965). Stuttard (1979) has shown that a temperate actinophage, ϕSV1, is capable of transduction. This should allow fine structure mapping of genes affecting chloramphenicol biosynthesis. Stuttard and Dwyer (1981) have described a second temperate phage, ϕSV2, that has double-stranded DNA with a molecular weight of 24×10^6. These phages may be developed as vectors for cloning genes but have narrow host specificity for *S. venezuelae*.

C. Plasmid Involvement in Chloramphenicol Production

Like many other antibiotic producers, *S. venezuelae* loses its productivity on repeated subculturing. The loss occurs in a stepwise manner, suggesting that high-level production could be polygenic in character. Genes responsible for chloramphenicol production are sensitive to treatment with intercalating agents (Akagawa et al., 1975; Michelson and Vining, 1978). Strains altered in productivity arose at frequencies as high as 12%, greatly exceeding the rate at which auxotrophic mutations appeared. The response of cultures varied with age, and at least three strain types—high, low, and nonproducers—could be distinguished. In routine screening, the true nonproducer class could not be distinguished from some very low ($<4 mg \cdot L^{-1}$) producers. The lesions in these strains were at first attributed to elimination of a plasmid carrying genes coding for antibiotic synthesis. However, intercalating agents not only cure plasmids, they are also known to cause rearrangements and alterations in structure of the extrachromosomal elements (Kelly and Catley, 1977; Komatsu et al., 1981). This could account for the formation of strains with various levels of chloramphenicol production.

Akagawa et al. (1975, 1979) crossed an apparently nonproducing strain (SVM3) carrying *his, leu,* and *ade* auxotrophy with a chloramphenicol producer (SVM1) carrying *str, lys, met, ilv,* and *pro* markers. These strains were derived from *S. venezuelae* strain ISP 5230. The nonproducer (Cpp^-) was prepared by acriflavine treatment and subsequently proved to be merely a very low producer. Recombinants were selected using all possible pairs of markers except Cpp and the genotypes were examined. When the Cpp marker was ignored, a circular linkage of *his, ade, str, leu, lys, met, ilv,* and *pro* could be obtained. However, the Cpp marker could not be placed between any pair of chromosomal markers, a finding suggesting that it was not linked to them and existed on a separate replicon. While the genetic evidence does not distinguish between plasmid loss or plasmid rearrangement, the observation that the Cpp^- strains produced a small amount of chloramphenicol (Akagawa et al., 1975) and evidence that it could be mutated to produce increased amounts (Malik, 1980) argue strongly against the first possibility.

Akagawa et al. (1979) have mapped other Cpp^- mutants, which were obtained from a producing strain, SVM2, by high-temperature incubation, UV irradiation, or nitrosoguanidine treatment. One mutant, SVM2-2A7, was clearly blocked in biosynthesis and produced 1-deoxychloramphenicol instead of chloramphenicol. Appropriate crosses with a producing strain carrying complementary auxotrophic markers indicated that these Cpp^- mutants were located between *met* and *ilv* on the chromosome of *S. venezuelae*. It was suggested that the plasmid might play a regulatory role in increasing chloramphenicol production.

A plasmid, pUC3, has been isolated from *S. venezuelae* and mapped with restriction enzymes (Malik and Reusser, 1979). Advances in plasmid isolation methodology now allow single clones or a few milliliters of culture not only to be screened for the presence of plasmids but to be analyzed with restriction enzymes (Birnboim and Doly, 1979; Davis et al., 1980; Schottel et al., 1981; Malik, 1981; Marko et al., 1982). With these techniques, it has been possible to establish that plasmid pUC3 is still intact in strains that have lost the capacity to produce chloramphenicol after treatment with intercalating dyes. Moreover, the plasmid could be eliminated by the protoplast regeneration procedure (see Chapter 1) without affecting chloramphenicol production. Hybridization of a radioactively labeled probe of purified plasmid pUC3 DNA to chromosomal DNA that had been cleaved by restriction enzymes and separated by agarose gel electrophoresis verified that the plasmid had not been integrated into the chromosome (Ahmed and Vining, 1983).

The DNA of two strains of *Streptomyces glaucescens,* one the wild type and the other an ethidium bromide–induced mutant that had become sensitive to 1 mg/l streptomycin and had lost the ability to form melanin, were labeled with ^{14}C-thymidine and ^{3}H-thymidine, respectively. They were then purified, mixed, digested with the restriction endonuclease *Bam*H1, and fractionated by agarose gel electrophoresis. The gel was then cut into small slices, and for each slice the ^{14}C cpm to ^{3}H cpm ratio was determined. Two ^{3}H peaks representing DNA fragments of approximately 6 and 7 kb were found in DNA from the mutant strain only. Visual examination of photographs of ethidium bromide–stained gels revealed a reproducible pattern of about 80 bands. Two independently obtained ethidium bromide–induced mutants lacked a band in the 13.5-kb range that was reproducibly present in preparations from wild-type strains (Hintermann et al., 1981). The banding pattern of seven other strains that differed by several mutagenic steps was identical. That the pattern was insensitive to point mutations and even to substantial DNA alterations is not unexpected, since the approximately 80 bands seen on the photographs were composed of about 2000 different restriction fragments. DNA from *Streptomyces scabies, S. coelicolor* A3(2), and *N. mediterranea* isolated and restricted by the same method gave banding patterns that were distinctive for each species. It thus seems that restriction patterns can be used as fingerprints for the identification of *Streptomyces* species. The results also indicate that ethidium bromide alters the DNA of *S. glaucescens* sufficiently to cause a distinct change in restriction pattern (Hintermann et al., 1981, 1982). Thus the effect of such dyes on chloramphenicol biosynthesis in *S. venezuelae* may well be due to chromosomal rearrangements.

VII. REGULATION

A. Lack of Feedback Controls

Gibson and Pittard (1968) have reviewed the regulatory mechanisms that control the synthesis of aromatic amino acids in various organisms. The essence of these is that enzyme synthesis is repressed by end products of the pathway and that the activity of the first and branch point enzymes of the pathway is inhibited by feedback inhibition from these compounds (Umbarger, 1978). In *S. venezuelae,* none of the enzymes involved in the biosynthesis of aromatic amino acids is repressed. The first enzyme of the pathway, DAHP synthetase, is also free from control by feedback inhibition. This leaves the pathway without restriction up to the synthesis of chorismate. Conversion of chorismate to the aromatic protein amino acids is under feedback control. Anthranilate synthetase and prephenate dehydratase are feedback-inhibited by tryptophan and phenylalanine, respectively. Thus when growth slows down and aromatic amino acids are not fully utilized, an increase in their pool size would be expected to inhibit the branching enzymes, thereby enlarging the chorismate pool. At this point, the excess chorismate can be channeled toward chloramphenicol synthesis (Malik, 1980).

B. Repression of Antibiotic Synthesis

Chloramphenicol is toxic to the producing organism, and mechanisms to prevent suicide by overproduction have evolved. Endogenous synthesis of chloramphenicol by *S. venezuelae* is inhibited by chloramphenicol in the medium. If the maximum quantity of chloramphenicol normally produced is added to the culture at the time of inoculation, no additional antibiotic is synthesized. Cultures with smaller initial supplements produce no more than the normal yield (Malik, 1979a). Addition of the antibiotic at any point during growth curtails the yield of biosynthetic chloramphenicol (Malik and Vining, 1972a). When the chloramphenicol concentration reaches a metabolically embarrassing level, it appears to repress its own biosynthesis.

VIII. FUTURE PROSPECTS

Chorismate is the branch-point intermediate for the synthesis of numerous secondary metabolites (Leisinger and Margraff, 1979). Antibiotics such as novobiocin (Kominek and Sebek, 1974), lincomycin (Eble, 1978), polyenes (Martín, 1977), and rifamycins (Ghisalba and Nüesch, 1978) contain significant components that also originate with chorismate. Organisms that produce these latter substances have been developed for high productivity and are potentially capable of generating high cellular levels of chorismate; thus it may be possible to manipulate them for commercial production of tryptophan, which awaits a good microbial process.

Streptomyces venezuelae has a manageable genetic system in at least one strain and harbors a plasmid (Malik and Reusser, 1979). It is also susceptible to phages (Stuttard and Dwyer, 1981; Stuttard, 1979). These extrachromosomal elements could be developed as vectors for introducing genes into *S. venezuelae*. Application of recombinant DNA technology will make possible an understanding of the many phenomena observed during improvement and preservation of antibiotic-producing strains. This could provide a scientific basis for planning strain improvement programs for the production of useful secondary metabolites.

REFERENCES

Ahmed, Z. U., and Vining, L. C. (1983) *J. Bacteriol.* (in press).
Akagawa, H., Okanishi, M., and Umezawa, H. (1975) *J. Gen. Microbiol. 90*, 336–346.
Akagawa, H., Okanishi, M., and Umezawa, H. (1979) *J. Antibiot. 32*, 610–620.
Bewick, M. W. M., and Williams, S. T. (1978) *Microbios 19*, 27–35.
Bewick, M. W. M., Williams, S. T., and Veltkamp, C. (1976) *Microbios 16*, 191–199.
Birnboim, H. C., and Doly, J. (1979) *Nucleic Acid Res. 7*, 1513–1523.
Christ, M., Gygax, D., and Ghisalba, O. (1981) *Experientia 37*, 1222.
Davies, J., and Smith, D. I. (1978) *Ann. Rev. Microbiol. 32*, 469–518.
Davis, R. W., Thomas, M., Cameron, J., John, T. P., Scherer, S., and Padgett, R. A. (1980) *Meth. Enzymol. 65*, 404–411.
Eble, T. E. (1978) *J. Chromatogr. Libr. 15*, 231–271.
Ehrlich, J., Gottlieb, D., Burkholder, P. R., Anderson, L. E., and Pridham, T. G. (1948) *J. Bacteriol. 56*, 467–477.
Elkersch, T. A., and Plouride, J. R. (1976) *J. Antibiot. 24*, 292–302.
Francis, M. M., Cella, R., and Vining, L. C. (1975) *Can. J. Microbiol. 21*, 1151–1159.
Francis, M., Vining, L. C., and Westlake, D. W. S. (1978) *J. Bacteriol. 134*, 10–16.
Freeman, R. F., and Hopwood, D. A. (1978) *J. Gen. Microbiol. 106*, 337–381.
Freeman, R. F., Bibb, M. J., and Hopwood, D. A. (1977) *J. Gen. Microbiol. 98*, 453–465.
Ghisalba, O., and Nüesch, J. (1978) *J. Antibiot. 31*, 215–225.
Gibson, F., and Pittard, J. (1968) *Bacteriol. Rev. 32*, 465–492.
Görisch, H., and Lingens, F. (1971) *Biochim. Biophys. Acta. 242*, 617–629.
Görisch, H., and Lingens, F. (1974) *Biochemistry 13*, 3790–3794.
Görisch, H., and Lingens, F. (1972) *Arch. Mikrobiol. 82*, 147–154.
Gottlieb, D., Battacharya, P. K., Anderson, H. W., and Carter, H. E. (1948) *J. Bacteriol. 55*, 409–417.
Hintermann, G., Crameri, R., Kieser, T., and Hütter, R. (1981) *Arch. Microbiol. 130*, 218–222.
Hintermann, G., Crameri, R., Kieser, T., and Hutter, R. (1982) *Anal. Biochem. 121*, 327–330.
Huang, M., and Gibson, F. (1970) *J. Bacteriol. 102*, 767–773.
Imagawa, Y., Shima, S., Hirota, A., and Sakai, H. (1979) *Agr. Biol. Chem. 43*, 2437–2440.
Irvin, J. E., and Ingram, J. M. (1980) *Can. J. Biochem. 58*, 1165–1171.
Jensen, R. A., and Stenmark, S. L. (1970) *J. Bacteriol. 101*, 763–769.
Jones, A., and Westlake, D. W. S. (1974) *Can. J. Microbiol. 20*, 1599–1611.
Jones, A., Francis, M. M., Vining, L. C., and Westlake, D. W. S. (1978) *Can. J. Microbiol. 24*, 238–244.

Jones, L. A., and Bradley, S. G. (1965) *J. Gen. Microbiol. 40*, 191–198.
Kane, J. F., Holmes, W. M., and Jensen, R. A. (1972) *J. Biol. Chem. 247*, 1587–1596.
Kelly, P. J., and Catley, B. J. (1977) *J. Gen. Microbiol. 102*, 249–254.
Komatsu, K., Leboul, J., Harford, S., and Davies, J. (1981) in *Microbiology 1981*, (Schlessinger, D., ed.) pp. 384–387, American Society for Microbiology, Washington, D.C.
Kominek, L. A., and Sebek, O. K. (1974) *Dev. Ind. Microbiol. 15*, 60–69.
Kono, M., O'Hara, K., Nagawa, M., and Mitsuhashi, S. (1972) *Jap. J. Microbiol. 16*, 461–467.
Lawrence, J., Cox, G. B., and Gibson, F. (1974) *J. Bacteriol. 118*, 41–45.
Leisinger, T., and Margraff, R. (1979) *Microbiol. Rev. 43*, 422–442.
Lowe, D. A., and Westlake, D. W. S. (1971) *Can. J. Biochem. 49*, 448–455.
Lowe, D. A., and Westlake, D. W. S. (1972) *Can. J. Biochem. 50*, 1064–1073.
Malik, V. S. (1972) *Adv. Appl. Microbiol. 15*, 297–336.
Malik, V. S. (1979a) *Adv. Appl. Microbiol. 25*, 75–93.
Malik, V. S. (1979b) *Adv. Genet. 20*, 37–126.
Malik, V. S. (1980) *Biotechnol. Lett. 2*, 455–457.
Malik, V. S. (1981) *Adv. Appl. Microbiol. 27*, 1–84.
Malik, V. S. (1982) *Adv. Appl. Microbiol. 28*, 27–115.
Malik, V. S., and Reusser, F. (1979) *Plasmid 2*, 627–631.
Malik, V. S., and Vining, L. C. (1970) *Can. J. Microbiol. 16*, 173–179.
Malik, V. S., and Vining, L. C. (1971) *Can. J. Microbiol. 17*, 1287–1290.
Malik, V. S., and Vining, L. C. (1972a) *Can. J. Microbiol. 18*, 137–143.
Malik, V. S., and Vining, L. C. (1972b) *Can. J. Microbiol. 18*, 583–590.
Marko, M. A., Chipperfield, R., and Birnboim, H. C. (1982) *Anal. Biochem. 121*, 382–387.
Martín, J. F. (1977) *Ann. Rev. Microbiol. 31*, 13–38.
Mavrides, C., and Orr, W. (1974) *Biochim. Biophys. Acta 336*, 70–78.
Mavrides, C., and Orr, W. (1975) *Biol. Chem. 250*, 4128–4133.
Michelson, A. M., and Vining, L. C. (1978) *Can. J. Microbiol. 24*, 662–669.
Mlynarczyk, G., Mlynarczyk, A., and Osowiecki, H. (1979) *Acta Microbiol. Pol. 28*, 101–109.
Nakano, H., Tomita, F., and Suzuki, T. (1974) *Agr. Biol. Chem. 38*, 2471–2475.
Nakano, H., Tomita, F., and Suzuki, T. (1976) *Agr. Biol. Chem. 40*, 207–212.
Nakano, H., Matsuhashi, Y., Takeuchi, T., and Umezawa, H. (1977a) *J. Antibiot. 30*, 76–82.
Nakano, H., Tomita, F., Yamaguchi, K., Nagashima, K., and Suzuki, T. (1977b) *Biotechnol. Bioeng. 19*, 1009–1018.
O'Neill, W. P., Nystrom, R. F., Rinehart, K. L., and Gottlieb, D. (1973) *Biochemistry 13*, 4775–4784.
Price, C. A., Lynch, E. M., Bowie, B. A., and Newman, D. J. (1981) *J. Antibiot. 34*, 118–119.
Sawula, R. V., and Crawford, I. P. (1973) *J. Biol. Chem. 248*, 3573–3581.
Schottel, J. L., Bibb, M. J., and Cohen, S. N. (1981) *J. Bacteriol. 146*, 360–368.
Sermonti, G., Petris, A., Micheli, M., and Lanfaloni, L. (1977) *J. Gen. Microbiol. 100*, 347–353.
Sermonti, G., Lanfaloni, L., and Micheli, M. R. (1980) *Mol. Gen. Genet. 177*, 453–458.

Shaw, W. V., and Hopwood, D. A. (1976) *J. Gen. Microbiol. 94*, 159–166.
Stuttard, C. (1979) *J. Gen. Microbiol. 110*, 479–482.
Stuttard, C., and Dwyer, M. (1981) *Can. J. Microbiol. 27*, 496–499.
Süssmuth, R., Haag, R., and Lingens, F. (1979) *J. Antibiot. 32*, 1293–1302.
Suzuki, T., Tomita, F., Nakano, H., and Honda, H. (1974) U.S. Patent 3,847,747.
Tamura, A., Takeda, I., Naruto, S., and Yoshimura, Y. (1971) *J. Antibiot. 24*, 270.
Umbarger, H. E. (1978) *Ann. Rev. Biochem. 47*, 533–606.
Umezawa, H., Tazaki, T., Kaneri, H., Okami, Y., and Fukuyama, S. (1948) *Jap. Med. J. 1*, 358–363.
Westlake, D. W. S., Sala, F., McGrath, R., and Vining, L. C. (1968) *Can. J. Microbiol. 14*, 587–593.
Wright, H. M., and Hopwood, D. A. (1977) *J. Gen. Microbiol. 102*, 417–421.
Zalkin, H. (1973) *Adv. Enzymol. Relat. Areas Mol. Biol. 38*, 1–39.

CHAPTER 13

The Lincomycin–Celesticetin–Anthramycin Group

Jeffrey L. C. Wright

I. INTRODUCTION

It is useful to group natural products according to their known or suspected biogenesis, but it is sometimes difficult to select the appropriate group for metabolites of mixed biogenetic origin. The lincomycin–celesticetin group and the anthramycin group of antibiotics are not closely related in chemical structure and display different biological activity. Nevertheless, they share a novel route for creating the five-membered heterocyclic ring present in each structure; for this reason, they are considered together in this chapter. The biosynthetic evidence that justifies this grouping has been reviewed by Hurley (1977, 1980) and by Hurley and Speedie (1980).

II. LINCOMYCINS AND CELESTICETINS

Lincomycin was first found in cultures of *Streptomyces lincolnensis* var. *lincolnensis* (Mason et al., 1963; Herr and Bergy, 1963). Chemical studies showed it to be a hygric acid derivative condensed through an amide link to methyl thiooctapyranoside (Hoeksema et al., 1964; Schroeder et al.,

1967; Slomp and MacKellar, 1967; Magerlein et al., 1967a). Later, Argoudelis et al. (1964, 1965) isolated the 3′-ethyl analog, lincomycin B, from the culture. The structures of these and other lincomycin derivatives that are the products of altered growth conditions and that will be discussed later are given in Figure 13.1.

Before lincomycin was discovered, Hoeksema et al. (1955) had described an antibiotic, celesticetin, from *Streptomyces caelestis*. Although active against gram-positive and gram-negative organisms, it did not display enough potential for development past the laboratory stage (DeBoer et al., 1955). A partial structure for celesticetin was reported by Hoeksema and Hinman (1964); and later (Hoeksema, 1964, 1968) its close similarity with lincomycin was established. Subsequently, a group at the Upjohn Company (Argoudelis and Brodasky, 1972; Argoudelis et al., 1972a) discovered desalicetin as well as celesticetins B, C, and D in fermentations of *S. caelestis* and isolated 7-O-demethylcelesticetin from a mutant (Fig. 13.2).

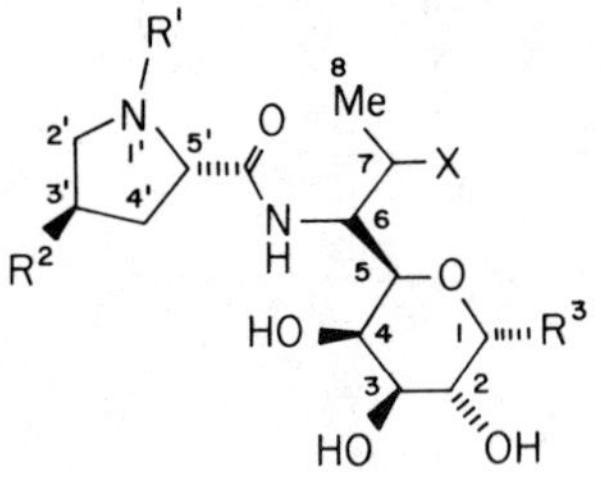

	R^1	R^2	R^3	X
Lincomycin	Me	Pr	SMe	OH
Lincomycin B	Me	Et	SMe	OH
Lincomycin sulfoxide	Me	Pr	SMe→O	OH
Demethylthio-1-hydroxylincomycin	Me	Pr	OH	OH
Lincomycin C	Me	Pr	SEt	OH
Lincomycin S	Et	Pr	SEt	OH
Lincomycin K	H	Pr	SEt	OH
Lincomycin D	H	Pr	SMe	OH
Clindamycin	Me	Pr	SMe	Cl
Clindamycin sulfoxide	Me	Pr	SMe→O	Cl
N-Demethylclindamycin	H	Pr	SMe	Cl

FIGURE 13.1 Structures of the lincomycins.

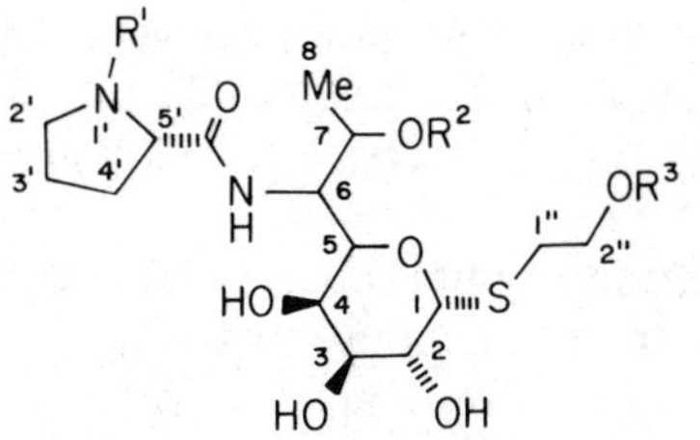

	R^1	R^2	R^3
Celesticetin	Me	Me	salicylyl
Desalicetin	Me	Me	H
Celesticetin B	Me	Me	isobutyryl
Celesticetin C	Me	Me	anthranilyl
Celesticetin D	Me	Me	acetyl
7-O-Demethylcelesticetin	Me	H	salicylyl
N-Demethylcelesticetin	H	Me	salicylyl
N-Demethyl-7-O-demethylcelesticetin	H	H	salicylyl
7-O-Demethyldesalicetin	Me	H	H
Desalicetin-2″-(*p*-amino)salicylate	Me	Me	*p*-aminosalicylyl

FIGURE 13.2 Structures of the celesticetin group.

A. Biological Activity

Lincomycin and the more potent semisynthetic offshoot, clindamycin (Magerlein et al., 1967b), are the only commercially established and clinically used antibiotics in this group (Magerlein, 1971; Leigh, 1981; Phillips, 1981). Both display considerable activity in vitro and in vivo toward gram-positive bacteria, including penicillinase-producing strains of *Staphylococcus aureus* and aerobic streptococci with the exception of *Streptococcus faecalis* and *Streptococcus faecium*. Anaerobic gram-positive bacilli are also susceptible, but resistant subspecies of *Clostridium* occur. Aerobic gram-negative bacteria are nearly all resistant to lincomycin or clindamycin; yeasts and animal cells grown in culture are insensitive to lincomycin at concentrations up to 100 μg/ml (Chang and Weisblum, 1967). Unlike sensitive bacteria, rat liver mitochondria continue to synthesize protein in the presence of lincomycin (Towers et al., 1972).

Lincomycin was introduced into therapy in the early 1960's. The chloro derivative, clindamycin, soon replaced it because of superior pharmacological properties; it is particularly effective in the treatment of superficial sepsis, often caused by anaerobes as well as by *S. aureus* and *Streptococcus pyogenes* (Phillips, 1981). Many serious pyogenic infections have the same staphylococcal, streptococcal, and anaerobic etiology as minor superficial infections and can be treated by intravenous and intramuscular injection of clindamycin phosphate. This biologically inactive derivative is rapidly hydrolyzed in vivo to the active base (Shoeman et al., 1972; DeHaan et al., 1973).

Association of pseudomembranous colitis with clindamycin and lincomycin therapy led to a decline in use that accelerated with the introduction of metronidazole as a highly effective alternative. The etiology of pseudomembranous colitis has since been traced to a toxin produced by *Clostridium difficile,* and the condition can be controlled by combined therapy with an antibiotic such as vancomycin (Leigh, 1981). Clindamycin is still the most effective treatment for certain diseases, including osteomyelitis (Phillips, 1981).

B. Producing Organisms

The original lincomycin producer was named *S. lincolnensis* var. *lincolnensis,* NRRL 2936 (Mason et al., 1963). Lincomycin is also produced by *Streptomyces umbrinus* var. *nova* (Patterson et al., 1964); by *Streptomyces espinosus* Dietz, sp. *nova* (Argoudelis et al., 1972b; Reusser and Argoudelis, 1974); by *Streptomyces pseudogriseolus* chemovar. *linmyceticus* Dietz var. *nova* (Argoudelis and Coats, 1973); by *Streptomyces variabolis* chemovar. *liniabilis* Dietz var. *nova* (Argoudelis and Coats, 1974a); by *Actinomyces roseolus* strain 1147 (Gause et al., 1975); and by *Streptomyces vellosus* sp. *nova* (Bergy et al., 1975). The lincomycin derivatives listed in Figure 13.1 are reported only from *S. lincolnensis* except for lincomycin C which is also produced by *S. umbrinus* var. *cyaneoniger* var. *nova* (Argoudelis et al., 1964). Celesticetin and its derivatives are all produced by *S. caelestis* sp. *nova* NRRL 2418 (Hoeksema et al., 1955; DeBoer et al., 1955; Argoudelis and Brodasky, 1972).

C. Fermentation

Few details of lincomycin production have been published. Lincomycin and the minor component, lincomycin B, are produced by standard methods of submerged culture. Typically, seed cultures of *S. lincolnensis* var. *lincolnensis* are incubated in a complex medium and used to inoculate complex or defined production media. Complex media have contained starch and glucose supplemented with cottonseed meal and blackstrap molasses. The defined media contained glucose, sodium citrate, and NH_4NO_3. The normal period of growth is 6 days; extension of the fermentation period to 12 days allows production of lincomycin sulfoxide and the 1-demethylthio-1-hydroxy compound (Argoudelis and Mason, 1969). Lincomycins C, D, and S are produced by adding inhibitors such as DL-ethionine, sulfonamides, and sulfanilamides (Argoudelis et al., 1970, 1973b). The effect of these inhibitors depends on the medium used; DL-ethionine with *S. lincolnensis* grown in complex media gives lincomycin C, but in a synthetic medium, lincomycin C and S are formed. Sulfonamides and sulfanilamides added to *S. lincolnensis* cultures give lincomycin D.

Fermentation conditions for producing celesticetin closely follow those for lincomycin. A variety of complex media using glucose, starch, peptone, yeast extract, cornsteep liquor, or molasses have been reported. Celesticetin, celesticetins B, C, and D, and desalicetin are all produced by *S. caelestis* under normal conditions. The demethylcelesticetins are produced by mutant strains.

D. Mode of Action

The mode of action of lincomycin has been reviewed by Chang and Weisblum (1967) and by Chang (1979). The antibiotic inhibits protein synthesis by binding hydrophobically to ribosomes of sensitive microorganisms. Lincomycin acts on the 50S ribosomal subunit, since it competes effectively with chloramphenicol for binding to the ribosome. It inhibits the peptidyl transferase reaction by interfering primarily with the binding of aminoacyl-tRNA to the A site.

No similar studies have been reported for the celesticetins.

E. Biosynthesis

Initial radiotracer experiments showed that several potential precursors of the propylhygric acid moiety of lincomycin, such as aminolevulinic acid, glutamate, hydroxyproline, and proline, were not incorporated to a significant extent. (Eble, 1967). Further biosynthetic studies (Argoudelis et al., 1969b) showed that both the C-methyl and N-methyl groups of the propylhygric acid moiety of lincomycin were derived from methionine by transmethylation.

Witz et al. (1971) observed that *S. lincolnensis* accumulated propylproline and ethylproline in sulfur-limited media. It was known that, in media containing tyrosine, *S. lincolnensis* produces melanin-like pigments that are biosynthesized from tyrosine via substituted proline intermediates. Since addition of tyrosine and L-dihydroxyphenylalanine (L-dopa) stimulated production of propylproline and ethylproline, it was suggested that the hygric acid moieties of lincomycins are derived from intermediates on the pathway between tyrosine and melanin (Fig. 13.3). As supporting evidence, [U-^{14}C]tyrosine was found to heavily label the amino acid portion of the antibiotic, while the label from [1-^{14}C]tyrosine and [^{15}N]tyrosine was exclusively in the amide carbonyl and proline nitrogen, respectively.

Cleavage of an aromatic ring, although unexpected, is certainly not unknown (Thomas, 1965), and the existence of this novel biosynthetic route has been confirmed during the studies on anthramycin biosynthesis discussed later in this chapter.

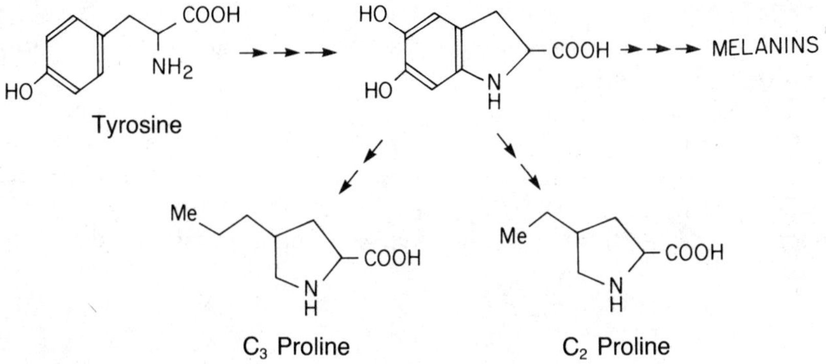

FIGURE 13.3 Suggested relationship between melanin biosynthesis and formation of substituted proline intermediates.

F. Biotransformations and Directed Biosynthesis

Biosynthetic modification of the lincomycin or clindamycin molecule has been achieved by using three general approaches. The first relied on the addition of metabolic inhibitors, the second on directed biosynthesis, and the third on biomodification by actinomycetes that are different from the producer.

The formation of lincomycins C and S in *S. lincolnensis* cultures when ethionine is added to the fermentation medium has already been noted. Antimetabolites such as sulfa drugs appear to inhibit N-methylation rather selectively, resulting in the production of lincomycin D (Argoudelis et al., 1973b). However, *p*-aminosalicylic acid, which should compete as a substrate in tetrahydrofolic acid biosynthesis so that modified folic acids are formed, neither inhibited growth of *S. lincolnensis* nor affected the antibiotic profile. Also, folic acid antagonists, such as aminopterin or amethopterin, at concentrations that did not suppress the growth of *S. lincolnensis*, gave only lincomycin. Interestingly, *p*-nitrobenzoic acid in appropriate concentrations completely inhibited antibiotic production without affecting growth.

Lincomycin is formed from two distinct biosynthetic units, and the effect of adding each component to *S. lincolnensis* fermentations has been investigated by Argoudelis et al. (1965). No increase in lincomycin production was observed with supplements of the C_2 or C_3 proline units, but adding the sugar caused the additional production of the N-demethyl derivative, lincomycin D. The corresponding S-thioethyl compound, lincomycin K, was produced by adding ethylthiolincosamide, the aminosugar of lincomycin C.

Streptomyces caelestis can produce a variety of celestosaminide antibiotics differing from celesticetin in the acid group that forms an ester bond with the primary hydroxyl of desalicetin. Argoudelis and Coats (1974b) showed that [carboxyl-^{14}C]salicylic acid was efficiently incorporated into celesticetin, offering an opportunity for directed biosynthesis of new celestosaminide antibiotics. Thus the addition of *p*-aminosalicylic acid to the usual glucose–starch medium steered production to a new antibiotic, desalicetin 2″-(*p*-amino)salicylate.

Several actinomycetes transform lincomycin and clindamycin; lincomycin is completely converted by *Streptomyces rochei* to the inactive 3-phosphate ester in less than 48 h (Argoudelis and Coats, 1969). The same result was obtained with lincomycin B, C, and D as well as with celesticetin. Clindamycin is converted to its inactive 3-phosphate by both whole cells and cell-free extracts of *Streptomyces coelicolor* Müller (Coats and Argoudelis, 1971). However, clindamycin added to cultures of *Streptomyces punipalus* is converted within 24 h to the more powerful antibiotic N-demethylclindamycin (Argoudelis et al., 1969a). Interestingly, this antibiotic is also active against *Plasmodium berghei* infections in mice. The observation has prompted the chemical synthesis of antimalarials based on the clindamycin skeleton but containing a modified heterocyclic ring (Lewis and Birkenmyer, 1981). When clindamycin was added to *Streptomyces armentosus* cultures, the main product was a less potent derivative, clindamycin sulfoxide. Several streptomycetes and some fungi convert clindamycin to either the N-demethyl compound or the sulfoxide or a mixture of both (Argoudelis et al., 1969a).

G. Mutant Strains

Two mutant strains of *S. caelestis,* induced by exposure to N-methyl-N′-nitro-N-nitrosoguanidine (Argoudelis et al., 1972, 1973a), appear to have impaired transmethylation ability; strain 22227a produces 7-O-demethylcelesticetin, which lacks a methyl group in lincosamine, the aminosugar present in lincomycin antibiotics. Strain 22218a produces 7-O-demethylcelesticetin and, in addition, the N-demethyl compound.

III. ANTHRAMYCIN AND RELATED ANTIBIOTICS

The pyrrolo[1,4]benzodiazepine antibiotics are formed by condensing an anthranilate moiety with the appropriate proline unit. The three main members of the group are anthramycin, tomaymycin, and sibiromycin; recently, the three related antibiotics, neothramycins A and B and mazethramycin, have been reported (Fig. 13.4). Several recent reviews have dealt mainly with the biosynthesis of these antibiotics (Hurley, 1977, 1980; Hurley and Speedie, 1980).

Anthramycin was first isolated in the 1950's from the fermentation broth of a thermophilic actinomycete, *Streptomyces refuineus* var. *thermolactans* (Tendler and Korman, 1963). Its structure was reported later (Leimgruber et al., 1965a, b). Following this, tomaymycin was isolated by a Japanese group from *Streptomyces achromogenes* var. *tomaymyceticus* (Arima et

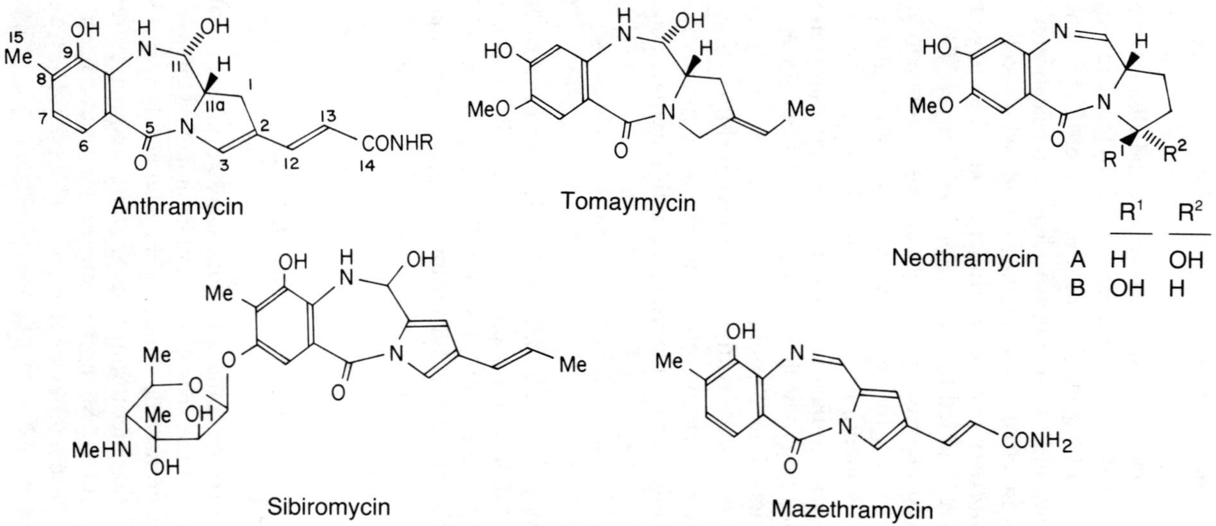

FIGURE 13.4 Structures of pyrrolo[1,4]benzodiazepine antibiotics.

al., 1972). The complete structure of this antibiotic and that of the related but inactive oxotomaymycin were deduced by Kariyone et al. (1971). Sibiromycin is a product of the actinomycete *Streptosporangium sibiricum* and was discovered by Gause et al. (1969) at the Moscow Institute for New Antibiotics. Details of its structure were disclosed later (Brazhnikova et al., 1972; Mesentsev et al., 1974). The isomeric and interconvertible neothramycins A and B (Takeuchi et al., 1976) and mazethramycin (Kunimoto et al., 1980) have also been isolated from the culture broths of actinomycetes.

These antibiotics differ mainly in their aromatic substitution patterns, the nature of the proline side chain, and the degree of unsaturation in the proline ring. Sibiromycin is unique in containing a sugar residue, sibirosaminide. In both anthramycin and tomaymycin, the absolute configuration at C-11 and C-11a is *R* and *S*, respectively. No information has been published for sibiromycin. It is worth noting that the Japanese workers who first described tomaymycin used the name to refer to the 11-methoxy compound, which, in fact, is formed only when the 11-hydroxy compound is heated in methanol. Since the names of the other antibiotics in the series refer to the 11-hydroxy forms, we will adopt the suggestion of Hurley (1977) and use the name tomaymycin for the 11-hydroxy compound.

A. Biological Activity

All the antibiotics in this group have antimicrobial, antiviral, and antitumor activity. In vitro, anthramycin has a wide antimicrobial spectrum in the dose range 0.1–5 μg/ml. However, in vivo it has no chemotherapeutic activity with bacterial, fungal, viral, and most protozoan and helminthic infections (Grunberg et al., 1966). Tomaymycin is mainly active against gram-positive bacteria but is weakly active against yeasts and fungi (Arima et al., 1972). Sibiromycin is also active against gram-positive bacteria (Brazhnikova et al., 1970). Both neothramycins A and B have weak antibacterial and antifungal activity in vitro (Takeuchi et al., 1976). Mazethramycin has more potent antibacterial activity (Kunimoto et al., 1980).

The antitumor activity of all pyrrolo(1,4)benzodiazepine antibiotics is most striking. Anthramycin and sibiromycin have activity against a variety of transplanted tumors. Tomaymycin has a marked inhibitory effect on leukemia L-1210 cells (Nishioka et al., 1972), and neothramycins A and B and mazethramycin have produced up to a 200% increase in the survival period of mice implanted with mouse leukemia L-1210 or Ehrlich ascites carcinoma cells (Takeuchi et al., 1976; Kunimoto et al., 1980).

Both anthramycin and sibiromycin have been used clinically. In trials with a small group of terminal cancer patients having solid tumors, lymphomas, or Hodgkins disease, about half responded favorably to treatment with pure anthramycin or fractions from the fermentation broth of *S. refuineus*. The tumors most responsive to the drug included gastrointestinal and breast neoplasms as well as lymphomas and sarcomas; ovarian and pancreatic carcinomas and malignant melanomas were the least responsive (Korman and Tendler, 1965; Hurley, 1977). Despite these encouraging results, clinical use of anthramycin has so far been severely restricted by its cardiotoxicity, though laboratory tests indicate that this may be alleviated by pretreatment with coenzyme Q_{10} or α-tocopherol (Lubawy et al., 1979, 1980).

B. Producing organisms

Streptomyces refuineus var. *thermotolerans,* which produces anthramycin, is a thermophilic organism; optimal growth and production occur in the temperature range from 40° to about 50°C (Berger et al., 1971). The other pyrrolo(1,4)benzodiazepine antibiotic producers all grow in the temperature range 27–32°C. Anthramycin has recently been isolated from *Streptomyces spadicogriseus* Komatsu (Komatsu et al., 1980). Tomaymycin is produced by *Streptomyces achromogenes* var. *tomaymyceticus* (Arima et al., 1972) and *Nocardia* sp. No. C-15003 (N-1 and N-37) (Tanida et al., 1980); sibiromycin by *Streptosporangium sibiricum* sp. *nova* (Gause et al., 1969); neothramycins from *Streptomyces* No. MC 916-C4 (Takeuchi et al., 1976); and mazethramycin from *Streptomyces thioluteus* ME 561-L4 (Kunimoto et al., 1980).

C. Fermentation

These antibiotics are produced by the usual methods of submerged culture, but few reports on fermentation conditions have been published. Most of these deal with the production of anthramycin by *S. refuineus* (Hurley and Speedie, 1980). Here, cultures grown in a complex seed medium for 24 h provided the cells that were used to inoculate complex or defined production media (Hurley et al., 1975; Gairola and Hurley, 1976). Anthramycin is produced during trophophase growth. In a defined (mannitol–sucrose–glycerol/ $NaNO_3$/salts) medium, production begins immediately, whereas in complex media, there is a lag of 8 or more hours followed by a surge in antibiotic synthesis during late growth (Hurley et al., 1975). Published production figures for anthramycin are low, in the range of 3–20 μg/ml. The titer reaches a maximum and then decreases, often quite markedly. The reason for this fall is unclear, but it may be due to changes in pH; incubation of exogenous anthramycin with filtered medium after various periods of fermentation indicates that antibiotic decomposition increases substantially in the period following the normal peak of anthramycin production.

Cultures of *S. refuineus* supplemented with methionine give significantly lower yields of anthramycin. This has been explained by interaction of anthramycin with a metabolic product of methionine, possibly SO_2 (Gairola and Hurley, 1976). When the three precursor amino acids, tryptophan, tyrosine, and methionine, were added to *S. refuineus* cultures at the start of incubation, antibiotic titres improved. A maximum threefold increase was obtained by raising the level of exogenous amino acids to 0.5 mM (Speedie and Forbes, 1977). It was subsequently discovered that tryptophan alone accounted for this observed increase. The addition of tryptophan did not alter the time course but increased the rate of antibiotic synthesis. Additions later in the fermentation had less effect, and the response was not strictly proportional to the tryptophan concentration at any given time. Methionine and tyrosine at concentrations greater than 0.5 mM suppressed the stimulatory effect of tryptophan. Recent studies on the regulation of tyrosine biosynthesis in *S. refuineus* using tyrosine auxotrophs showed that phenylalanine can feedback inhibit prephenate dehydratase, increasing NAD-dependent prephenate dehydrogenase activity and steering the metabolism of prephenic acid toward tyrosine (Speedie and Park, 1980).

All other antibiotics in this group have been produced in complex media; tomaymycin is formed near the end of the growth phase in yields of about 40 μg/ml (Arima et al., 1972; Hurley et al., 1976a). Sibiromycin production begins about 24 h after inoculation and peaks at about 100 μg/ml some 30 h later (Hurley et al., 1979).

The biologically inactive metabolite oxotomaymycin is produced along with tomaymycin in cultures of *S. achromogenes*. Labeling experiments have shown that tomaymycin is converted to oxotomaymycin by an intracellular enzyme that is present in both the trophophase and the idiophase (Hurley et al., 1976a). Whether this protects the cells from toxic effects of the antibiotic or is merely a fortuitous side reaction is not known. The neothramycin isomers are isolated from cultures in yields ranging from 80 to 800 μg/ml (Takeuchi et al., 1976).

D. Mode of Action

With DNA, these antibiotics form a nonintercalative labile covalent adduct that strongly inhibits DNA and RNA synthesis but has little or no effect on protein synthesis. The carbinolamine function of the antibiotics is essential for activity. Generally speaking, their reactivity with DNA is slow, shows a high specificity for deoxyguanosine in a double-stranded template, and is stable only as long as the secondary structure of DNA is preserved.

Although the mechanistic details are still debatable (Petrusek et al., 1981; Lown and Joshua, 1979), there is considerable evidence in the case of the anthramycin–DNA adduct that a covalent bond is established between C-11 of the drug and the 2-amino group of guanine as the drug lies deep within the narrow groove of DNA (Hurley and Petrusek, 1979). Detailed reviews of the mechanism of action of individual pyrrolo[1,4]benzodiazepine antibiotics are available (Kohn, 1975; Hurley, 1977; Krey, 1980; Gause and Dudnik, 1980).

E. Biosynthesis

Tryptophan, tyrosine, and methionine are the principal precursors of anthramycin, tomaymycin, and sibiromycin (Fig. 13.5). Biosynthetic studies with these three antibiotics indicate that the anthranilate portion of all the

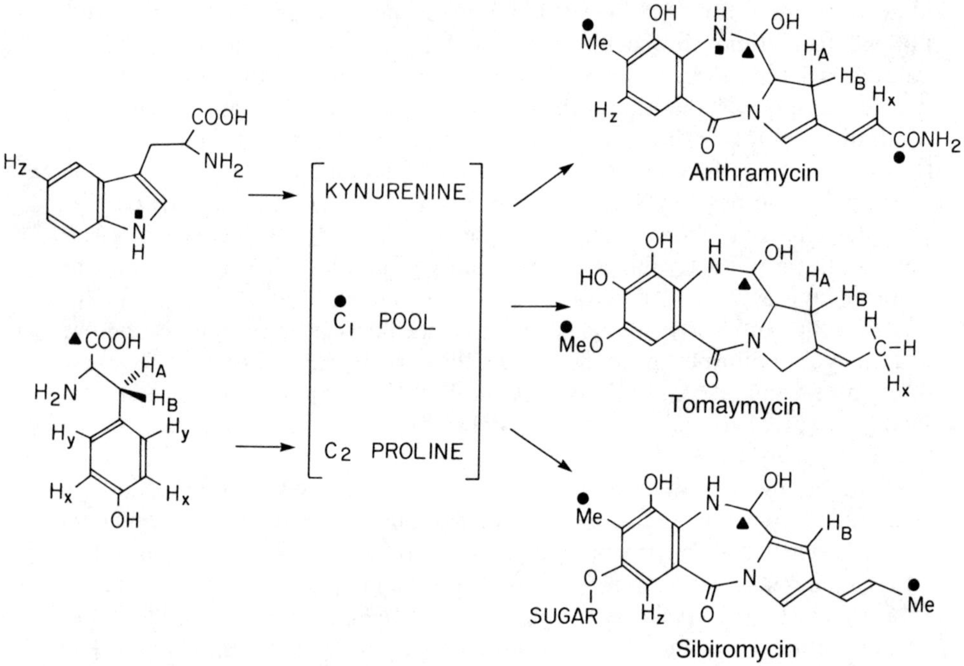

FIGURE 13.5 Biosynthetic precursors of the pyrrolo[1,4]benzodiazepine antibiotics.

antibiotics in this group is derived from tryptophan via the kynurenine pathway. Isotope from ^{15}N-labeled tryptophan was found exclusively at N-10 of anthramycin (Ostrander et al., 1980). Tritium from D,L-[5-^{3}H]tryptophan was completely retained in anthramycin (Hurley et al., 1976b) and almost completely retained (93%) at C-6 in sibiromycin (Hurley and Gairola, 1979). The latter result indicates that an NIH shift has occurred and that the hydroxylation substrate possesses a less activated para substituent such as an amide. Furthermore, the efficient incorporation of 3-hydroxy-4-methyl-[2-^{14}C]anthranilic acid into sibiromycin indicates that hydroxylation at C-9 and methylation at C-8 precede this NIH shift (Hurley and Gairola, 1979). On the other hand, there was very low retention of tritium in tomaymycin labeled from D,L-[5-^{3}H]tryptophan. This suggests that hydroxylation at C-8 precedes hydroxylation at C-7, since by NIH rules, hydroxylation ortho to an existing hydroxyl group would lead to loss of tritium (Hurley et al., 1976b). These data have led Hurley (1977) to propose the general route outlined in Figure 13.6 for biosynthesis of the anthranilate moieties of anthramycin, sibiromycin, and tomaymycin.

As was established by Witz et al. (1971) during their studies on lincomycin biosynthesis, the C_2 and C_3 proline units are derived from tyrosine. All seven carbon atoms of the C_2 proline moiety and seven of the eight carbon

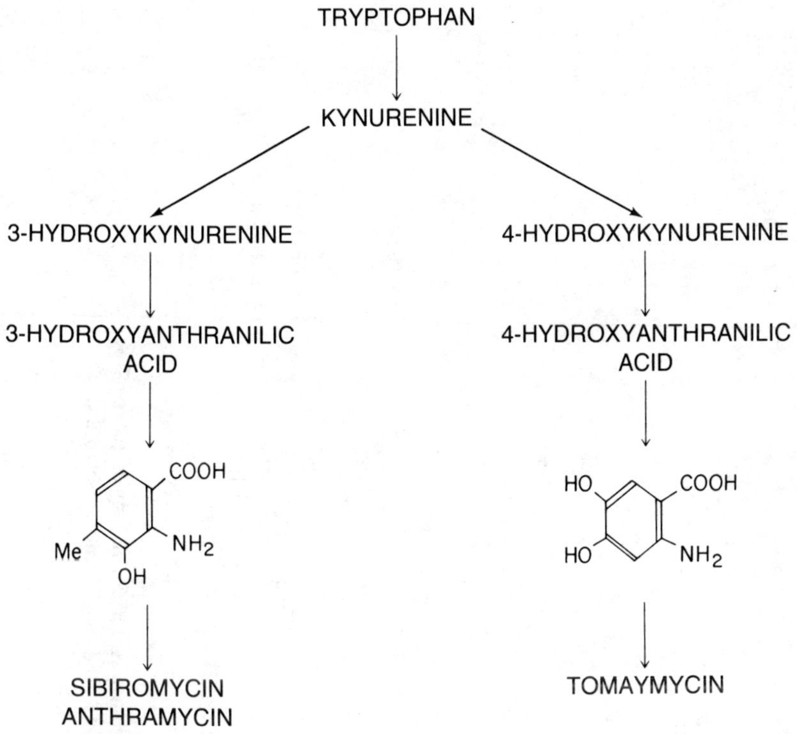

FIGURE 13.6 Proposed pathways from tryptophan to the anthranilate moieties of anthramycin, sibiromycin and tomaymycin.

atoms of the C_3 proline moieties in the pyrrolo(1,4)diazepine antibiotics come from the amino acid. The additional carbon is provided by the methyl group of methionine; ^{13}C-NMR analysis of anthramycin and sibiromycin enriched from [methyl-^{13}C]-methionine clearly showed that only the terminal carbon at position 14 (numbering according to Hurley (1980)) contained isotopes from the precursor (Hurley et al., 1975).

During incorporation of [3- or 5-^{3}H]tyrosine into anthramycin and tomaymycin, 50% of the tritium is lost. This, combined with the observation that fermentation broths of all three antibiotic-producing strains contain a bright yellow acidic compound with spectral characteristics of the aldehydo oxo acid (Fig. 13.7), has persuaded Hurley (1980) that proximal–distal cleavage of dopa, followed by ring closure (without rotation around C-1–C-3′) to form a Schiff base, is the pathway to these antibiotics. In the case of sibiromycin, all the tritium from [3- or 5-^{3}H]tyrosine is lost, but in the plausible expectation that a common overall pathway exists, Hurley (1980) has suggested that the tritium loss occurs at a later biosynthetic stage and is not due to a different cyclization process following ring cleavage.

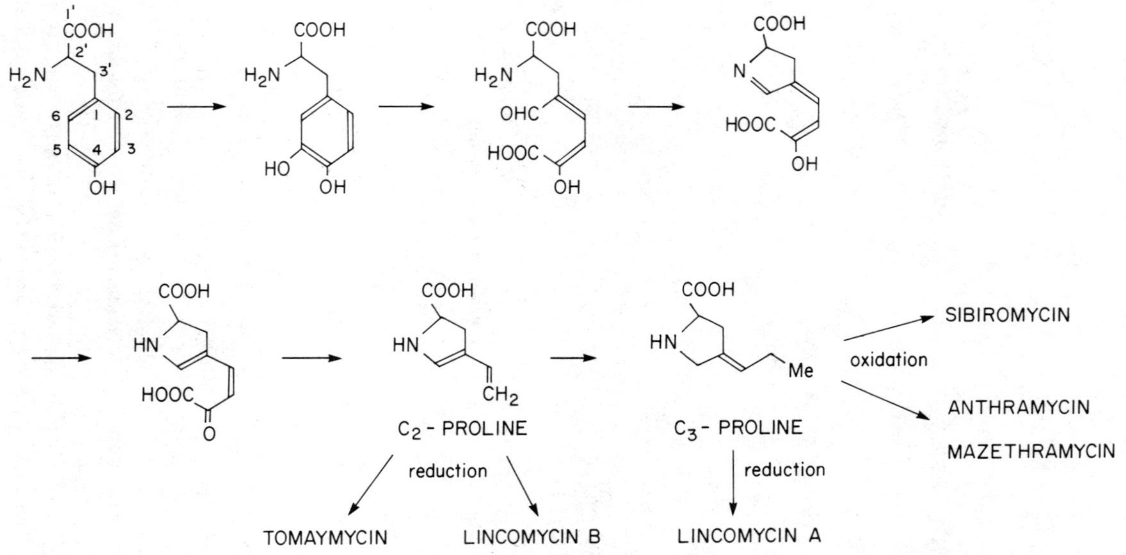

FIGURE 13.7 Proposed general pathway for formation of the proline moieties of lincomycins and pyrrolo[1,4]benzodiazepine antibiotics.

It follows from the biosynthetic details outlined in Figures 13.5 and 13.7 that tritium from D,L-[3′-*RS*-(3′-^{3}H)]tyrosine would label the hydrogen atoms at C-1 of these antibiotics. This indeed occurs with anthramycin and tomaymycin, and as was predicted, chemical conversion of labeled anthramycin to the didehydro derivative results in a 50% loss of isotope (Malhotra et al., 1980). In sibiromycin there is a stereospecific loss (66%) of the precursor 3′-(*S*)-hydrogen, presumably during formation of the C-1, C-11a double bond (Hurley et al., 1979).

A remaining puzzle concerns the fate of the C-2′ hydrogen of tyrosine. The C-2′ carbon finally appears at C-11a in the antibiotics, and although no proton retention is possible at this position in sibiromycin, it is theoretically possible in anthramycin and tomaymycin. Moreover, the stereochemistry at C-11a in the latter two antibiotics is the same as that at C-2′ of tyrosine. There is complete loss of the C-2′ hydrogen from L-[1-^{14}C; 2′-^{3}H]tyrosine during its utilization in antibiotic synthesis (Hurley et al., 1979). Experiments with ^{15}N-labeled tyrosine indicate that 70-80% of the isotope can be lost through transamination but this is not enough to account for the complete loss of tritium.

The biosynthesis of neothramycins is more straightforward; as usual, the anthranilate portion is derived directly from tryptophan via the kynurenine pathway, and the proline moiety, which contains no additional side chain, arises directly from proline (Miyamoto et al., 1978).

REFERENCES

Argoudelis, A. D., and Brodasky, T. F. (1972) *J. Antibiot.* 25, 194–196.

Argoudelis, A. D., and Coats, J. H. (1969) *J. Antibiot.* 22, 341–343.

Argoudelis, A. D., and Coats, J. H. (1973) U.S. Patent 3,726,766.

Argoudelis, A. D., and Coats, J. H. (1974a) U.S. Patent 3,812,014.

Argoudelis, A. D., and Coats, J. H. (1974b) *J. Antibiot.* 27, 674–676.

Argoudelis, A. D., and Mason, D. J. (1969) *J. Antibiot.* 22, 289–291.

Argoudelis, A. D., Fox, J. A., Mason, D. J., and Eble, T. E. (1964) *J. Amer. Chem. Soc.* 86, 5044–5045.

Argoudelis, A. D., Fox, J. A., and Eble, T. E. (1965) *Biochemistry* 4, 698–703.

Argoudelis, A. D., Coats, J. H., Mason, D. J., and Sebek, O. K. (1969a) *J. Antibiot.* 22, 309–314.

Argoudelis, A. D., Eble, T. E., Fox, J. A., and Mason, D. J. (1969b) *Biochemistry* 8, 3408–3411.

Argoudelis, A. D., Eble, T. E., and Mason, D. J. (1970) *J. Antibiot.* 23, 1–8.

Argoudelis, A. D., Coats, J. H., Lemaux, P. G., and Sebek, O. K. (1972a) *J. Antibiot.* 25, 445–455.

Argoudelis, A. D., Coats, J. H., and Pyke, T. R. (1972b) U.S. Patent 3,697,380.

Argoudelis, A. D., Coats, J. H., and Lemaux, P. G., and Sebek, O. K. (1973a) *J. Antibiot.* 26, 7–14.

Argoudelis, A. D., Johnson, L. E., and Pyke, T. R. (1973b) *J. Antibiot.* 26, 429–436.

Arima, K., Kohsaka, M., Tamura, G., Imanaka, H., and Sakai, H. (1972) *J. Antibiot. 24*, 437–444.

Berger, J., Karr, A. E., Leimgruber, W., Tabenkin, B., Schocher, A. J., and Stevanovic, V. (1971) U.S. Patent 3,619,374.

Bergy, M. E., Coats, J. H., and Malik, V. S. (1975) Belgium Patent 830,068.

Brazhnikova, M. G., Konstantinova, I. N., Mesentsev, A. S., Proshlyakova, V. V., and Tolstykh, I. V. (1970) *Antibiotiki 15*, 297–300.

Brazhnikova, M. G., Konstantinova, H. V., and Mesentsev, A. S. (1972) *J. Antibiot. 25*, 668–673.

Chang, F. N. (1979) in *Antibiotics* Vol. V (Part 1) (Hahn, F. E., ed.) pp. 127–134, Springer, New York.

Chang, F. N., and Weisblum, B. (1967) in *Antibiotics* Vol. I (Gottleib, D., and Shaw, P. D., eds.) pp. 440–445, Springer, New York.

Coats, J. H., and Argoudelis, A. D. (1971) *J. Bacteriol. 108*, 459–464.

DeBoer, C., Dietz, A., Wilkins, J. R., Lewis, C., and Savage, G. M. (1955) *Antibiotics Ann. 2*, 831–836.

DeHaan, R. M., Metzler, C. M., and Schellenberg, D. (1973) *J. Clin. Pharmacol. 13*, 190–209.

Eble, T. E. (1967) in *Antibiotics* Vol. I (Gottleib, D., and Shaw, P. D., eds.) pp. 353–358, Springer, New York.

Gairola, C., and Hurley, L. H. (1976) *Eur. J. Appl. Microbiol. 2*, 1129–1149.

Gause, G. F., and Dudnik, Y. V. (1980) *Antibiot. Chemother. 28*, 102–108.

Gause, G. F., Preobrazhenskaya, T. P., Ivanitskaya, L. P., and Sveshnikova, M. A. (1969) *Antibiotiki 14*, 963–969.

Gause, G. F., Lapchinskaya, O. A., Sveshnikova, M. A., Preobrazhenskaya, T. P., Ukholina, R. S., Nechaeva, N. P., Pogozheva, V. V., Korobkova, T. P., and Trenina, G. A. (1975) U.S.S.R. Patent 287,743.

Grunberg, E., Price, H. N., Titsworth, E., Buskid, G., and Tendler, M. D. (1966) *Chemotherapia 11*, 249–260.

Herr, R. R., and Bergy, M. E. (1963) *Antimicrob. Agents Chemother. 1962*, 560–564.

Hoeksema, H. (1964) *J. Amer. Chem. Soc. 86*, 4224–4225.

Hoeksema, H. (1968) *J. Amer. Chem. Soc. 90*, 755–757.

Hoeksema, H., and Hinman, J. W. (1964) *J. Amer. Chem. Soc. 86*, 4979–4982.

Hoeksema, H., Crum, G. F., and DeVries, W. H. (1955) *Antibiot. Ann. 2*, 837–841.

Hoeksema, H., Bannister, B., Birkenmeyer, R. D., Mackellar, F. A., Kagan, F., Magerlein, B. J., Schroeder, W., Slomp, G., and Herr, R. R. (1964) *J. Amer. Chem. Soc. 86*, 4223–4224.

Hurley, L. H. (1977) *J. Antibiot. 30*, 349–370.

Hurley, L. H. (1980) *Acc. Chem. Res. 13*, 263–269.

Hurley, L. H., and Gairola, C. (1979) *Antimicrob. Agents Chemother. 15*, 42–46.

Hurley, L. H., and Petrusek, R. (1979) *Nature* (London) *282*, 529–531.

Hurley, L. H., and Speedie, M. K. (1980) in *Antibiotics* Vol. IV (Corcoran, J. W., ed.) pp. 262–294, Springer, New York.

Hurley, L. H., Zmijewski, M., and Chang, C. J. (1975) *J. Amer. Chem. Soc. 97*, 4372–4378.

Hurley, L. H., Gairola, C., and Das, N. (1967a) *Biochemistry 15*, 3760–3769.

Hurley, L. H., Gairola, C., Das, N., and Zmijewski, M. (1976b) *Tetrahedron Lett.* 1419–1423.
Hurley, L. H., Lasswell, W. L., Ostrander, J., and Parry, R. (1979) *Biochemistry 18*, 4230–4237.
Kariyone, K., Yazawa, H., and Kohsaka, M. (1971) *Chem. Pharm. Bull. 19*, 2289–2293.
Kohn, K. W. (1975) in *Antibiotics* Vol. III (Corcoran, J. W., and Hahn, F. E. eds.) pp. 3–11, Springer, New York.
Komatsu, N., Kimura, K., Shizuko, A., and Kagitani, Y. (1980) *J. Antibiot. 33*, 54–60.
Korman, S., and Tendler, M. D. (1965) *J. New Drugs 5*, 275–285.
Krey, A. K. (1980) *Progress Mol. Subcellular Biol. 7*, 43–87.
Kunimoto, S., Masuda, T., Kanbayashi, N., Hamada, M., Naganawa, H., Miyamoto, M., Takeuchi, T., and Umezawa, H. (1980) *J. Antibiot. 33*, 665–667.
Leigh, D. A. (1981) *J. Antimicrob. Chemother. 7* (Suppl. A), 3–9.
Leimgruber, W., Batcho, A. D., and Schenker, F., (1965a) *J. Amer. Chem. Soc. 87*, 5793–5795.
Leimgruber, W., Batcho, A. D., Schenker, F., Karr, A., and Berger, J. (1965b) *J. Amer. Chem. Soc. 87*, 5791–5793.
Lewis, C., and Birkenmeyer, R. D. (1981) U.S. Patent 4,293,547.
Lown, J. L., and Joshua, A. V. (1979) *Biochem. Pharmacol. 28*, 2017–2026.
Lubawy, W. C., Whaley, J., and Hurley, L. H. (1979) *Res. Comm. Chem. Pathol. Pharmacol. 24*, 401–404.
Lubawy, W. C., Dallam, R. A., and Hurley, L. H. (1980) *J. Nat. Cancer Inst. 64*, 105–109.
Magerlein, B. J. (1971) *Adv. Appl. Microbiol. 14*, 185–229.
Magerlein, B. J., Birkenmeyer, R. D., Herr, R. R., and Kagan, F. (1967a) *J. Amer. Chem. Soc. 89*, 2459–2464.
Magerlein, B. J., Birkenmeyer, R. D., and Kagan, F. (1967b) *Antimicrob. Agents Chemother. 1966*, 727–736.
Malhotra, R. K., Ostrander, J. M., Hurley, L. H., McInnes, A. G., Smith, D. G., Walter, J. A., and Wright, J. L. C. (1980) *J. Nat. Prod. 44*, 38–44.
Mason, D. J., Dietz, A., and DeBoer, C. (1963) *Antimicrob. Agents Chemother. 1962*, 554–559.
Mesentsev, A. S., Kuljaeva, V. V., and Rubasheva, L. M. (1974) *J. Antibiot. 27*, 866–873.
Miyamoto, M., Sawa, T., Kondo, S., Takeuchi, T., and Umezawa, H. (1978) *J. Ferment. Technol. 56*, 329–333.
Nishioka, Y., Beppu, T., Kohsaka, M., and Arima, K. (1972) *J. Antibiot. 25*, 660–667.
Ostrander, J. M., Hurley, L. H., McInnes, A. G., Smith, D. G., Walter, J. A., and Wright, J. L. C. (1980) *J. Antibiot. 33*, 1167–1171.
Patterson, E. L., Hash, J. H., Lincks, M., Miller, P. A., and Bohonos, N. (1964) *Science 146*, 1691–1692.
Petrusek, R. L., Anderson, G. L., Garner, T. F., Fannin, Q. L., Kaplan, D. J., Zimmer, S. G., and Hurley, L. H. (1981) *Biochemistry 20*, 1111–1119.
Phillips, I. (1981) *J. Antimicrob. Chemother. 7* (Suppl. A), 11–18.
Reusser, F., and Argoudelis, A. D. (1974) U.S. Patent 3,833,475.

Schroeder, W., Bannister, B., and Hoeksema, H. (1967) *J. Amer. Chem. Soc. 89*, 2448–2453.
Shoeman, W. D., Wan, S. H., and Azarnoff, D. L. (1972) *Clin. Pharmacol. Therapeut. 13*, 704–709.
Slomp, G., and MacKellar, F. A. (1967) *J. Amer. Chem. Soc. 89*, 2454–2459.
Speedie, M. K., and Forbes, M. B. (1977) *Lloydia 40*, 612–613.
Speedie, M. K., and Park, M. O. (1980) *J. Antibiot. 33*, 579–584.
Takeuchi, T., Miyamoto, M., Ishizuka, M., Naganawa, H., Kondo, S., Hamada, M., and Umezawa, H. (1976) *J. Antibiot. 29*, 93–96.
Tanida, S., Haibara, K., and Asai, M. (1980) *J. Takeda Res. Lab. 39*, 138–139.
Tendler, M. D., and Korman, S. (1963) *Nature* (London) *199*, 501.
Thomas, R. (1965) in *Biogenesis of Antibiotic Substances* (Vaněk, Z., and Hošťálek, Z., eds.) pp. 155–169, Academic Press, New York.
Towers, N. R., Dixon, H., Kellermann, G. M., and Linnane, A. W. (1972) *Arch. Biochem. Biophys. 151*, 361–369.
Witz, D. F., Hessler, E. J., and Miller, T. L. (1971) *Biochemistry 10*, 1128–1133.

CHAPTER 14

The Aminocyclitol Glycosides (Aminoglycosides)

Julian E. Davies
Morimasa Yagisawa

I. INTRODUCTION

The aminocyclitol glycosides constitute a large group of related antibiotics, all containing aminosugar residues (Rinehart and Suami, 1980). The major groups and their producing organisms are listed in Table 14.1, and some representative structures are given in Fig. 14.1. There has been some confusion in naming the compounds; they are often given the collective name "aminoglycoside," which is, in fact, incorrect. While the 2-deoxystreptamine antibiotics (neomycins, kanamycins, etc.) and the streptamine antibiotics (streptomycins) are aminoglycosides, the actinamine antibiotics (spectinomycin) are not. To avoid further confusion and in the interests of accuracy in nomenclature, it would perhaps be better if the word "aminoglycoside" was quietly retired and only "aminocyclitol" used.

The biological activity of this group continues to excite interest, and additional members have been isolated with regularity (Table 14.1); this is in spite of the known ototoxicity and nephrotoxicity of these compounds. They remain one of the most potent groups of agents for nosocomial gram-negative infections (Phillips, 1982). The third- and fourth-generation ceph-

TABLE 14.1 Major Aminocyclitols and Producing Organisms

Structure[1]	*Generic Name*	*Trivial Name*	*Origin (first reported producer)*	*Discovery*
A1	Neomycin A	Neamine	*Streptomyces fradiae* ATCC 10745	1949
B1	Ribostamycin		*S. ribosidificus* ATCC 21294	1970
B2	Xylostasin		*Bacillus* sp. ATCC 21932	1974
B3	Butirosin A	Butirosin	*B. circulans* NRRL B-3312	1971
B4	Putirosin B			
C1	Neomycin B	Neomycin	*S. fradiae* ATCC 10745	1949
C2	Neomycin C			
C3	Paromomycin I	Paromomycin	*S. rimosus* f. *paromomycinus* NRRL 2455	1958
C4	Paromomycin II			
C5	Lividomycin A	Lividomycin	*S. lividus* ATCC 21178	1971
C6	Lividomycin B			
D1	Kanamycin A	Kanamycin	*S. kanamyceticus* ATCC 12853	1957
D2	Amikacin		Synthesis from kanamycin A	1972
D3	Kanamycin B	Bekanamycin	*S. kanamyceticus* ATCC 21259	1957
D4	Tobramycin		*S. tenebrarius* ATCC 17920	1967
D5	Dibekacin		Synthesis from kanamycin B	1971
D6	Kanamycin C		*S. kanamyceticus* ATCC 12853	1957
E1	Gentamicin C_{1a}	Gentamicin	*Micromonospora purpurea* NRRL 2953	1963
E2	Gentamicin C_2			
E3	Gentamicin C_1			
E4	Gentamicin C_{2b}	Micronomycin or Sagamycin	*M. sagamiensis* ATCC 21826	1974
E5	G-418		*M. grisea* NRRL 3800	1974
F1	Sisomicin		*M. inyoensis* NRRL 3292	1970
F2	Netilmicin		Synthesis from sisomicin	1976
F3	Verdamicin		*M. grisea* NRRL 3800	1975
F4	G-52		*M. zionensis* NRRL 5466	1974

TABLE 14.1 (Continued)

Structure[1]	*Generic Name*	*Trivial Name*	*Origin (first reported producer)*	*Discovery*
G1	Fortimicin A	Fortimicin	*M. olivoasterospora* ATCC 21819	1976
G2	Sporaricin A		*Saccharopolyspora hirsuta* subsp. *kobensis* ATCC 20501	1979
G3	Istamycin A		*Streptomyces tenjimariensis* SS 939	1979
G4	Istamycin B			
G5	Dactimicin		*Dactinosporangium matsuzakiensis* ATCC 31570	1979
H1	Streptomycin		*Streptomyces griseus*	1944
H2	Dihydrostreptomycin		*S. humidus* ATCC 12760	1957
I1	Spectinomycin		*S. spectabilis* NRRL 2494	1961
I2	Dihydrospectinomycin			1975
J1	Destomycin A		*S. rimofaciens* ATCC 21066	1965
J2	Hygromycin B		*S. hygroscopicus*	1958
K	Validamycin A	Validamycin	*S. hygroscopicus* subsp. *limoneus* ATCC 21431	1970
L	Apramycin		*S. tenebrarius* ATCC 17920	1968

[1] See Fig. 14.1.

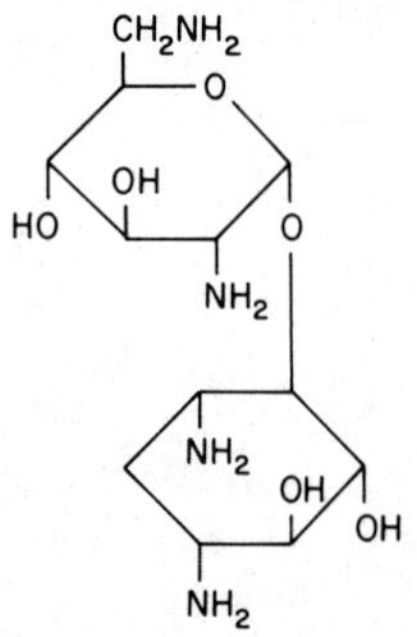

A. Neomycin A
(Neamine)

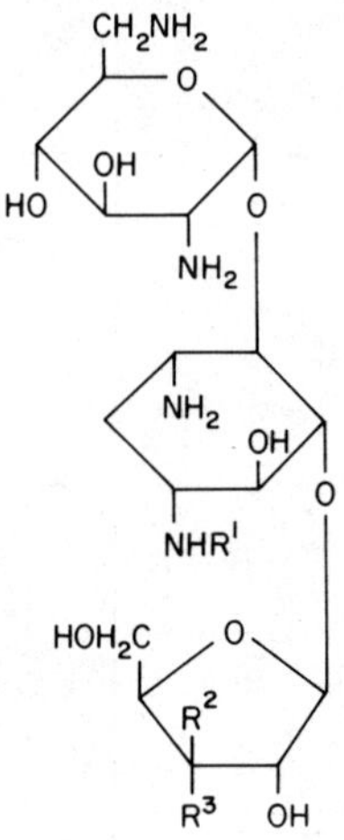

	R^1	R^2	R^3
B1. Ribostamycin	H	OH	H
B2. Xylostasin	H	H	OH
B3. Butirosin A	*	H	OH
B4. Butirosin B	*	OH	H

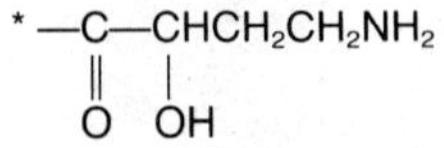

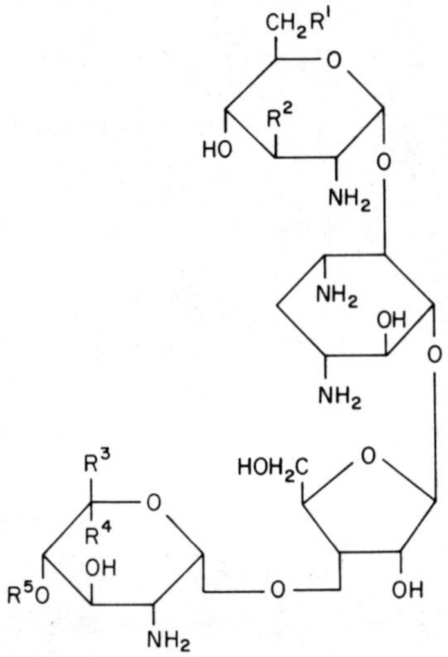

	R^1	R^2	R^3	R^4	R^5
C1. Neomycin B	NH_2	OH	H	CH_2NH_2	H
C2. Neomycin C	NH_2	OH	CH_2NH_2	H	H
C3. Paromomycin I	OH	OH	H	CH_2NH_2	H
C4. Paromomycin II	OH	OH	CH_2NH_2	H	H
C5. Lividomycin A	OH	H	CH_2NH_2	H	‡
C6. Lividomycin B	OH	H	CH_2NH_2	H	H

‡ Mannose

FIGURE 14.1 Structures of the major classes of aminocyclitols.

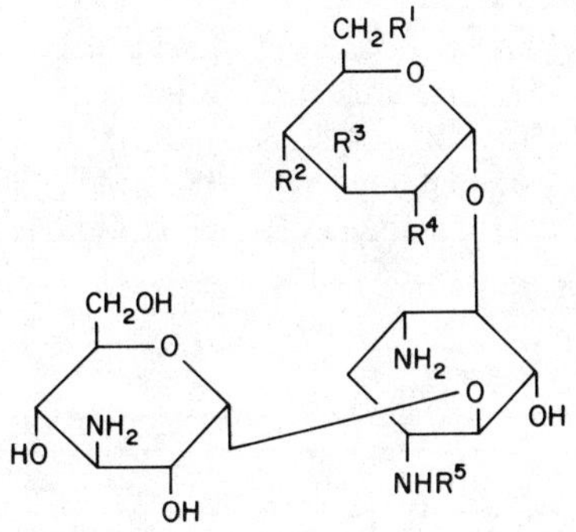

		R^1	R^2	R^3	R^4	R^5
D1.	Kanamycin A	NH_2	OH	OH	OH	H
D2.	Amikacin	NH_2	OH	OH	OH	*
D3.	Kanamycin B	NH_2	OH	OH	NH_2	H
D4.	Tobramycin	NH_2	OH	H	NH_2	H
D5.	Dibekacin	NH_2	H	H	NH_2	H
D6.	Kanamycin C	OH	OH	OH	NH_2	H

* $-C(=O)-CH(OH)CH_2CH_2NH_2$

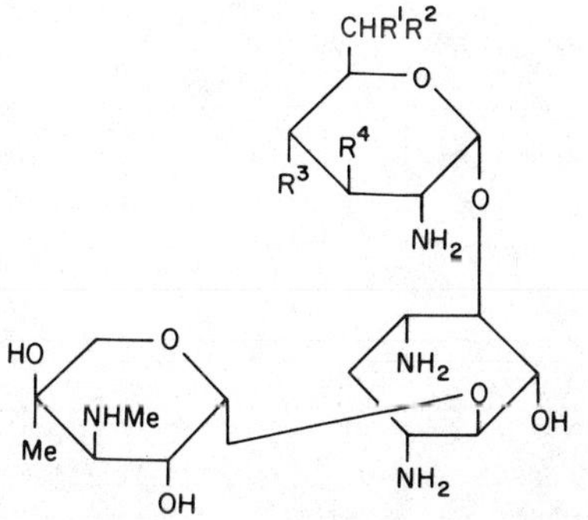

		R^1	R^2	R^3	R^4
E1.	Gentamicin C_{1a}	H	NH_2	H	H
E2.	Gentamicin C_2	Me	NH_2	H	H
E3.	Gentamicin C_{2b}	H	NHMe	H	H
E4.	Gentamicin C_1	Me	NHMe	H	H
E5.	G-418	Me	OH	OH	OH

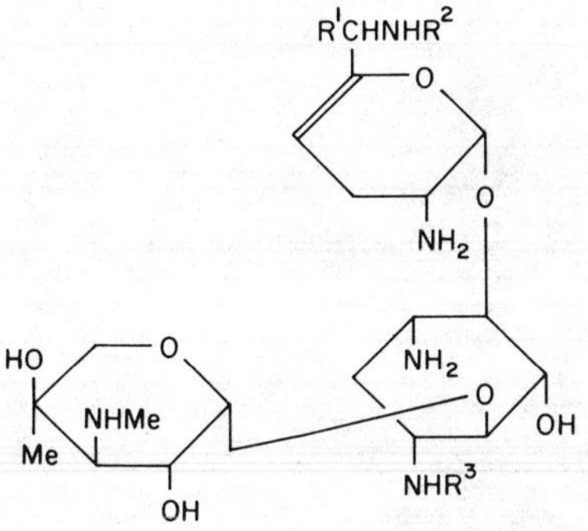

		R^1	R^2	R^3
F1.	Sisomicin	H	H	H
F2.	Netilmicin	H	H	Et
F3.	Verdamicin	Me	H	H
F4.	G-52	H	Me	N

FIGURE 14.1 (*continued*)

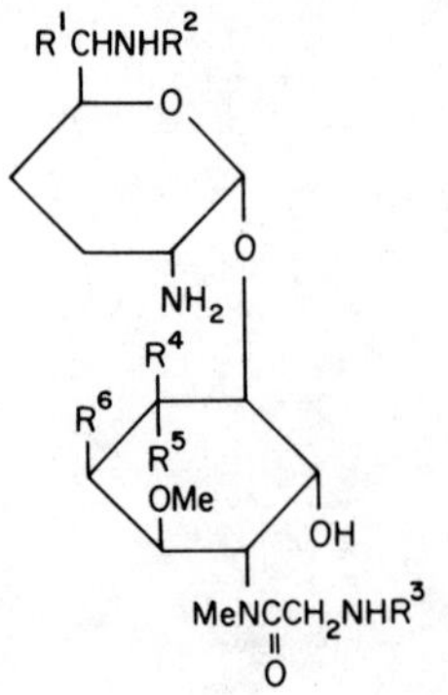

		R^1	R^2	R^3	R^4	R^5	R^6
G1.	Fortimicin A	Me	H	H	H	NH_2	OH
G2.	Sporaricin A	Me	H	H	NH_2	H	H
G3.	Istamycin A	H	Me	H	H	NH_2	H
G4.	Istamycin B	H	Me	H	NH_2	H	H
G5.	Dactimicin	Me	H	CH=NH	H	NH_2	OH

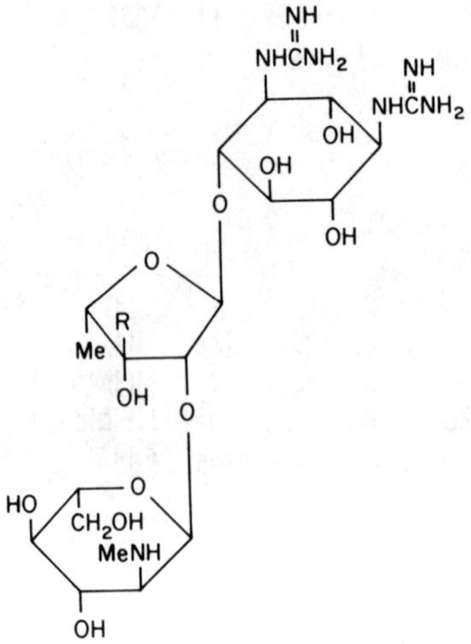

		R
H1.	Streptomycin	CHO
H2.	Dihydrostreptomycin	CH_2OH

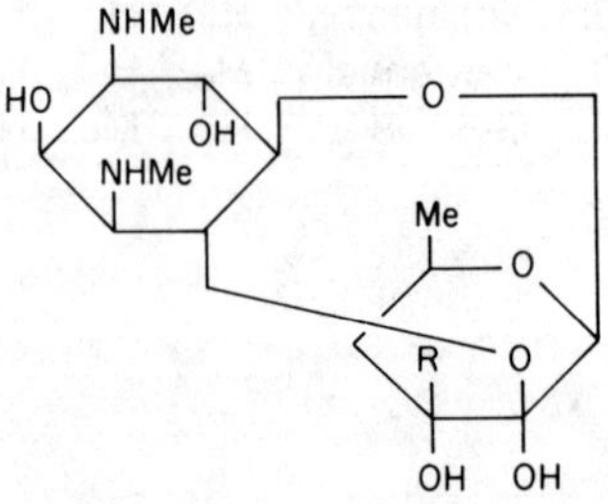

		R
I1.	Spectinomycin	OH
I2.	Dihydrospectinomycin	H

FIGURE 14.1 *(continued)*

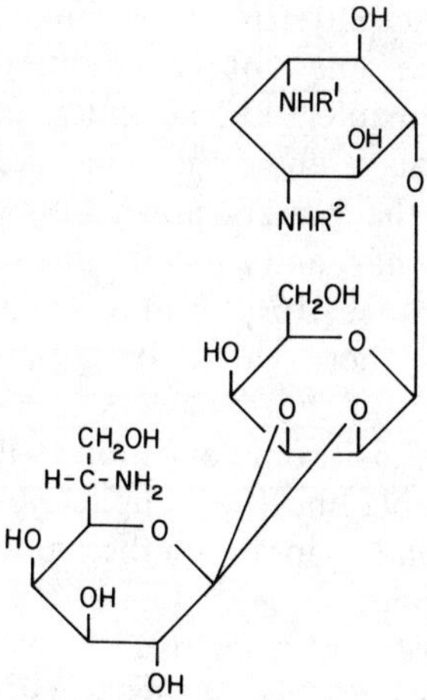

		R^1	R^2
J1.	Destomycin A	H	Me
J2.	Hygromycin B	Me	H

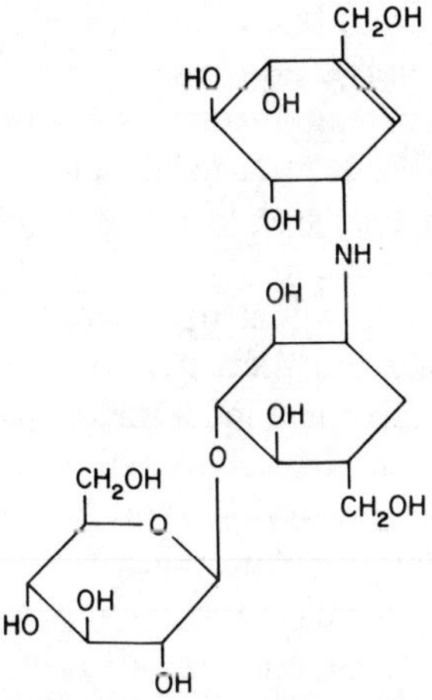

K. Validamycin A

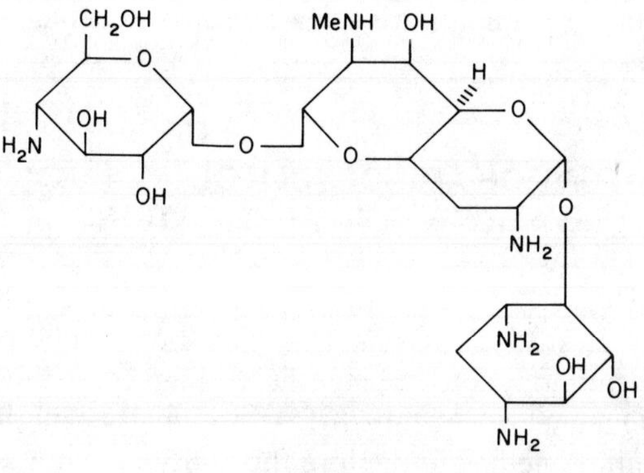

L. Apramycin

FIGURE 14.1 ***(continued)***

alosporins are making inroads into the use patterns of the aminocyclitols; but, provided they are used with proper awareness of the toxicity problems, the aminocyclitols will continue to be a sensible and economical alternative to the cephalosporins (or even remain the drug of choice for some indications).

The chemistry and biosynthesis of the better-known aminocyclitols have been reviewed with authority by Rinehart and his colleagues in several recent publications (Rinehart and Stroshane, 1976; Pearce and Rinehart, 1981), and repeating this source material here would be superfluous. It should be consulted if a detailed treatment of these aspects is needed; our aim is to present more recent work and to raise some questions with respect to biosynthetic studies of the aminocyclitols and their applications.

It is worth noting that only the macrolides approach the aminocyclitols in the complexity and diffuseness of the biological and biosynthetic problems they present for study. The aminocyclitols are composed of a variety of identical or related sugar molecules, with a variety of positional variations and secondary modifications (C-methylation, N-methylation, etc.). They are produced by a wide variety of different bacterial genera (*Streptomyces, Micromonospora, Bacillus, Pseudomonas*), and it is reasonable to expect that many common functions and genes related to aminocyclitol biosynthesis would be distributed among these different microorganisms. The degree of commonality is particulary evident in the case of the modifying enzymes; however, the entire range of interwoven functions provides interesting fodder for the researcher who undertakes a biological investigation of antibiotic biosynthetic pathways, their evolution, and their genetic distribution. Are 2-deoxystreptamine-producing enzymes identical in all microorganisms making this unusual aminocyclitol? What genetic relationships exist between 2-deoxystreptamine, streptamine, and actinamine producers?

Without discussing the point in detail, we note that relationships between structure and toxicity in the aminocyclitols have been virtually ignored, with the consequence that the utility of these compounds remains threatened by properties that are unknown in structural terms and for which not even empirical attempts have been made at solutions. This is undoubtedly due to the difficulty in obtaining quantitative assays for ototoxicity and nephrotoxicity.

II. METHODS FOR THE STUDY OF AMINOCYCLITOL BIOSYNTHESIS: RESULTS, VALIDITY, AND APPLICATIONS

A. Nonproducing Mutants

One classic approach to pathway analysis that has been only partially successful in elucidating aminocyclitol biosynthesis has been the use of nonproducing mutants (Queener et al., 1978). Although this method has provided some useful results, one has to deplore the lack of genetic analysis of the mutants. Thus we have no idea if the strains used have single or multiple mutations, whether the mutations are in structural genes or in regulatory genes, and whether or not the mutant genes are clustered; mutagenesis has often involved the use of such powerful agents as nitrosoguanidine, so the probability of multiple hits on the chromosome is very high (Singer and Kusmierek, 1982). This sorry situation places us in a position where mutants are available and are known to do useful things but no interrelationships are known. A partial list of some of the useful mutants is given in Table 14.2. It is easy to find fault with these studies. Almost no cross-feeding tests have been done with the mutants, which means that valuable information on the possible existence of complementation groups has not been obtained. No reversion studies have been reported; thus one does not know if point mutants or deletions may be involved. In addition, the interesting and potentially useful temperature-conditional class of mutant has been ignored completely. Another class of mutants derived by treatment with plasmid-curing agents has been misunderstood.

In spite of these deficiencies, nonproducing mutants of aminocyclitol-producing organisms have been extremely useful in biosynthetic studies. By precursor feeding, or mutasynthesis, a number of probable precursors have been identified, and useful antibiotic derivatives (mutasynthons) have been obtained (Rinehart, 1977). Unfortunately these compounds are made in relatively low yield, and this approach is not generally suitable for producing new compounds. However, it allows the identification and testing of interesting derivatives that can then be obtained by more practical methods. As an example, we can note the 5-*epi* derivatives of the gentamicin antibiotics, which were first obtained by feeding a blocked mutant with 5-*epi*-streptamine (Daniels et al., 1980) but are now prepared chemically (Daniels, 1977). Other such derivatives are known and may prove to be useful; thus discovery of the first mutasynthons, the aminocyclitol hybrimycins, by Shier et al. (1969) has opened up a profitable field of study applicable to all antibiotics.

The main criticism of the mutasynthesis approach is that the mutants are totally uncharacterized and may often be multiple; perhaps most unfortunately, they have not been characterized biochemically, so the approach is entirely empirical. One interesting observation can be made from aminocyclitol mutasynthesis studies: it is somewhat surprising that so many of the mutants respond to feeding by 2-deoxystreptamine (DOS) or its derivatives. Does this mean that DOS lies at a key regulatory step in the biosynthetic pathway, that DOS$^-$ mutants are the only ones that were screened, or that DOS is the only intermediate that gets into the cell easily? Many such questions have remained unanswered in the mutasynthetic approach. Most studies have taken a DOS$^-$ mutant and used this for subsequent feeding experiments. As far as we know, a large batch of nonproducing mutants has never been screened with a variety of possible precursors. While this should be done, it might also be profitable to take a DOS$^-$ mutant and remutagenize it, aiming to improve its properties with respect to DOS uptake and the assimilation of other probable precursor molecules.

TABLE 14.2 Some Mutants of Aminocyclitol-Producing Organisms

Mutant Strain	*Mutation*	*Application*
Structures A–C6[1]		
Streptomyces fradiae 3535X M68 ATCC 21401	2-Deoxystreptamine(DOS) idiotroph	Hybrimycin A and B production
S. fradiae N4C2	Final addition of neosamine	Ribostamycin production
S. ribosidificus AF-1	DOSidiotroph	Ribostamycin analogs from neamine & DOS
S. rimosus f. *paromomycinus* UC57 A67M17	Paromomycin low producer	Neomycin production by interspecific fusion
S. rimosus f. *paromomycinus* r24s ATCC 21484	DOSidiotroph	Hybrimycin C production
Bacillus circulans 301-2B	Glucose-resistant, high-temp (37°C)-resistant	Butirosin production
B. circulans S-11	DOSidiotroph	Biosynthesis of DOS
B. circulans B15M305	N^1-Acylation	Xylostasin production
B. circulans B15M306	N^1-Acylation	Ribostamycin production
B. circulans MCRL5004	Neamine$^-$ (D-glu ⇸ DOS)	Butirosin analogs
B. circulans MCRL5003	Neamine$^-$, DOS$^-$ (DOS ⇸ paromamine)	Butirosin analogs
Structure D1[1]		
S. kanamyceticus 12-48 ATCC 21257	3″-Amination	NK-1001 production
S. kanamyceticus 21-16	3″-Amination	NK-1012-1 production
S. kanamyceticus K-8 ATCC 21486	DOSidiotroph	Kanamycin analogs
Structures E1–F4[1]		
Micromonospora purpurea VIb-3P ATCC 31164	DOSidiotroph	Gentamicin analogs

TABLE 14.2 (Continued)

Mutant Strain	*Mutation*	*Application*
M. purpurea Paro346	Final addition of garosamine	Gentamicin C pathway
M. sagamiensis KY11533	C-3″ *N*-Methylation	Elucidation of biosynthetic pathway
M. sagamiensis KY11538	C-6′ *C*-Methylation	
M. sagamiensis KY11565	C-6′ Amino substitution	
M. sagamiensis KY11535	4′, 5′-Hydrogenation	
M. sagamiensis KY11509	C-1 Dehydrogenation	
M. sagamiensis G-726	C-1 & C-3 Amino transfer	
M. inyoensis 1550F	DOSidiotroph	Mutamicin production
M. sp. K-6993 Y41 ATCC 31348	Gentamicin nonproducer	Combimicin production
Structures H1 and H2[1]		
S. griseus SC2376 ATCC 27001	Phage-resistance, high mannosidostreptomycinase	Streptomycin hyperproducer
S. griseus MIT-A5 ATCC 31087	Streptidine idiotroph	Streptomutacin A
S. griseus FT-1	A-factor requirement	Regulation of streptomycin production
S. bikiniensis IFO13350 (A-factor$^-$)		
Structure L[1]		
S. tenebrarius No. 13	Nonproducer of other nebramycin factors	Apramycin production

1. See Fig. 14.1

B. Plasmid Curing

Another red herring surfaced when it was discovered that DOS^- strains could be obtained by treatment of producing organisms with plasmid-curing agents (Yagisawa et al., 1978). The suggestion was made that plasmids may be intimately involved in aminocyclitol biosynthesis and that a new class of DOS^- variants might be accessible by this route. It is surprising that such a high percentage of DOS^- variants were obtained in this way, and they may well prove useful in biosynthetic studies; however, the role of plasmids in aminocyclitol biosynthesis is by no means clear. A variety of plasmids may play a (presently) ill-defined regulatory role in antibiotic production, but it is now apparent that the effect of plasmid-curing conditions is to produce mutants, probably arising by enhanced recombination across repeated sequences or by induced transposition of these sequences (Hopwood, 1978; Davies et al., 1982). The so-called plasmid-cured variants are yet another class of aminocyclitol nonproducing mutants that deserve more intensive biochemical and genetic study. Whether or not they are altered in structural genes or in regulatory functions has yet to be determined.

C. In Vitro Studies

In studies on antibiotic biosynthesis, an approach that interfaces nicely with mutant analysis is the use of cell-free systems. This type of study has not yet received enough attention in the aminocyclitol series. Rinehart has taken the lead in clarifying the early steps in neomycin biosynthesis, and he is attempting to demonstrate some of the intermediates between nucleotidylated glucose precursors and neomycin. The enzyme reactions that lead to formation of the aminocyclitol moieties in bluensomycin and streptomycin have been explored in depth by J. B. Walker (1975) and parallel experiments to investigate steps in the biosynthesis of streptomycin have been undertaken by Kniep and Grisebach (1980). Such analyses, combined with the use of well-characterized mutants, could prove especially valuable in the future, when the role of intermediates or enzymes in regulating aminocyclitol biosynthesis is studied.

D. Labeled Precursors

By far the most information characterizing aminocyclitol biosynthetic pathways has come from studies of the metabolic fate of labeled intermediates. This approach has been applied successfully by J. B. Walker and M. S. Walker (1969) and by Grisebach (1978) to streptomycin, and by a variety of other workers to the 2-deoxystreptamine antibiotics. Although ^{14}C-labeling studies were used formerly, they have been replaced by the more efficient and revealing use of ^{13}C, which is amenable to NMR analysis (Rinehart et al., 1974). Some of the pathways indicated by such analyses are outlined in Fig. 14.2; although ^{13}C NMR analyses provide more information than

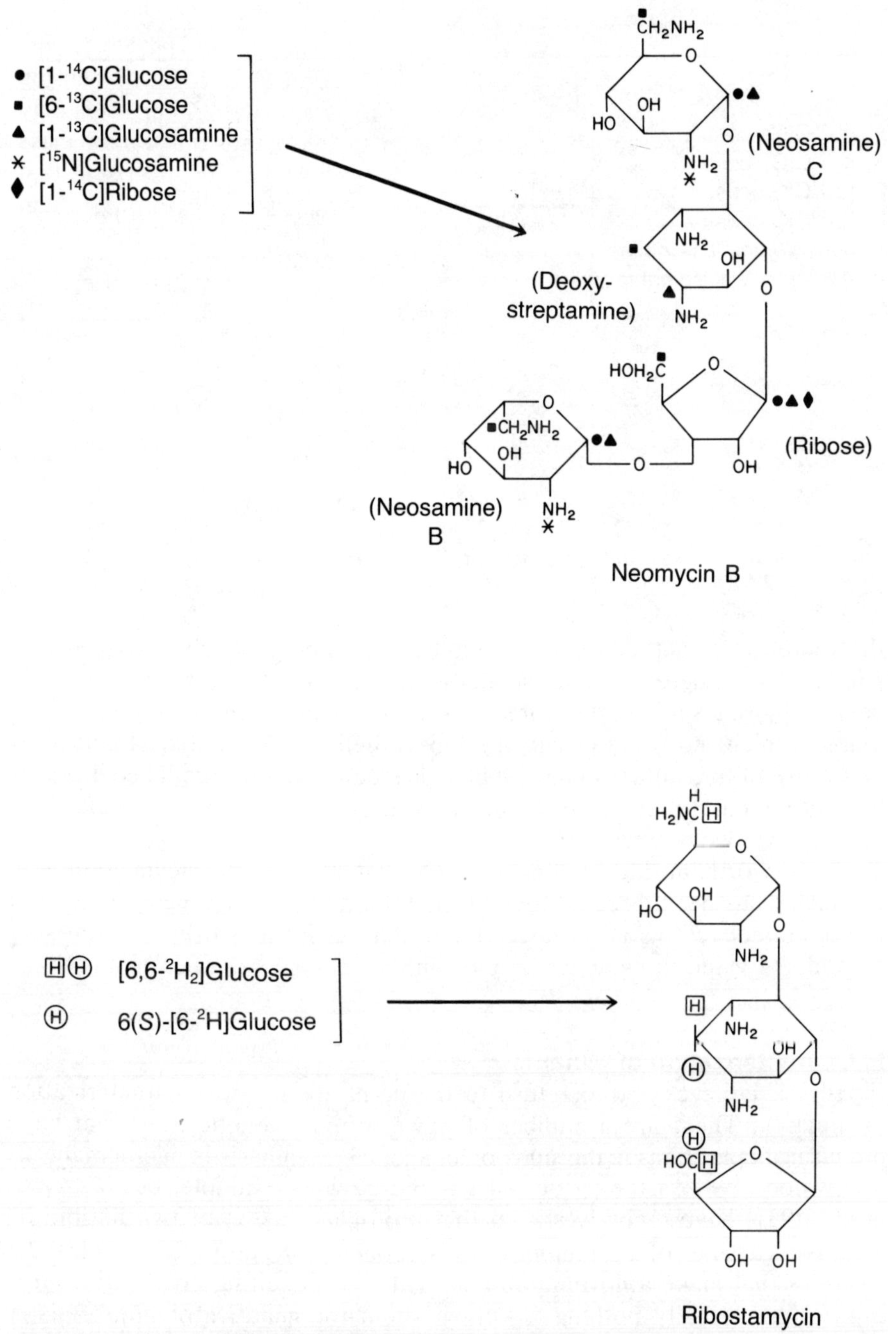

FIGURE 14.2 Labelling of representative aminocyclitols by isotopically enriched precursors.

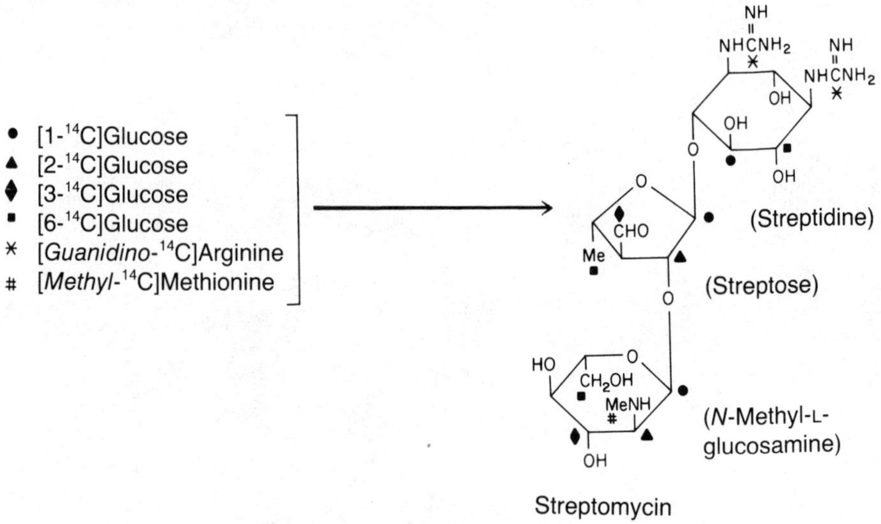

FIGURE 14.2 *(continued)*

the radiotracer methods used previously, a number of key intermediates still have to be detected (and isolated), and it is not clear that mutants have been explored sufficiently. Closer examination of a few commonly used mutants might not only clarify the biosynthetic pathway, but also lead to better use of the mutant in mutasynthesis studies. Deuterium-labeled sugars have proved useful in studies of the biosynthesis of ribostamycin (Kakinuma et al., 1981; Kakinuma, 1982).

It is worth noting at this juncture that some very clever synthetic chemistry has been directed toward the total or partial synthesis of several of the aminocyclitols (Umezawa, 1980). Although this work has been successful, its value in establishing biosynthetic routes has been minimal.

E. Characterization of Mutants

Finally, it is necessary to return to the point about poor characterization of mutants. There are a number of powerful new genetic tools that have promising applications in the study of antibiotic biosynthesis in microorganisms (Hopwood, 1981). Cloning of DNA is the obvious example, but Mazieres et al. (1981) have recently shown that protoplast fusion of two antibiotic-negative variants of an aminocyclitol-producing streptomycete yields derivatives that make a *novel* antibiotic. This is an exciting observation; it is a pity that virtually nothing is known about the genetics or biochemistry of the parent nonproducing organisms.

It is highly likely that fusion is going to be more and more useful in the production of new and modified antibiotics and in obtaining increased yields. However, if the parent strains are not characterized, we can draw no useful predictive conclusions for future work. Are the altered properties of the "fusants" due to biochemical alterations, to regulatory changes, or simply to the release from repression of a "hidden" biosynthetic pathway? An empirical approach may provide results, but only rarely does it generate useful predictions, so this kind of blind searching is not going to be very helpful as we move to apply newer methods of gene manipulation.

III. REGULATION OF THE EXPRESSION OF AMINOCYCLITOL BIOSYNTHESIS

A. Environmental Effects

This is another grey area. What little work has been done is difficult to interpret, mainly because pleiotropic effects may follow the imposition of environmental changes. For example, carbon and phosphate starvation, which are known to affect the production of some aminocyclitols (Table 14.3), are totally nonspecific, since the synthesis of many antibiotics and metabolic products are drastically affected by such catabolic variations (Martin and Demain, 1980). One problem in this respect, which is well recognized, is that the biosynthesis of antibiotics involves a carefully modulated interaction between primary and secondary metabolism. Starvation conditions (including late logarithmic and stationary phases of growth) involve shifts in metabolism and degradation that must be channelled for the benefit of cell survival; only when we isolate overproducing mutants under extreme metabolic conditions will we begin to understand the control of antibiotic biosynthesis. Once again, the past neglect in analyzing aminocyclitol non-producing mutants for regulatory effects leaves us in a position where even speculation is useless. It is not known whether the mutants affect transcription (gene expression by promoter or repressor/activator alterations), translation (changes in enzyme activity due to control of mRNA processing), or post-translational events (modification of enzymes by proteases or by phosphorylation, etc., or feedback inhibition of enzyme activity). It is quite clear that

TABLE 14.3 Carbon Catabolite Effects on Aminocyclitol Production with Glucose as the Carbon Source

Strain	*Antibiotic Produced*	*Effect*
S. kanamyceticus K133-28	Kanamycin	Represses *N*-acetylkanamycin amidohydrolase
S. griseus	Streptomycin	Represses mannosidostreptomycinase
S. griseus ATCC 12475	Streptomycin	Represses streptomycin phosphatase
S. fradiae ATCC 10745	Neomycin	Represses alkaline phosphatase
B. circulans 301	Butirosins	Catabolite represses

any of these activities can influence both primary and secondary metabolism (or indeed product secretion); until mutants are analyzed directly or by recombinant DNA approaches, our knowledge and application of this subject will remain, at best, empirical.

This is especially true in the case of the aminocyclitol antibiotics, because resistance to these antibiotics in the producing organism almost always involves antibiotic modification by *O*-phosphorylation or *N*-acetylation (Davies et al., 1978). Since the other substrates for these reactions are acetyl-coenzyme A and adenosine triphosphate, it is not surprising that variations in carbon metabolites and phosphate should affect aminocyclitol biosynthesis; the producing cell presumably coordinates antibiotic synthesis and defense mechanisms to guarantee survival. In another example, we take note of the evidence that aminocyclitols are made in the cell in a protected form (*O*-phosphate or *N*-acetate). The protecting group is cleaved during export of the antibiotic (but whether this cleavage is essential for export is another complication). As shown by M. Walker and J. B. Walker (1970) and by others (Nimi et al., 1981), streptomycin is made in the cell as streptomycin-6-phosphate and cleaved by a phosphatase on secretion so that the main medium component is unmodified streptomycin. The same deprotection mechanism (probably allied to secretion) may be expected for other aminocyclitols (Satoh et al., 1976). One can easily imagine that changes in essential metabolites (especially phosphate) will affect the kinetic constants for these deblocking enzymes and cause corresponding increases or decreases in apparent antibiotic yield.

B. Mutants

To support the examination of laboratory-derived mutants referred to earlier, a study of natural variants of antibiotic-producing microorganisms may prove educational. For example, there are a number of different species of streptomycetes that produce streptomycin and many that can produce neomycin. Assuming that the biosynthetic pathways are the same in the different producers, one can ask why different species produce different amounts of the antibiotics. Can useful regulatory information be obtained in this way?

We may have been overly pessimistic in this appraisal of aminocyclitol biosynthesis and its regulation. The pathways are largely known (although not in detail), and many potentially interesting mutants have been isolated. There also exist gene transfer systems for a number of aminocyclitol producers (either transformation, conjugation, or transduction). Thus the tools are available, and much useful knowledge awaits discovery and interpretation; the study of antibiotic production must pass from commercial to academic exploitation if the necessary information is to become available. The end results will surely be of biological interest and commercial application.

Considering that most "strain improvement" procedures used by industry for many years involved the isolation of mutant producers that make substantially more antibiotic, it is somewhat depressing to realize that we know so little of what has happend in these mutations, many of which must involve control steps.

C. Regulatory Factors

To complete this section on aminocyclitol biosynthesis, we must mention two unusual findings concerning the effects of small molecules on the production of aminocyclitols. In 1967, Khokhlov and his colleagues first described the isolation of a so-called "A-factor," subsequently identified as 2(*S*)-isocapryloyl-3(*S*)-hydroxymethyl-γ-butyrolactone (Fig. 14.3). This compound is excreted by streptomycin-producing strains of *Streptomyces griseus* and *Streptomyces bikiniensis* (Khokhlov et al., 1967). If concentrated and returned to the medium, it stimulates both streptomycin production and spore formation. It was reported that A-factor stimulates NADP glycohydrolase and NADP-linked glucose-6-phosphate dehydrogenase. On this basis, these enzymes were assumed to take part in regulating streptomycin biosynthesis (Voronina et al., 1978). Hara and Beppu (1982a, b) isolated *S. griseus* mutants that were deficient in streptomycin and A-factor production (not all streptomycin nonproducing mutants are A-factor–deficient—many are blocked in the antibiotic biosynthetic pathway). When A-factor was added back to the streptomycin nonproducing, A-factor–negative mutants, streptomycin production could be restored, but there was no significant change in activity of the two enzymes studied by Voronina et al.

Hara and Beppu (1982b) noted that A-factor–responsive nonproducers of *S. griseus* were restored in their levels of streptomycin resistance as well as in streptomycin production. Resistance could be correlated with an increase in streptomycin-6-phosphotranferase activity. Hara and Beppu (1982a), along with others, have produced evidence that genetic control of A-factor lies on a plasmid or some readily mutable element, but no convincing genetics has yet been done. We are left with the paradox that A-factor has a number of pleiotropic effects—on streptomycin biosynthesis, streptomycin resistance, and spore formation. Does A-factor have direct or secondary

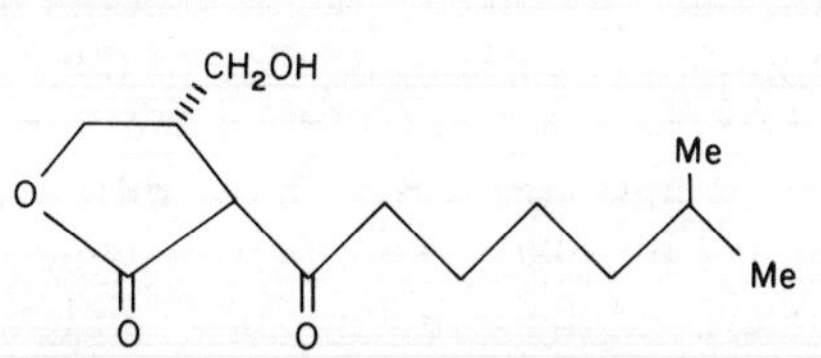

FIGURE 14.3 The structure of A-factor.

effects, and at what level do they occur? Is it a true regulatory effector? Since *S. griseus* produces not only streptomycin, responsive to the factor, but a variety of antibiotics including the antifungal cycloheximide, yet another question can be asked: does A-factor have any effect on the production of these other antibiotics? The question is pertinent because A-factor is now known to be produced by a number of streptomycetes that are not streptomycin producers.

Another somewhat unusual and even less well characterized effect is that of sclerin, a substance that stimulates oxidative phosphorylation in microorganisms, plants, and animals. When sclerin was added to cultures of *S. griseus, Streptomyces ribosidificus,* and *Streptomyces kanamyceticus,* production of streptomycin, ribostamycin, and kanamycin, respectively, was stimulated about twofold (Satoh et al., 1975). Sclerin also increased the levels of some of the enzymes known to be associated with resistance to these antibiotics. These unusual experiments have (apparently) not been continued, and the results presented do not lend themselves to any simple interpretation.

IV. BIOCHEMISTRY OF AMINOCYCLITOLS

A. Mode of Action

The mechanisms of action of aminocyclitols have been studied for more than 20 years. Although many different effects have been reliably reported, there now seems little reason to doubt that the primary effect of these drugs is on the ribosome, resulting in inhibition of protein synthesis. In spite of this consensus, the detailed biochemical mechanism by which protein synthesis is inhibited and the relationship to a variety of other effects on the cell (including death) remain incompletely characterized. The best review of the current state of affairs has been assembled by Cundliffe (1981). It should be noted that aminocyclitols vary in their effects on translation; thus kasugamycin and spectinomycin have somewhat different modes of action from gentamicin or streptomycin, which, in turn, differ substantially from one another in their biochemical effects. Some of the aminocyclitols cause extensive mistranslation in vitro and are bacteriocidal (streptomycin, gentamicin), while others do not derange the decoding mechanism and are not bacteriocidal antibiotics (kasugamycin, spectinomycin). Between these two extremes are numerous variations. One of the clearest indications that such differences exist comes from the recognition that members of the aminocyclitol group of antibiotics may have different binding sites on the ribosome (Davis and Tai, 1980). This fact became clear when target site mutants were characterized; these mutants have undergone specific alterations of the ribosome affecting the primary structure of the ribosomal RNA or ribosomal proteins (Table 14.4); in general, no broad cross-resistance patterns were detected.

TABLE 14.4 Aminocyclitol-Resistant Mutants of Bacteria[1]

Antibiotic	*Ribosomal Component Altered*
Kasugamycin	16S RNA[2]; mutants fail to N^6-dimethylate two adjacent adenine residues near the 3′ terminus
Spectinomycin	Protein S5 of the 30S subunit; mutation characterized in *E. coli* and *B. subtilis*
Neamine	Proteins S17, S5, and S12 of the 30S subunit
Streptomycin	Protein S12 of the 30S subunit[3]; characterized in *E. coli* and *B. subtilis*
Gentamicin	Protein L6 of the 50S subunit

[1] Many nonribosomal mutations to aminocyclitol resistance are known. These usually involve alterations in drug uptake (see Hancock, 1981).
[2] Several altered ribosomal proteins have been found in kasugamycin-dependent mutants.
[3] Alterations in this protein lead to resistance or dependence.

B. Uptake

Bacterial mutants with broad aminocyclitol cross-resistance properties have been isolated. Detailed studies have illustrated another interesting aspect of the biology of aminocyclitol antibiotics: the mechanism of uptake into susceptible microorganisms. Largely from the work of Bryan and co-workers (e.g., Bryan and van den Elzen, 1977), which has been reviewed in detail by Hancock (1981), it has been established that there are no specific transport systems for aminocyclitol uptake into bacteria but that the drugs enter the cell across a proton gradient generated by oxidative phosphorylation. Anaerobes are generally not susceptible to aminocyclitols, since the drugs do not enter the cell in the absence of the required redox-established charge differential across the cell membrane. All of the variations in bacterial antibiotic susceptibility have not yet been completely explained in these terms—for example, *Pseudomonas aeruginosa* is much more susceptible to gentamicin-like antibiotics than is *Escherichia coli*, and in general, gram-positive organisms require lower drug concentrations for inhibition. Some of these variations may involve different membrane structures and perhaps different ribosome sensitivity, but by and large, the susceptibility of microorganisms to aminocyclitols is based on the activity of the required transport mechanism.

Several aminocyclitol antibiotics show a very broad toxicity range outside the bacteria. This probably says something about the action of these compounds, although as was mentioned earlier, the highly specific toxic side effects of gentamicins, tobramycins, kanamycins, and streptomycin in humans have not been explained on a rational basis. Aminocyclitols with broad inhibition spectra have promise as cross-species selective agents in cloning experiments involving bacteria, plant, and mammalian cells. Of particular interest are the kanamicin-gentamicin-related antibiotic G418 (Jimenez and Davies, 1980) and hygromycin B (Leboul and Davies, 1982), which, in combination with the appropriate inactivating enzymes, can be used to select for intergeneric gene transfer. Antibiotic G418 has seen use for gene transfer in mammalian cells (Canaani and Berg, 1982).

C. Mechanisms of Resistance

We have mentioned resistant mutants in the foregoing discussion. Of greater interest from a clinical and biochemical standpoint are the enzymatic modifications of aminocyclitols (Table 14.5). The clinical aspects of plasmid-determined resistance to aminocyclitols have been fully discussed (Davies, 1982) and do not need additional review in this chapter. However, the genes for aminocyclitol inactivation have different significance when their involvement in antibiotic biosynthesis and in protecting the producing organism against an aminocyclitol product are being considered. J. B. Walker and Skorvaga (1973) suggested that the streptomycin-modifying enzymes of *S. griseus* and *S. bikiniensis* may bear a relationship to the enzymes produced by bacteria containing R-plasmids. Subsequently, Benveniste and Davies (1973) proposed that antibiotic resistance determinants encoded by extrachromosomal elements of clinical isolates had their origins in antibiotic-producing organisms. It has been established that the mechanisms of resistance are biochemically homologous, not only for the aminocyclitols (Table 14.5), but for several other important classes of antibiotics. Extensive studies have so far failed to reveal homologies at the nucleic acid level for antibiotic resistance determinants in various organisms; however, recent DNA sequencing studies (C. Thompson and G. S. Gray, private communication) have indicated strong homologies at the protein sequence level in several domains of the aminoglycoside-3′-phosphotransferases (Fig. 14.4). These enzymes have not been shown to be immunologically cross-reactive. The functions of the domains (aminocyclitol-substrate binding, ATP binding, etc.) have not been identified, but there is a reasonable expectation that, when the genes for additional aminocyclitol-modifying enzymes have been sequenced, some of these features may become apparent.

The precise role of the aminocyclitol-modifying enzymes in producing organisms is not known. A number of antibiotic nonproducing derivatives were found not to have the enzymes, so there is at least a causal relationship (Shaw and Piwowarski, 1977; Yagisawa et al., 1978). Several groups of workers have noted that increased antibiotic production may be accompanied by enhanced production (or activity) of the modifying enzyme (Komatsu et al., 1981). However, aminocyclitol-modifying enzymes are not the only means of self-protection for the producing organisms; several aminocyclitol-producing strains have ribosomes resistant to the inhibitory effects of the drugs in vitro (Yamamoto et al., 1982). It will be interesting to see if there are any rRNA or r-protein alterations in these strains.

N-Acylated and *O*-phosphorylated aminocyclitols have been isolated from several producing organisms, albeit in small quantity (Rinehart, 1964). It is not known if production of these compounds indicates a role for the

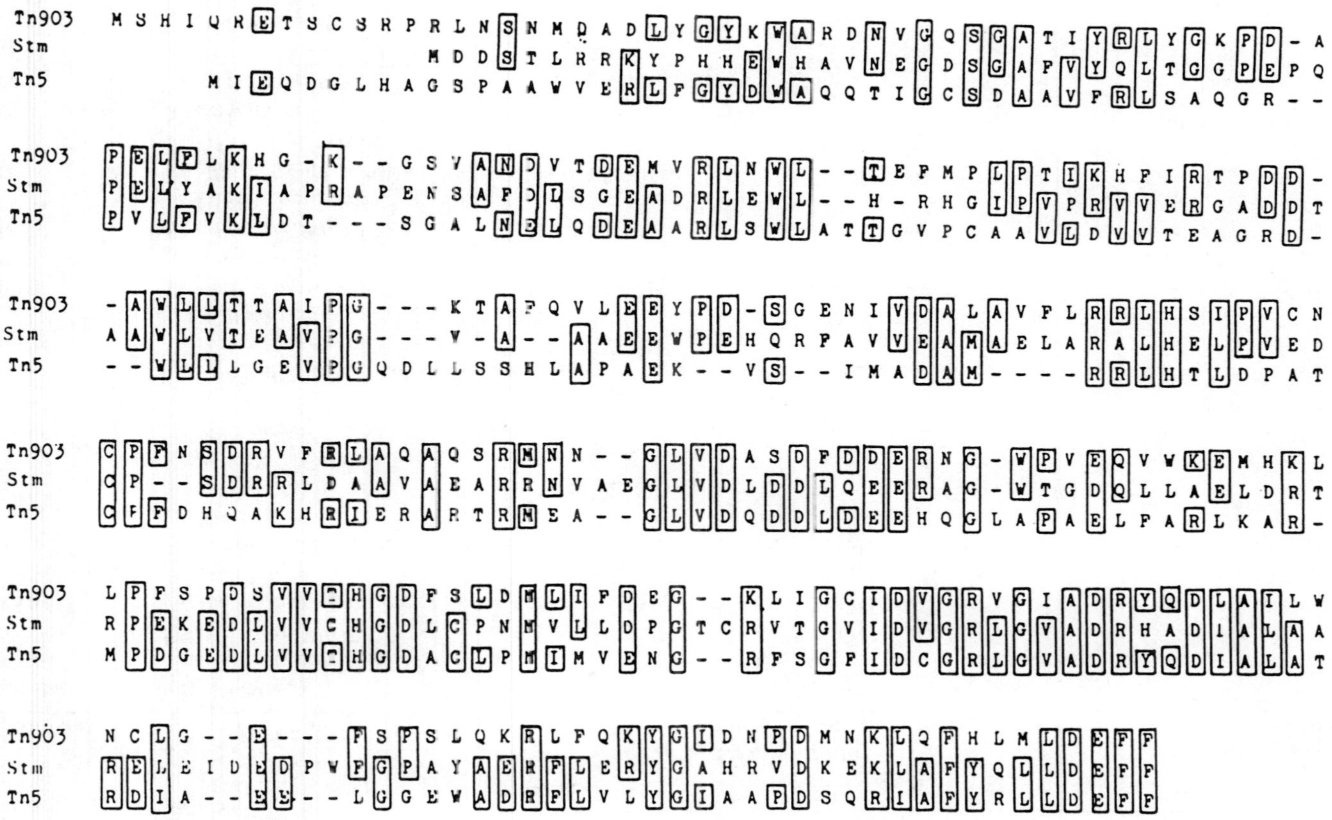

FIGURE 14.4 **Comparison of amino acid sequence homologies in 3′-phosphotransferases from *Streptomyces fradiae* (Stm) and two enterobacteriaceae (Tn903 and Tn5) (Source: C. Thompson and G. S. Gray, private communication).**

TABLE 14.5 Enzymatic Mechanisms of Resistance to Aminocyclitols in Clinical Isolates and Producers

		Enzymatic Modification							
Mechanism	*Strain*	*Neomycin*	*Kanamycin B*	*Lividomycin*	*Butirosin*	*Tobramycin*	*Gentamycin* C_{1a}	*Streptomycin*	*Spectinomycin*
Phosphorylation	Clinical isolate								
APH(3′)-I	*Escherichia coli* JR35/W677	3′-OH	3′-OH	5″-OH	—	—	—	—	—
APH(3′)-II	*E. coli* JR66/W677	3′-OH	3′-OH	—	3′-OH	—	—	—	—
APH(3′)-III	*Pseudomonas aeruginosa* 21-75	3′-OH	3′-OH	5″-OH	3′-OH	—	—	—	—
APH(3′)-III	*Staphylococcus aureus* 20240	3′-OH	3′-OH	5″-OH	3′-OH	—	—	—	—
APH(2″)	*S. aureus* Palm	—	(2″-OH)	—	—	(2″-OH)	2″-OH	—	—
APH(3″)	*E. coli* JR35/C600	—	—	—	—	—	—	3″-OH	—
APH(6)	*P. aeruginosa* GN573	—	—	—	—	—	—	6-OH	—
	Producer (antibiotic produced)								
APH(3′)	*Streptomyces fradiae* UC2046 (neomycin)	3′-OH	(3′-OH)	—	—	—	—	—	—
APH(3′)	*S. ribosidificus* ATCC 21294 (ribostamycin)	3′-OH	3′-OH	—	—	—	—	—	—
APH(3′)	*S. albogriseolus* NRRL B-1305 (neomycin)	3′-OH	3′-OH	—	—	—	—	—	—
APH(3′)	*Bacillus circulans* NRRL B-3312 (butirosin)	3′-OH	3′-OH	—	3′-OH	—	—	—	—
APH(6)	*S. griseus* HUT 6037 (streptomycin)	—	—	—	—	—	—	6-OH	—
APH(3″)	*S. griseus* ATCC 10971 (none)	—	—	—	—	—	—	3″-OH	—

TABLE 14.5 (Continued)

Mechanism	Strain	Enzymatic Modification							
		Neomycin	*Kanamycin B*	*Lividomycin*	*Butirosin*	*Tobramycin*	*Gentamycin* C_{1a}	*Streptomycin*	*Spectinomycin*
Acetylation	Clinical isolate								
AAC(6′)	*E. coli* R5/W677	6′-NH_2	6′-NH_2	—	6′-NH_2	6′-NH_2	6′-NH_2	—	—
AAC(3)-I	*P. aeruginosa* 130	—	(3-NH_2)	—	—	(3-NH_2)	3-NH_2	—	—
AAC(3)-II	*P. aeruginosa* PST	3-NH_2	3-NH_2	3-NH_2	?	3-NH_2	3-NH_2	—	—
AAC(2′)	*Proteus inconstance* 164	(2′-NH_2)	(2′-NH_2)	2′-NH_2	2′-NH_2	2′-NH_2	2′-NH_2	—	—
	Producer (antibiotic produced)								
AAC(2′)	*S. spectabilis* UC 2072 (spectinomycin)	2′-NH_2	2′-NH_2				2′-NH_2		—
AAC(2′)	*S. tenebrarius* (tobramycin)	2′-NH_2	2′-NH_2						
AAC(3)	*S. fradiae* UC 2064 (neomycin)	3-NH_2	3-NH_2			3-NH_2	3-NH_2		
AAC(6′)	*S. kanamyceticus* ATCC 12853 (kanamycin)	6′-NH_2	6′-NH_2				6′-NH_2		
Nucleotidylation	Clinical isolate								
AAD(3″)	*E. coli* NR73/W677	—	—	—	—	—	—	3″-OH	9-OH
ANT(2″)	*E. coli* JR66/W677	—	2″-OH	—	—	2″-OH	2″-OH	—	—
ANT(4′)	*S. aureus* Ap01	4′-OH	4′-OH	4′-OH	4′-OH	4′-OH	—		
AAD(6)	*S. aureus* MS27							6-OH	—

APH: ATP:aminoglycoside phosphotransferase [E.C. 2.7.1.72 2.7.1.87, 2.7.1.88, 2.7.1.95].
AAC: Acetyl-CoA:aminoglycoside acetyltransferase [E.C. 2.3 1.55, 2.3.1.59, 2.3.1.60].
AAD: ATP:aminoglycoside adenylyltransferase [E.C. 2.7.7.47].
ANT: NTP:aminoglycoside nucleotidyltransferase [E.C. 2.7.7.46].

modifying enzymes in the biosynthetic pathway to produce suitable protected or blocked intermediates or whether they are noncleaved inactivated products—or both. The modifying enzymes, their production, and control of their expression constitute an interesting and as yet poorly studied aspect of aminocyclitol biosynthesis.

The natural resistance of aminocyclitol-producing organisms to the antibiotics they produce, and the genes encoding these enzymes, are interesting in other respects. First, it is likely that increased production of these enzymes leading to increased levels of resistance to the drug may provide a means of augmenting production of the drug. Second, several of the genes for aminocyclitol-modifying enzymes have been cloned and analyzed in structural detail. They show promise as selective and insertional inactivation markers for actinomycete and, possibly, fungal cloning vectors (Jimenez and Davies, 1980). Sequence analysis of the regulatory regions of the antibiotic-modifying genes of streptomycetes provides the most direct approach to the identification of operator, promoter, ribosome-binding, and other regulatory regions that will undoubtedly prove useful in gene cloning to enhance specific steps in antibiotic biosynthetic pathways.

Finally, among the many other interesting aspects of aminocyclitol biosynthesis, one in particular is worthy of comment. Butirosin and related antibiotics possess a 4-amino-2-hydroxybutyryl (AHB) side chain (Claridge et al., 1974); discovery of this interesting blocking group provided the impetus for the synthesis of a variety of 1-substituted derivatives, the most notable of which, amikacin (Kawaguchi et al., 1972) and netilmicin (Miller et al., 1976), have excellent therapeutic potential, since they are inert to a number of aminoglycoside modifications that determine resistance to these antibiotics in clinical isolates. However, neither the biosynthetic origin of the AHB substituent nor the timing of its addition is presently known. Better knowledge of these events might point the way to the preparation of such 1-substituted derivatives in vivo.

ACKNOWLEDGMENTS

We wish to thank Jenny Bassin for her steadfast efforts in the preparation of this manuscript and Leo Vining for his patience.

REFERENCES

Benveniste, R., and Davies, J. (1973) *Proc. Nat. Acad. Sci. U.S.A. 70,* 2276–2279.

Bryan, L. E., and van den Elzen, H. M. (1977) *Antimicrob. Agents Chemother. 12,* 163–177.

Canaani, D., and Berg, P. (1982) *Proc. Nat. Acad. Sci. U.S.A. 79,* 5166–5170.

INDEX